Nanotechnology in the Life Sciences

Series Editor

Ram Prasad
Department of Botany
Mahatma Gandhi Central University, Motihari, Bihar, India

Nano and biotechnology are two of the 21st century's most promising technologies. Nanotechnology is demarcated as the design, development, and application of materials and devices whose least functional make up is on a nanometer scale (1 to 100 nm). Meanwhile, biotechnology deals with metabolic and other physiological developments of biological subjects including microorganisms. These microbial processes have opened up new opportunities to explore novel applications, for example, the biosynthesis of metal nanomaterials, with the implication that these two technologies (i.e., thus nanobiotechnology) can play a vital role in developing and executing many valuable tools in the study of life. Nanotechnology is very diverse, ranging from extensions of conventional device physics to completely new approaches based upon molecular self-assembly, from developing new materials with dimensions on the nanoscale, to investigating whether we can directly control matters on/in the atomic scale level. This idea entails its application to diverse fields of science such as plant biology, organic chemistry, agriculture, the food industry, and more.

Nanobiotechnology offers a wide range of uses in medicine, agriculture, and the environment. Many diseases that do not have cures today may be cured by nanotechnology in the future. Use of nanotechnology in medical therapeutics needs adequate evaluation of its risk and safety factors. Scientists who are against the use of nanotechnology also agree that advancement in nanotechnology should continue because this field promises great benefits, but testing should be carried out to ensure its safety in people. It is possible that nanomedicine in the future will play a crucial role in the treatment of human and plant diseases, and also in the enhancement of normal human physiology and plant systems, respectively. If everything proceeds as expected, nanobiotechnology will, one day, become an inevitable part of our everyday life and will help save many lives.

More information about this series at http://www.springer.com/series/15921

Ram Prasad
Editor

Microbial Nanobionics

Volume 1, State-of-the-Art

Springer

Editor
Ram Prasad
Department of Botany
Mahatma Gandhi Central University
Motihari, Bihar, India

ISSN 2523-8027 ISSN 2523-8035 (electronic)
Nanotechnology in the Life Sciences
ISBN 978-3-030-16385-3 ISBN 978-3-030-16383-9 (eBook)
https://doi.org/10.1007/978-3-030-16383-9

This Springer imprint is published by the registered company Springer Nature Switzerland AG
The registered company address is: Gewerbestrasse 11, 6330 Cham, Switzerland

Foreword

Bionics is a discipline focused on the application of knowledge gained by studying living organisms and their structures in the development of new technologies. Although officially this is a young interdisciplinary field of science that originated in the 1960s, the principles of this idea were used, for example, by Leonardo da Vinci in his flying machine inspired by a bat at the early sixteenth century. Nanobionics is an even younger field; its beginnings can be traced back to the first decade of the new millennium. Nanotechnology has become a phenomenon of the twenty-first century, and bionics, under the name "nanobionics," aims to prepare a variety of nanomaterials by so-called green synthesis using nature-inspired resources. Nanoparticles can be prepared using "heavy" chemistry, plant extracts, or using a variety of microorganisms, which can be natural, bred, or genetically modified as in many cases when microorganisms are used in the industrial biotechnology.

Microorganisms (bacteria, fungi) can be found without exaggeration anywhere on our planet, are able to survive extreme conditions, and provide remarkable products. Microorganisms are an essential part of biota, including human. Some are beneficial, e.g., for digestion, for the production of secondary metabolites for industrial and biomedical purposes, and for good soil quality. It should be noted that soil microorganisms are the most represented of all soil biota; they are responsible for the creation, circulation of nutrients and other compounds in the soil, fertility, soil regeneration, overall plant health, and sustainable soil ecosystem. Useful soil microorganisms include especially those that are symbiotic with plant roots (i.e., mycorrhizal fungi, rhizobia, actinomycetes, diazotrophic bacteria), ensure nutrient availability, promote mineralization, produce plant growth hormones, and help plants to fight diseases, pests, and parasites (so-called biocontrol agents). On the other hand, microorganisms can have adverse impact on living organisms, which is reflected, for example, in various diseases of plants, animals, and humans. The increase in the number of drug-resistant and multidrug-resistant strains/species of microorganisms causing both major damage to agriculture and the food industry and disproportionately increasing the cost of human treatment poses a major global threat. With climate change, other pathogenic agents are emerging at new latitudes, and the reduced immunity of the human population contributes to the increase in life-threatening infections caused by originally nonpathogenic microorganisms.

This book series deals with the use of microorganisms, i.e., a wide range of different bacteria, yeasts, and filamentous fungi, to produce various nanomaterials for the application in a wide range of human activities and purposes. It provides valuable modern knowledge on how to choose a suitable microorganism, how to adjust/prepare it, what to avoid, and what is the negative impact of nanoparticles on "living synthesizers." The current overview of applications of microorganisms for the biosynthesis of nanomaterials for environmental, energy, or biomedical purposes is the matter of course. I am convinced that both volumes of the book series *Microbial Nanobionics* will become an indispensable resource and reference for any engineers and researchers interested or engaged in the biosynthesis of nanomaterials using microorganisms, the use of microbial photosynthetic reaction centers combined with nanoparticles for photocurrent generation, or the application of nanobiosensors for microbial growth and diagnostics.

Faculty of Natural Sciences, Comenius University Josef Jampílek
Bratislava, Slovak Republic
Regional Centre of Advanced Technologies
and Materials, Palacký University
Olomouc, Czech Republic

Biographical Sketch of Josef Jampilek

Josef Jampílek completed his Ph.D. degree in Medicinal Chemistry at the Faculty of Pharmacy of the Charles University (Czech Republic) in 2004. During 2004–2011, he worked in expert and managerial posts in the R&D Division of the pharmaceutical company Zentiva (Czech Republic). Prof. Jampilek deepened his professional knowledge at the Medicinal Chemistry Institute of the Heidelberg University (Germany) and at multiple specialized courses. In 2009, he became an Associate Professor of Medicinal Chemistry at the Faculty of Pharmacy of the University of Veterinary and Pharmaceutical Sciences Brno (Czech Republic). In 2017, he was designated as a Full Professor of Medicinal Chemistry at the Comenius University in Bratislava (Slovakia). He is currently working at the Faculty of Natural Sciences, Comenius University in

Bratislava (Slovakia) and at the Regional Centre of Advanced Technologies and Materials, Palacky University Olomouc (Czech Republic). He is an author/co-author of more than 30 patents/patent applications, almost 200 peer-reviewed scientific publications, 7 university textbooks, 20 chapters in monographs, and many invited lectures at international conferences and workshops. He also received several awards for his scientific results, e.g., from Aventis, Elsevier, Willey, Sanofi, and FDA. His research interests include design, synthesis, and structure-activity relationships of heterocyclic compounds as anti-invasive and anti-inflammatory agents as well as photosynthesis inhibitors. He is also interested in ADME, drug bioavailability, nanosystems, and pharmaceutical analysis, especially solid-state analytical techniques.

Preface

Microbes are widely used in the chemical and food sectors, biomedicine and in various areas such as environmental remediation, green chemistry, sustainable manufacturing, biomass energy and resources. The use of microbes to synthesize biogenic nanoparticles has been of great interest. Microorganisms can change the oxidation state of metals, and these microbial processes have opened up new opportunities for us to explore novel applications, for example, the biosynthesis of metal nanomaterials. In contrast to chemical and physical methods, microbial processes for synthesizing nanomaterials can be achieved in aqueous phase under gentle and eco-friendly benign conditions. This approach has become an attractive focus in microbial nanotechnology research toward resource-efficient and sustainable development. This book covers the synthesis of nanoparticles by microbes (bacteria, fungi, actinomycetes, and so on), and the mechanisms involved in such biosynthesis, and a unique template for synthesis of tailored nanoparticles targeted at therapeutics, medicine, agriculture, biofuel and toward new applications that integrate microbes with nanomaterials to produce biohybrids and the next generation of bionic architectures.

This book should be immensely useful to nanoscience scholars, especially microbiologists, nanotechnologists, researchers, technocrats, and scientists of microbial nanobiotechnology. I am honoured that the leading scientists who have extensive, in-depth experience and expertise in microbial systems and nanobiotechnology took the time and effort to develop these outstanding chapters. Each chapter was written by internationally recognized researchers and scientists so the reader is given an up-to-date and detailed account of our knowledge of nanobiotechnology and innumerable applications of microbes.

I wish to thank Dr. Eric Stannard, Senior Editor, Springer; Mr. Rahul Sharma, Project Coordinator, Springer Nature, for generous assistance, constant support, and patience in initializing the volume. I give special thanks to my exquisite wife Dr. Avita for her constant support and motivation in putting everything together.

I am in particular very thankful to Honorable Vice Chancellor Professor Dr. Sanjeev Kumar Sharma, Mahatma Gandhi Central University, Bihar for constant encouragement. I am also grateful to my esteemed friends and well-wishers and all faculty colleagues of the School of Life Sciences and Department of Botany and Biotechnology, India.

Motihari, Bihar, India Ram Prasad

Contents

Contributors

Shagufta Afreen CAS Key Laboratory of Bio-based materials, Qingdao Institute of Bioenergy and Bioprocess Technology, Chinese Academy of Sciences, Qingdao, China

Mehrdad Alizadeh Department of Plant Pathology, College of Agriculture, Tarbiat Modares University, Tehran, Iran

Serap Altindag Aksaray University, Graduate School of Science, Aksaray, Turkey

Elif Apohan Inonu University, Faculty of Science and Letters, Department of Biology, Malatya, Turkey

Mohammad Ashfaq School of Life Science, BS Abdur Rahaman Institute of Science and Technology, Chennai, India

Department of Mechanical Engineering, Faculty of Engineering, Chulalongkorn University, Bangkok, Thailand

Mehmet Dogan Asik Ankara Yildirim Beyazit University, Musculoskeletal Regenerative Medicine, Ankara, Turkey

T. Aswani School of Biotechnology, National Institute of Technology Calicut, Kozhikode, Kerala, India

Emre Birhanli Inonu University, Faculty of Science and Letters, Department of Biology, Malatya, Turkey

Juan Bueno Research Center of Bioprospecting and Biotechnology for Biodiversity Foundation (BIOLABB), Bogotá, Colombia

Monica Butnariu Banat's University of Agricultural Sciences and Veterinary Medicine, "*King Michael I of Romania*" from Timisoara, Timis, Romania

Alina Butu National Institute of Research and Development for Biological Sciences, Bucharest, Romania

Divya Chauhan Department of Chemistry, Punjab University, Chandigarh, India

Bharti P. Dave Department of Life Sciences, Maharaja Krishnakumarsinhji Bhavnagar University, Bhavnagar, Gujarat, India

Karuna Dhiman Department of Biotechnology, Dr YS Parmar University of Horticulture and Forestry, Himachal Pradesh, India

Hesham Ali El Enshasy Institute of Bioproduct Development (IBD), Universiti Teknologi Malaysia (UTM), Johor Bahru, Johor, Malaysia

Faculty of Chemical and Energy Engineering, Universiti Teknologi Malaysia (UTM), Johor Bahru, Johor, Malaysia

City of Scientific Research and Technology Application, New Burg Al Arab, Alexandria, Egypt

Elsayed Ahmed Elsayed Bioproducts Research Chair, Zoology Department, Faculty of Science, King Saud University, Riyadh, Kingdom of Saudi Arabia

Chemistry of Natural and Microbial Products Department, National Research Centre, Cairo, Egypt

Ezgi Emul Hacettepe University, Nanotechnology and Nanomedicine Division, Ankara, Turkey

Jiya George Nanotechnology and Biotechnology, Nabio Consulting, Pretoria, South Africa

Ekrem Gurel Abant Izzet Baysal University, Faculty of Science and Literature, Department of Biology, Bolu, Turkey

Siti Zulaiha Hanapi Institute of Bioproduct Development (IBD), Universiti Teknologi Malaysia (UTM), Johor Bahru, Johor, Malaysia

Mehrnaz Hatami Department of Medicinal Plants, Faculty of Agriculture and Natural Resources, Arak University, Arak, Iran

Institute of Nanoscience and Nanotechnology, Arak University, Arak, Iran

Sedef İlk Nigde Omer Halisdemir University, Faculty of Medicine, Department of Immunology, Nigde, Turkey

Asim K. Jana Department of Biotechnology, Dr BR Ambedkar National Institute of Technology, Jalandhar, India

Daniel Joel Institute of Bioproduct Development (IBD), Universiti Teknologi Malaysia (UTM), Johor Bahru, Johor, Malaysia

Faculty of Chemical and Energy Engineering, Universiti Teknologi Malaysia (UTM), Johor Bahru, Johor, Malaysia

Puja Khare Agronomy and Soil Science Division, CSIR-Central Institute of Medicinal and Aromatic Plants, Lucknow, Uttar Pradesh, India

Kugen Kumar Institute of Bioproduct Development (IBD), Universiti Teknologi Malaysia (UTM), Johor Bahru, Johor, Malaysia

Behnam Asgari Lajayer Department of Soil Science, Faculty of Agriculture, University of Tabriz, Tabriz, Iran

Roslinda Abd Malek Institute of Bioproduct Development (IBD), Universiti Teknologi Malaysia (UTM), Johor Bahru, Johor, Malaysia

Nariman Marooufpour Department of Plant Protection, Faculty of Agriculture, University of Tabriz, Tabriz, Iran

Zamanzima Mazibuko Knowledge Economy and Scientific Advancement, Mapungubwe Institute for Strategic Reflection (MISTRA), Johannesburg, South Africa

Disha Mishra Agronomy and Soil Science Division, CSIR-Central Institute of Medicinal and Aromatic Plants, Lucknow, Uttar Pradesh, India

Maluta Steven Mufamadi Nanotechnology and Biotechnology, Nabio Consulting, Pretoria, South Africa

Rishabh Anand Omar Centre for Environmental Science and Engineering, Indian Institute of Technology Kanpur, Kanpur, India

Necdet Saglam Hacettepe University, Nanotechnology and Nanomedicine Division, Ankara, Turkey

Marcia Regina Salvadori Department of Microbiology, Biomedical Institute-II, University of São Paulo, São Paulo, São Paulo, Brazil

Mesut Sam Aksaray University, Faculty of Science and Letters, Department of Biology, Aksaray, Turkey

Sasi Reshmi School of Biotechnology, National Institute of Technology Calicut, Kozhikode, Kerala, India

Sneh Sharma Department of Biotechnology, Dr YS Parmar University of Horticulture and Forestry, Himachal Pradesh, India

Dhananjaya P. Singh ICAR-National Bureau of Agriculturally Important Microorganisms, Kushmaur, Uttar Pradesh, India

Werayut Srituravanich Department of Mechanical Engineering, Faculty of Engineering, Chulalongkorn University, Bangkok, Thailand

Biomedical Engineering Research Center, Faculty of Engineering, Chulalongkorn University, Bangkok, Thailand

T. V. Suchithra School of Biotechnology, National Institute of Technology Calicut, Kozhikode, Kerala, India

Neetu Talreja Department of Bio-nanotechnology, Gachon University, Incheon, South Korea

Gamze Tan Aksaray University, Faculty of Science and Letters, Department of Biology, Aksaray, Turkey

Theivasanthi Thirugnanasambandan International Research Centre, Kalasalingam Academy of Research and Education (Deemed University), Krishnankoil, TN, India

Thilivhali Emmanuel Tshikalange Department of Plant and Soil Sciences, University of Pretoria, Pretoria, South Africa

Anjana K. Vala Department of Life Sciences, Maharaja Krishnakumarsinhji Bhavnagar University, Bhavnagar, Gujarat, India

Madan L. Verma Centre for Chemistry and Biotechnology, Deakin University, Geelong, VIC, Australia

Department of Biotechnology, Dr YS Parmar University of Horticulture and Forestry, Himachal Pradesh, India

Sandeep Kumar Verma Abant Izzet Baysal University, Faculty of Science and Literature, Department of Biology, Bolu, Turkey

Ozfer Yesilada Inonu University, Faculty of Science and Letters, Department of Biology, Malatya, Turkey

About the Editor

Ram Prasad, Ph.D. has been associated with Mahatma Gandhi Central University, Motihari, Bihar, India. His research interests include applied microbiology, plant–microbe interactions, sustainable agriculture and nanobiotechnology. Dr. Prasad has more than one hundred fifty publications to his credit, including research papers, review articles, book chapters, and five patents issued or pending, and has edited or authored several books. Dr. Prasad has 12 years of teaching experience and has been awarded the Young Scientist Award (2007) and the Prof. J.S. Datta Munshi Gold Medal (2009) by the International Society for Ecological Communications; FSAB fellowship (2010) by the Society for Applied Biotechnology; the American Cancer Society UICC International Fellowship for Beginning Investigators, USA (2014); Outstanding Scientist Award (2015) in the field of microbiology by Venus International Foundation; BRICPL Science Investigator Award (ICAABT-2017) and Research Excellence Award (2018). He serves as editorial board member for *Frontiers in Microbiology*, *Frontiers in Nutrition*, *Academia Journal of Biotechnology* and has been series editor of *Nanotechnology in the Life Sciences*, Springer Nature, USA. Previously, Dr. Prasad served as Assistant Professor, Amity University Uttar Pradesh, India; Visiting Assistant Professor, Whiting School of Engineering, Department of Mechanical Engineering at Johns Hopkins University, USA and Research Associate Professor at School of Environmental Science and Engineering, Sun Yat-sen University, Guangzhou, China.

Chapter 1
Processing of Nanoparticles by Biomatrices in a Green Approach

Marcia Regina Salvadori

Contents

1.1 Introduction

The definition of nanotechnology was first described by physicist Richard Feynman in 1959 (Bhattacharya et al. 2009). This science is considered to have the ability to control and manipulate matter at the atomic and molecular levels (Balasooriya et al. 2017). The synthesis of nanoparticles (NPs) has emerged as a

M. R. Salvadori (✉)
Department of Microbiology, Biomedical Institute-II, University of São Paulo, São Paulo, São Paulo, Brazil
e-mail: mrsal@usp.br

R. Prasad (ed.), *Microbial Nanobionics*, Nanotechnology in the Life Sciences,
https://doi.org/10.1007/978-3-030-16383-9_1

promising area in nanotechnology with exponential progress in many fields such as chemical industries, biomedical, electronics, drug–gene delivery, catalysis, agriculture, cosmetics, optics, mechanics, food and feed, and the environment (Sreejivungsa et al. 2016; Zhao et al. 2015a, b; Itohara et al. 2016; Lohani et al. 2014; Prasad 2014; Prasad et al. 2014, 2016, 2017a; Pei et al. 2015; Zhang et al. 2016; Yuan et al. 2015).

Chemical and physical methods have been used for the synthesis of NPs; however, they present disadvantages such as the use of hazardous products, toxic solvents, and high-energy consumption (Azandehi and Moghaddam 2015; Yu et al. 2016). The production of NPs can be divided into bottom-up or top-down approaches (Murphy 2002). The bottom-up approach involves microemulsion (Darbandi et al. 2005), chemical vapor deposition (Li et al. 2004), sol–gel processing (Yi et al. 1988), laser pyrolysis (Lacour et al. 2007), and plasma- or flame-spraying synthesis (Mädler et al. 2002). The top-down approach consists of a process of the breaking down of large structures to create small structures and uses physical methods such as vapor deposition (Kumar and Ando 2010), pulsed electrochemical etching (Nissinen et al. 2016), sputtering deposition (Bell et al. 2001), lithography (Tapasztó et al. 2008), and laser ablation (Barcikowski et al. 2009).

In the last few decades, green approaches have substituted some of the chemical and physical methods in NP synthesis using biomatrices (bacteria, viruses, fungi, yeasts, algae, protozoans, plants or plant extracts) from aqueous solutions of the corresponding salts (Klaus-Joerger et al. 2001; Ahmad et al. 2011; Ramezani et al. 2012; Alghuthaymi et al. 2015; Boroumand et al. 2015; Prasad et al. 2016; Dahoumane et al. 2017a, b). These techniques present the advantages of being low-cost and environmentally safe, as they are implemented at room temperature or under mild heating; in aqueous solution, atmospheric pressure and salts are used, as they are received without additional synthesis. These procedures fit in with the precepts of the green approach to nanotechnology (Dahoumane et al. 2017a, b; Anastas and Kirchhoff 2002).

The applications of NPs produced through green chemistry using biomatrices are rapidly increasing in various areas, such as antimicrobial activity, health care, drug delivery, cosmetics, biomedical, food and feed, high-temperature superconductors, wood preservatives, catalysis, light emitters, gas sensor, medical imaging devices, agriculture, and environmental clean-up (Khalil et al. 2013; Kaviya and Viswanathan 2011; Ahmed et al. 2016; Evans et al. 2008; Li et al. 2007; Guo et al. 2007; Salvadori et al. 2016; Bansal et al. 2014; Prasad et al. 2014, 2017a, b, c, 2018a; Aziz et al. 2015, 2016, 2019). This fact is emphasizing the responsibility of researchers to seek a synthetic route, which is not only profitable, but also environmentally safe; this has been shown through several reports in the literature on the green approach to the synthesis of NPs.

In this chapter, a green approach to NPs synthesis is discussed as an environmentally safer advancement in the synthesis process compared to chemical and physical processes. The main objective is a deeper understanding of the utilization of biomatrices as microorganisms and plants in NP production using green chemistry, and their variety of applications, such as in life science, industry, and environmental science.

1.2 Bacteria: Biomatrices for NP Synthesis

Bacteria are microorganisms compressing one of the three domains of living organisms. They are prokaryotic organisms, unicellular and either free living in water or soils or parasitizing animals or plants. Owing to their abundance in nature, their remarkable capacity to reduce metal ions, and the ability to adapt to extreme conditions, the bacteria are a good choice as candidates for biomatrices in NP production (Pantidos and Horsfall 2014; Prasad et al. 2016).

The bacteria synthesize NPs both intra- and extracellularly, according to the place where NPs are formed in the bacterial cells. Intracellular NP production by bacteria encompasses, for example, the synthesis of silver NPs (AgNPs), such as those produced by *Rhodococcus* sp. inside the cell and in the cytoplasm (Otari et al. 2015), *Vibrio alginolyticus* through reduction of silver nitrate (Rajeshkumar et al. 2013a), and *Pseudomonas stutzeri* AG 259 isolated from silver mine-produced NPs within the periplasmic space of a bacterium (Klaus et al. 1999). Gold NPs (AuNPs) were produced by *Shewanella algae* (Konishi et al. 2007), which are facultatively anaerobic bacteria, Gram-negative, found predominantly in marine sediments and in association with fish (Nealson and Scott 2006). The alkalotolerant actinomycetes *Rhodococcus* sp. (Ahmad et al. 2003) and *Brevibacterium casei* produced spherical AuNPs (Kalishwaralal et al. 2010). The bacteria *Escherichia coli* synthesized intracellularly cadmium sulfide (CdS) nanocrystals (Sweeney et al. 2004). The palladium NPs were synthesized by the bacteria *Desulfovibrio desulfuricans* and *Bacillus benzeovorans* (Omajali et al. 2015). Non-metallic nanomaterials such as selenium NPs were synthesized by the *Bacillus licheniformis* JS2, which was able to inhibit proliferation and induce caspase-independent necrosis in human prostate adenocarcinoma cells (Sonkusre et al. 2016).

Several bacteria have been studied for their potential for extracellular NP synthesis. Zonooz and Salouti (2011) reported the biosynthesis of AgNPs using supernatant of *Streptomyces* sp., *Bacillus megaterium* showed the ability to synthesize silver, lead, and cadmium NPs extracellularly (Prakash et al. 2010). AuNPs were synthesized by *Pseudomonas denitrificans* (Mewada et al. 2012). The bacteria *Bacillus subtilis* has been described as being capable of synthesizing iron oxide NPs extracellularly (Sundaram et al. 2012). Other examples of NPs produced by bacteria are described in Table 1.1.

The exact mechanism of NP production by bacteria has not yet been elucidated. Probable mechanisms that can be considered for NP biosynthesis by bacteria included extracellular complexion or precipitation of metals, efflux system, bioaccumulation, lack of specific metal transportation system, alteration of solubility, and toxicity via reduction or oxidation and bio-absorption (Beveridge and Murray 1980; Pantidos and Horsfall 2014; Prasad et al. 2016). Figure 1.1 illustrates the possible extra- or intracellular mechanisms of the synthesis of metallic NPs by bacteria, taking into account important physico-chemical parameters in NP production, such as pH, contact time, temperature, amount of bacteria, and concentration of the metal salts.

Table 1.1 Biosynthesis of nanoparticles (NPs) by bacteria

Bacteria species	Metal NPs	Location	Size (nm)	References
Shewanella oneidensis	U	Extracellular	150	Marshall et al. (2007)
Pseudomonas aeruginosa	Au	Extracellular	15–30	Husseiney et al. (2007)
Lactobacillus sp.	Au, Ag	Intracellular	20–50	Nair and Pradeep (2002)
Acetobacter xylinum	Ag	Extracellular	–	Barud et al. (2008)
Bacillus selenitireducens	Te	Extracellular	~10	Baesman et al. (2007)
Magnetospirillum magnetotacticum	Fe_3O_4	Intracellular	47.1	Philipse and Maas (2002)
Desulfovibrio desulfuricans NCIMB 8307	Pd	Intracellular	~50	Yong et al. (2002)
Aquaspirillum	Fe_3O_4	Intracellular	40–50	Mann et al. (1984)
Lactobacillus sp.	Ti	Extracellular	40–60	Prasad et al. (2007)
Rhodopseudomonas capsulate	Au	Extracellular	10–20	Shiying et al. (2007)
Plectonema boryanum UTEX485	Au	Intracellular	10	Lengke et al. (2006)
Enterobacter cloacae, Klebsiella pneumonia, E. coli	Ag	Extracellular	52.5	Shahverdi et al. (2007)
Gluconacetobacter xylinus	CdS	Extracellular	30	Li et al. (2009)
Actinobacter sp.	Magnetite	Extracellular	10–40	Bharde et al. (2005)

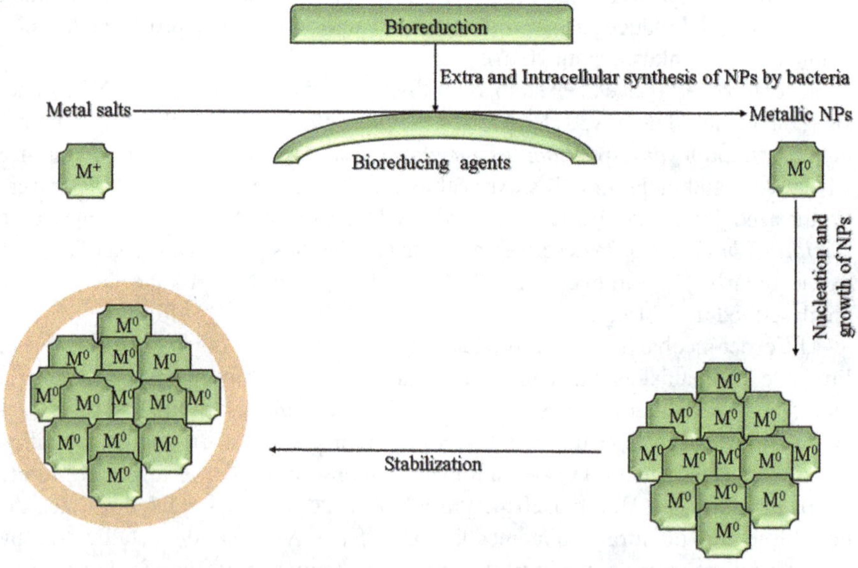

Fig. 1.1 Scheme of nanoparticle (NP) synthesis by bacteria

The extracellular biosynthesis of NPs is still the most promising, owing to easy scale-up processing and rapid processing, which is why this is the most frequently studied mechanism for NP-forming bacterial systems (Velusamy et al. 2016).

1.3 Protozoans: Biomatrices of NP Processing

Protozoans are microscopic eukaryotic organisms belonging to the Protista kingdom, and are primarily unicellular, living singly or in colonies. Usually, they are nonphotosynthetic, and are often classified further into phyla according to their capacity for motility, i.e., pseudopods, flagella or cilia, which have a relatively complex internal structure. Most species are free-living, but all higher animals are infected with one or more species of protozoans (Baron 1996).

There are few reports in the literature about the use of protozoans as biomatrices in NP synthesis. Juganson et al. (2013) described the extracellular synthesis of AgNPs by cell-free exudates of the ciliated protozoan *Tetrahymena thermophila*, with a mean hydrodynamic size of 70 nm. *Tetrahymena thermophila* has also been described as a synthesizer of amorphous spherical selenium NPs with a diameter between 50 and 500 nm. In experimental results, it is believed that in addition to glutathione, three more of the protozoan proteins are involved in the synthesis of selenium NPs (Cui et al. 2016). Ramezani et al. (2012) described the synthesis of Ag- and AuNPs using the protozoan *Leishmania* sp., with sizes ranging from 10 to 100 nm for silver NPs and 50 to 100 nm for AuNPs. The authors also observed the presence of proteins as stabilizing agents on the NP surface.

In view of the literature reports, protozoans may also offer a good choice for a green approach in NP production. A probable scheme of metallic NP synthesis by protozoa is illustrated in Fig. 1.2.

1.4 Viruses: Biomatrices of Nanomaterials

The virus is a small parasite, which, when it infects a susceptible cell, can direct the cell machinery to produce more viruses. Normally, the viruses have either RNA or DNA as their genetic material. The entire infectious virus particle, called a virion, consists of the nucleic acid and an outer shell of protein (Lodish et al. 2000).

The viruses have shown great promise as templates and scaffolds for NP synthesis (Lee et al. 2009), through the interconnection of nanometric components. They exhibit the characteristics of an ideal nano-building block, as their ability to associate into desired structures, monodispersity, morphology, and the variety of chemical groups available for modification and size (Douglas and Young 1998; Mao et al. 2004; Velusamy et al. 2016).

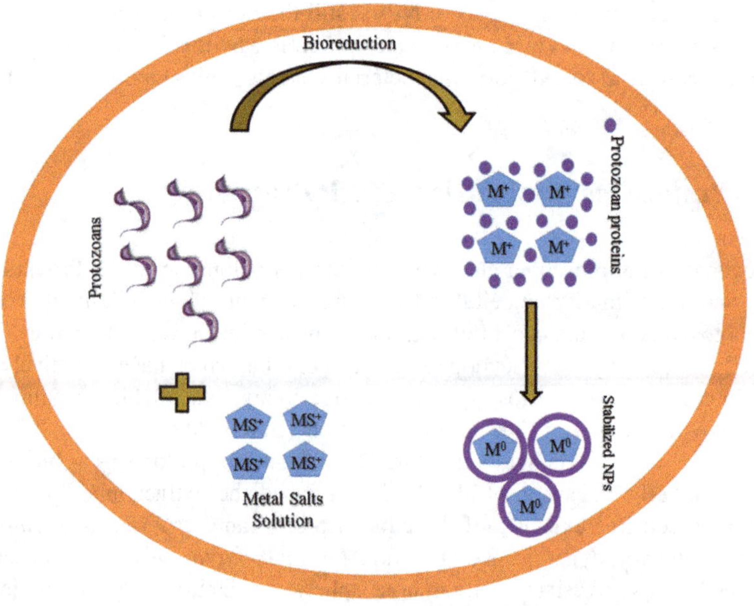

Fig. 1.2 Illustration of the probable scheme of production of metallic NPs by protozoans

The viruses have been exposed to various metal or other inorganic precursors and this exposure, through interactions with amino acids of the capsids, has resulted in the nucleation and formation of nanomaterials on their outer or inner surface (Pokorski and Steinmetz 2011).

Hou et al. (2014) reported the production of platinum NPs by a wild-type virion, bacteriophage T4. The NPs presented a uniform size of 3–4 nm and covered the whole viral capsid, in addition to presenting high electrocatalytic activity in electrochemical measurement. The rod-like viruses showed the ability to synthesize copper nanorods and nanowires (Zhou et al. 2012). Mao et al. (2003) described the production of ZnS and CdS nanocrystals with an average size of 50–100 nm by the M13 bacteriophage. Cowpea chlorotic mottle virus and cowpea mosaic virus have been used for the mineralization of inorganic materials (Love et al. 2014).

Tobacco mosaic virus (TMV) is a very versatile scaffold, because of its external surface and inner channel, which can be mineralized, the lysine-/arginine-rich exterior is positively charged, whereas the aspartate/glutamate-rich lumen is negatively charged (Namba and Stubbs 1986). These characteristics allow several materials to be deposited on TMV such as SiO_2, CdS, PbS, and Fe_2O_3 nanocrystals (Shenton et al. 1999). Cobalt, copper, gold, nickel, palladium, platinum, and silver can be deposited on the virus via electrolytic deposition (Balci et al. 2006; Dujardin et al. 2003; Bromley et al. 2008; Knez et al. 2002, 2003, 2004). The atomic layer deposition has been used to produce hybrid aluminum oxide and titanium oxide structures (Knez et al. 2006).

Therefore, nowadays the viruses are not only considered pathogens, but are seen as new biomatrices for the construction of nanomaterials, with applications in several areas.

1.5 Yeasts: Biomatrices That Mediate NP Production

The yeasts are eukaryotic microorganisms, classified in the fungi kingdom, with approximately 1500 species described (Lachance 2016). A new evolution in the use of yeasts is in the field of intracellular and extracellular NP synthesis through green routes (Salvadori et al. 2014a; Skalickova et al. 2017).

Over the last few decades, the yeasts have become important in nanotechnology because of their valuable attributes for NP production, such as the ease of control under laboratory conditions, growth under extreme conditions of temperature, pH, and nutrients, the rapid growth, the production of various enzymes, the simple scale-up, the cost-effectiveness, the easy processing, and biomass handling (Anand et al. 2006; Kumar et al. 2011; Varshney et al. 2012).

Several research projects have been conducted to obtain the biosynthesis of metal NPs using yeasts (Table 1.2). Jha et al. (2009) reported a green and low-cost synthesis of Sb_2O_3 NPs by *Saccharomyces cerevisiae*. The synthesis of Au- and AgNPs by *Candida guilliermondii* (Mishra et al. 2011) and *Saccharomyces cerevisiae*

Table 1.2 Biosynthesis of metal NPs by yeasts

Yeasts species	Metal NPs	Location	Size (nm)	References
Baker's yeast	Au	Extracellular	13	Attia et al. (2016)
Candida albicans	Au	Cell-free extract	5	Ahmad et al. (2013)
Yeast cells	Fe_3O_4	Extracellular	–	Zhou et al. (2009b)
Yeast cells	$FePO_4$	Extracellular	–	Zhou et al. (2009a)
Candida utilis	Au	Intracellular	–	Gericke and Pinches (2006)
Yeast	Zr	–	–	Tian et al. (2010)
Yeast	Au/Ag	Extracellular	9–25	Zheng et al. (2010)
Yarrowia lipolytica NCIM3589	Au	Cell surface	Varying	Agnihotri et al. (2009)
Candida glabrata	CdS	Intra and extracellular	20 Å, 29 Å	Dameron et al. (1989)
Yeast strain MKY3	Ag	Extracellular	2–5	Kowshik et al. (2003)
Saccharomyces cerevisiae	Sb_2O_3	Intracellular	2–10	Jha et al. (2009)
Schizosaccharomyces pombe	CdS	Intra and extracellular	18 Å, 29 Å	Dameron et al. (1989)
Schizosaccharomyces pombe	CdS	Intracellular	1–1.5	Kowshik et al. (2002)
Yeast	CdS	Intracellular	3.6	Prasad and Jha (2010)
Yeasts	$Zn_3(PO_4)_2$	Extracellular	10–80 × 80–200	Yan et al. (2009)
Rhodotorula mucilaginosa	Cu	Intracellular	10.5	Salvadori et al. (2014a)
Rhodotorula mucilaginosa	Ni/NiO	Extracellular	5.5	Salvadori et al. (2016)
Rhodotorula mucilaginosa	Ag	Intracellular	11	Salvadori et al. (2017)

(Lim et al. 2011) was also described. The ability of the tropical marine yeast *Yarrowia lipolytica* NCIM 3589 to produce AuNPs was reported by Agnihotri et al. (2009) and the yeast strain MKY3 was reported for the production of AgNPs by Kowshik et al. (2003). A green approach using dead biomass of the yeast *Rhodotorula mucilaginosa* as a biomatrix was described for the intracellular production of copper and AgNPs (Salvadori et al. 2014a, 2017) and the extracellular production of Ni/NiO magnetic NPs (Salvadori et al. 2016). The CdS NPs were produced by *Candida glabrata* and *Schizosaccharomyces pombe* (Dameron et al. 1989), the NPs produced by *Schizosaccharomyces pombe* being used for the production of a cadmium diode (Kowshik et al. 2002).

The mechanisms of NP synthesis by yeasts have not yet been elucidated, as future research into the understanding of this interesting green approach to nanotechnology requires further research.

Salvadori et al. (2016) proposed a natural protocol for extracellular metal NP synthesis using yeasts. This mechanism consists in the interaction between the metal cations and the amide groups located in the yeast cell wall and its subsequent bioreduction, possibly because of the presence of extracellular enzymes in the yeast cell wall; these same proteins serve as capping and stabilizing agents of the NPs. A probable mechanism of intracellular NP production comprises the electrostatic interaction between metal cations and amide groups found in yeast cell wall enzymes, followed by the bioreduction of the ions by enzymes located inside the cell wall, which results in the aggregation of the metal ions and the formation of NPs (Salvadori et al. 2017). Figure 1.3 schematizes a possible mechanism of intracellular NP synthesis by yeasts.

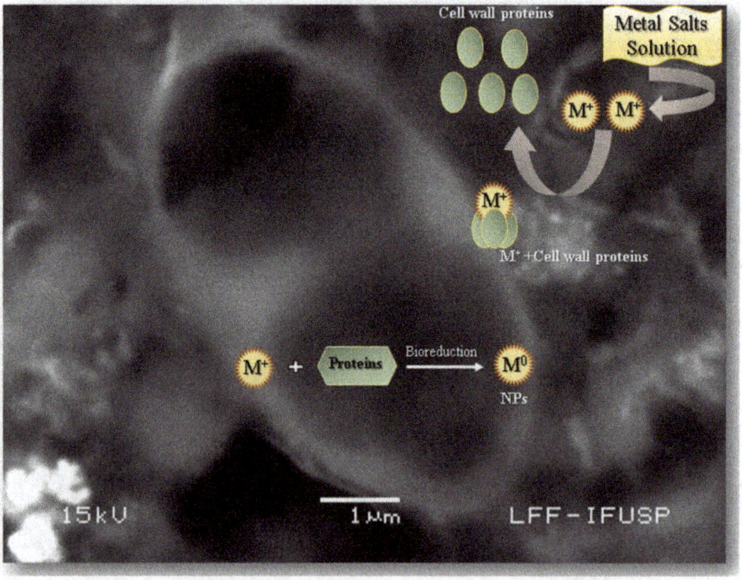

Fig. 1.3 Illustration of the mechanism of intracellular NP synthesis by yeasts

In view of several reports in the literature, yeasts are considered to be versatile biomatrices in NP construction, making a great contribution to green nanotechnology.

1.5.1 Filamentous Fungi: Biomatrices of Nanofactories

Filamentous fungi are ubiquitous eukaryotic organisms, multicellular, spore-producing, they reproduce both sexually and asexually, and show filamentous branched structures (Gow and Gadd 1994). The mycogenic route using filamentous fungi for NP production has been well accepted because this totipotent eukaryotic microorganism has various notable features, which have been well reported (Saxena et al. 2014), such as they are easy to culture (Rai et al. 2009), they have the ability to secrete extracellular enzymes (Kar et al. 2014), easy handling and growth control (Salvadori et al. 2015), extracellular NP synthesis (Gade et al. 2008), and biomolecules in the cell wall, which play an important role in the sorption of various metals (Srivastava and Thakur 2006; Prasad 2016, 2017; Abdel-Aziz et al. 2018; Prasad et al. 2018b).

The mycosynthesis of several metallic NPs by different filamentous fungi has been reported (Table 1.3). Filamentous fungi have the ability to produce metal NPs/meso- and nanostructured by reducing intra- or extracellular enzymes and through the biomimetic mineralization process (Ahmad et al. 2003).

Table 1.3 Metal NPs produced by filamentous fungi

Fungi species	Metal NPs	Location	Size (nm)	References
Aspergillus clavatus	Ag	Extracellular	10–25	Narayanan and Sakthivel (2010)
Aspergillus niger	Ag	Extracellular	3–30	Jaidev and Narasimha (2010)
Bipolaris nodulosa	Ag	Extracellular	10–60	Saha et al. (2010)
Phoma glomerata	Ag	Extracellular	60–80	Birla et al. (2009)
Mucor hiemalis	Ag	Extracellular	5–15	Aziz et al. (2016)
Pestalotia sp.	Ag	Extracellular	10–40	Raheman et al. (2011)
Aspergillus terreus	Mg	Extracellular	48–98	Raliya et al. (2013)
Aspergillus aculeatus	NiO	Extracellular	5.89	Salvadori et al. (2014c)
Aspergillus niger	CeO_2	Extracellular	5–20	Gopinath et al. (2015)
Hypocrea lixii	Cu	Extracellular	24.5	Salvadori et al. (2013)
Aspergillus fumigatus	ZnO	Extracellular	1.2–6.8	Raliya and Tarafdar (2013)
Trichoderma koningiopsis	Cu	Extracellular	87.5	Salvadori et al. (2014b)
Colletotrichum sp.	Au	Extracellular	8–40	Shankar et al. (2003)
Hypocrea lixii	NiO	Intra and extracellular	1.25 3.8	Salvadori et al. (2015)
Lecanicillium lecanii	Ag	Extracellular	45–100	Namasivayam and Avimanyu (2011)
Rhizopus oryzae	Au	Cell surface	10	Das et al. (2009)
Aureobasidium pullulans	Au	Intracellular	29	Zhang et al. (2011)

The fungus *Rhizopus stolonifer* has been reported for synthesizing AgNPs with antibacterial activity (Afreen and Ranganath 2011). The biosynthesis of AgNPs by the filamentous fungus *Verticillium* sp. has also been reported (Mukherjee et al. 2001). Zhang et al. (2009) described the use of the fungus *Penicillium* sp. for intracellular AuNP production. The extracellular biosynthesis of copper NPs was performed by dead biomass of the fungi *Hypocrea lixii* and *Trichoderma koningiopsis* (Salvadori et al. 2013; 2014a). Syed and Ahmad (2013) reported CdTe quantum dot biosynthesis by the fungus *Fusarium oxysporum*. Nickel oxide NPs in film form were produced by *Aspergillus aculeatus* with an average size of 5.89 nm (Salvadori et al. 2014c), and intra- and extracellular nickel oxide NPs were also synthesized by *Hypocrea lixii* with a spherical shape (Salvadori et al. 2015). Rajakumar et al. (2012) reported TiO_2 NP synthesis with a probable antibacterial property by the fungus *Aspergillus flavus*.

The exact mechanism by which filamentous fungi produce NPs intra- or extracellularly is not fully understood. Salvadori et al. (2015) described a probable mechanism of intracellular NP synthesis, where the metal binding to protein from the fungal cell wall by electrostatic interactions. Next, the metal ions are bioreduced by enzymes present inside the cell wall, which leads to the aggregation of the metal ions and the formation of NPs. The extracellular mechanism of NP synthesis consists in the interaction between the metal and enzymes located in the cell wall of the filamentous fungi and its subsequent reduction and NP formation (Salvadori et al. 2013). This synthesis has advantages because it does not require downstream processing for recovery of NPs and lysis of the fungal cell wall (Saxena et al. 2014). Figure 1.4 illustrates a possible mechanism of extracellular NP formation mediated by filamentous fungi.

Mycosynthesis of NPs using filamentous fungi as biomatrices constitutes a promising route in nanofactories, owing to their great versatility of structural and metabolic characteristics.

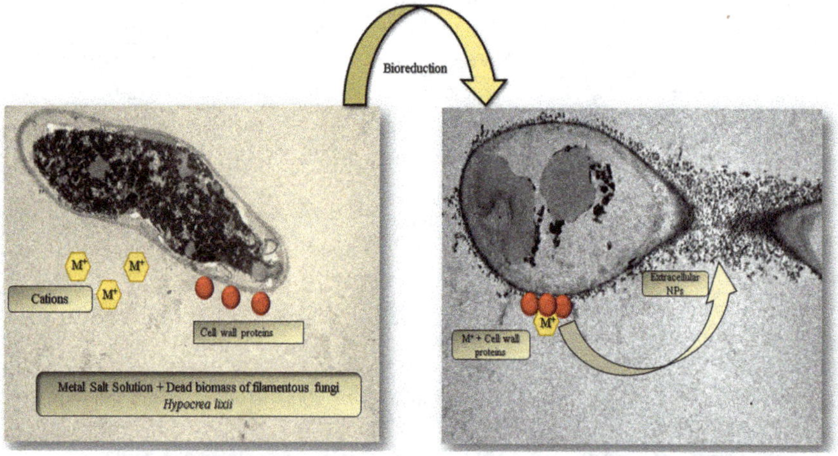

Fig. 1.4 Mechanism of extracellular synthesis of metal NPs by the dead biomass of the filamentous fungi *Hypocrea lixii.* (From Salvadori et al. (2013))

1.6 Algae: NP Biomatrices

Algae are uni- or multicellular organisms, have a nuclear envelope and membranous organelles, occupy all environments that offer them light and sufficient moisture, and are classified as microalgae and macroalgae. They are photosynthetic organisms and can inhabit the surface of moist rocks, freshwater or marine water, and even snow (Thajuddin and Subramanian 2005; Oscar et al. 2014).

Algae have the capacity to accumulate metals and reduce metal ions, making them good candidates for NP synthesis, and also present some advantages such as NP synthesis at low temperature, low toxicity and being easy to handle (Oscar et al. 2016). There are several reports of NP synthesis using groups of algae such as *Cyanophyceae*, *Phaeophyceae*, *Rhodophyceae*, *Chlorophyceae* (Sharma et al. 2015; Aziz et al. 2014, 2015).

Algae may synthesize NPs from various metal salts through functional groups and enzymes present in the cell wall; even edible forms of algae are used in metallic NP synthesis. The processing of metal and metal oxide NPs may be performed by several algae species, such as *Fucus vesiculosus*, *Pithophora oedogonia*, *Spirulina platensis*, *Chlorella vulgaris*, *Sargassum wightii* (Siddiqi and Husen 2016). The biomatrix of the freshwater, edible red alga *Lemanea fluviatilis* has been used to accomplish AuNP synthesis (Sharma et al. 2014). Abdel-Raouf et al. (2018) reported the AgNP synthesis using the marine brown alga *Padina pavonia*. Fucoidan is polysaccharide secreted from the cell walls of marine brown algae, with the ability to synthesize AuNPs (Lirdprapamongkol et al. 2014). Table 1.4 shows some examples of NPs synthesized by different algae species.

The marine macroalgae are organisms categorized according to their pigmentation into green (Chlorophyta), brown (Phaeophyta), and red (Rhodophyta), which

Table 1.4 Algae-synthesized metal NPs

Algae species	Metal NPs	Location	Size (nm)	References
Ulva fasciata	Ag	Extracellular	28–41	Rajesh et al. (2012)
Sargassum plagiophyllum	AgCl	Extracellular	18–42	Dhas et al. (2014)
Caulerpa racemose	Ag	Extracellular	5–25	Kathiraven et al. (2015)
Spirulina platensis	Au	Intracellular	5	Suganya et al. (2015)
Galaxaura elongata	Au	Extracellular	3.85–77.13	Abdel-Raouf et al. (2013)
Sargassum wightii Grevilli	Ag	Extracellular	8–27	Govindaraju et al. (2009)
Padina pavonica	Ag	Extracellular	54	Sahayaraj et al. (2012)
Parachlorella kessleri	Ag	Extracellular	–	Kaduková et al. (2014)
Chlorella pyrenoidosa	Ag	Extracellular	5–15	Aziz et al. (2015)
Scendesmus abundans	Ag	Extracellular	59–66	Aziz et al. (2014)
Chlorococcum humicola	Ag	Intracellular	4–6	Jena et al. (2013)
Bifurcaria bifurcate	CuO	Extracellular	5–45	Abboud et al. (2014)
Amphora-46	Ag	Extracellular	5–70	Jena et al. (2015)
Gracilaria dura	Ag	Extracellular	6	Shukla et al. (2012)
Turbinaria conoides	Au	Extracellular	60	Rajeshkumar et al. (2013b)

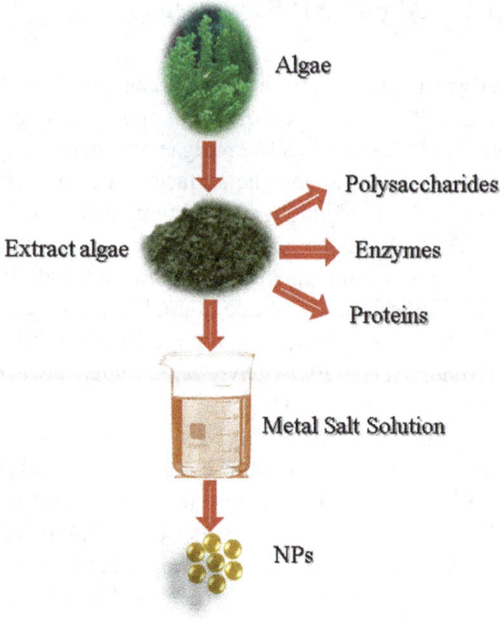

Fig. 1.5 Scheme of the mechanism of NP formation using seaweed extract

live either in a marine or in a brackish water environment. Studies have shown that these algae can synthesize Ag- and AuNPs (Fulekar and Pathak 2017).

Dahoumane et al. (2016) proposed four possible pathways for NP synthesis by microalgae. The first method comprises the use of biomolecules extracted from the rupture of microalgae cells; the second method uses supernatants produced from culture media from which the cells were removed by centrifugation or filtration; the third method uses whole microalgae cells resuspended in distilled water; and the fourth method proposes the use of live cells maintained in their normal culture. Figure 1.5 shows a possible mechanism for metal NP synthesis engineered by algae extract.

In relation to other organisms already studied as nanofactories, algae are also important in NP biosynthesis; this study is nowadays called phyconanotechnology.

1.7 Plants: Biomatrices for NP Building

Owing to the large amount of plant extracts and plant tissues currently available as sources of biomatrices for NP synthesis via the green approach, they have become one of the best NP biosynthesizers compared with other biological systems (Prasad 2014). The advantages of using plants in NP synthesis compared with other biological systems is that they support large-scale NP synthesis, do not require maintenance of cell culture (Chung et al. 2016), provide natural capping agents, possess a wider number of secondary metabolites, are low-cost, safe to handle, and easily available (Prasad 2014).

The plant systems may biosynthesize NPs of different metals intra- or extracellularly (Renugadevi and Aswini 2012; Hesgazy et al. 2015). Use various parts of plants for NP synthesis have been described, such as latex, bark, fruits, callus, seeds, and stems (Anjum et al. 2016; Prasad et al. 2011; Prasad 2014). Dubey et al. (2009) described AgNP synthesis by the extract of *Eucalyptus hybrida* (safeda) leaves. Sathishkumar et al. (2009a, b) described AgNP and palladium NP synthesis by *Cinnamomum zeylanicum* bark, whose components such as eugenol, linalool, and methyl chavicol were responsible for the synthesis. *Dioscorea bulbifera* has been described as being able to synthesize AuNPs (Ghosh et al. 2011). Song et al. (2010) reported platinum NP synthesis by the plant *Diospyrus kaki*. AgNPs were easily synthesized by Krishnaraj et al. (2010) using leaf extract of *Acalypha indica* and its production was observed within 30 min. Dwivedi and Gopal (2010) reported AgNP biosynthesis with sizes within range of 10–30 nm, from an obnoxious weed, *Chenopodium album*. Aljabali et al. (2018), described the production of AuNPs with antimicrobial activity using leaf extract of *Ziziphus zizyphus*; other examples of metallic NP synthesis by plants are described in Table 1.5.

Table 1.5 Synthesis of metal NPs by different plant extracts

Plant species	Plant parts	Metal NPs	Size (nm)	References
Acorus calamus	Rhizome	Ag	19	Nayak et al. (2015)
Allium cepa	Extract	Ag	–	Saxena et al. (2010)
Alstonia scholaris	Bark	Ag	50	Shetty et al. (2014)
Averrhoa carambola	Leaf	Ag	14	Mishra et al. (2015)
Nicotiana tobaccum	Leaf	Ag	8.43 ± 1.15	Prasad et al. (2011)
Azadirachta indica	Leaf	Ag	9.6–25.5	Bhuyan et al. (2015)
Syzygium cumini	Leaf	Ag	100–160	Prasad et al. (2012)
Santalum album	Leaf	Ag	80–200	Swamy and Prasad (2012)
Syzygium cumini	Bark	Ag	20–60	Prasad and Swamy (2013)
Carissa carandas	Berries	Ag	10–60	Joshi et al. (2018)
Avena sativa	Leaf	Au	25–85	Shankar et al. (2005)
Cucurbita maxima	Petals	Ag	19	Nayak et al. (2015)
Emblica officinalis	Leaf	Ag and Au	10–20 and 15–25	Ankamwar et al. (2005b)
Enteromorpha flexuosa	Seaweed	Ag	2–32	Yousefzadi et al. (2014)
Jatropha curcas	Latex	Ag	10–20	Bar et al. (2009)
Momordica cymbalaria	Fruit	Ag	15.5	Swamy et al. (2015)
Nelumbo nucifera	Root	Ag	16.7	Sreekanth et al. (2014)
Myrmecodia pendans	Whole plant	Ag	10–20	Zuas et al. (2014)
Onosma dichroantha	Root	Ag	5–65	Nezamdoost et al. (2014)
Piper longum	Fruit	Ag	46	Reddy et al. (2014)
Sinapis arvensis	Seed	Ag	14	Khatami et al. (2015)
Tephrosia tinctoria	Stem	Ag	73	Rajaram et al. (2015)

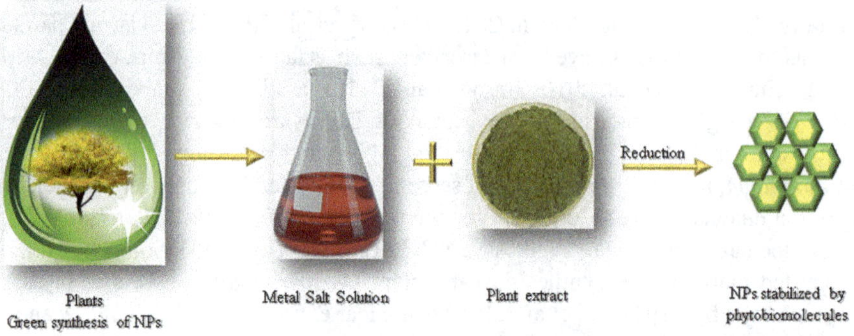

Fig. 1.6 Schematic representation of the mechanism of NP synthesis by plants

The intracellular processes of NP synthesis encompass plant growth in a hydroponic solution rich in metals, metal-rich media, metal-rich soils, among others (Haverkamp et al. 2006; Harris and Bali 2007). The extracellular NP synthesis comprises the use of leaves in water or ethanol and leaf extract; the use of the latter may be of low cost owing to easier downstream processing (Ankamwar et al. 2005a; Parashar et al. 2009).

In the NP synthesis, the plant extracts may act as stabilizing and reducing agents. Although the ability of plant extracts to reduce metals has been known since 1900, the performance of the reducing agents has not yet been fully elucidated (Mittal et al. 2013). Usually, the plant extract reduction process comprises mixing the aqueous solution of the metal salt with the extract at room temperature, and NP formation is finished within a few minutes (Mittal et al. 2013). Figure 1.6 exemplifies the green mechanism of metal NP biosynthesis through plants.

Synthesis of NPs using plant extracts is promising via nanotechnology because their special characteristics as easily scaled up, they are environmentally benign, and of low cost.

1.8 Applications of Nanomaterials Produced by the Green Approach

Over the last few years, NP synthesis has gained the attention of several scientists in the nano-universe of technology, because of its diverse applications in several industrial sectors. NP synthesis encompasses several fields of science, combining multiple scientific sectors, such as biotechnology, nanotechnology, material science, physics, and chemistry. In search for an environmentally safe nanotechnology, researchers have opted for the use of green approach processes, which involve the use of bacteria, protozoans, viruses, yeasts, filamentous fungi, algae, and plants in NP synthesis (Fig. 1.7).

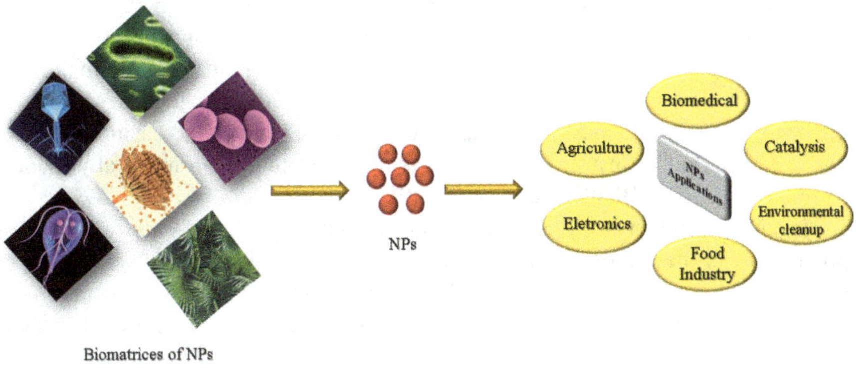

Fig. 1.7 Scheme of NP applications synthesized by the green approach

1.8.1 Biomedical Applications

Nanoparticles have many applications in the biomedical sector, such as biosensors, soaps, detergents, shampoos, toothpastes, cosmetics, and medical and therapeutic products (Banerjee et al. 2014). Biosynthesized AgNPs are employed in several health sectors such as bio-labeling, antimicrobial agents, mainly in the pharmaceutical industry, owing to resistance to antibiotics of pathogenic microorganisms, NPs being a new treatment option, in addition to anti-inflammatory activities, that can be incorporated into composite fibers. The broad spectrum activity of AgNPs make them relevant agents, not only in the fight against infections, but also in the treatment of tumors, and especially of multidrug-resistant tumor cells (Nazeruddin et al. 2014). NPs are also used in drug delivery systems, which is a very interesting area of biomedical application, as it is aimed at the safe delivery of drugs to specific sites. In recent years, the use of NPs as medical imaging devices has been quite widespread, for example, AgNPs synthesized by the fungus *Trichoderma viride*, which, when submitted for laser excitation, showed photoluminescence measurements that allowed its use for the labeling and generation of images (Sarkar et al. 2010). Nowadays, AgNPs are being used in modern medical practices for dressing various wounds (Vlachou et al. 2007).

1.8.2 Agricultural Applications

Precision agriculture is a relevant topic within the agrarian sciences, as it assists in the reduction of agricultural waste in addition to environmental pollution. Although not fully implemented, nanotechnology-enabled monitoring systems are of great importance for precision agriculture as they promote the accurate monitoring of crop growth and soil conditions, allowing for an early response to climate change (Elrahman and Mostafa 2015; Prasad et al. 2014, 2017a, b; Sangeetha et al. 2017).

The uncontrolled use of pesticides has caused serious human and animal health problems, and several adverse effects on ecosystems. However, the use of NPs and nanocapsules of pesticides has brought good results, such as increasing the effectiveness of pesticides in a smaller dose, in addition to being environmentally friendly (Sharon et al. 2010; Bhattacharyya et al. 2016).

Another important area of agriculture in which the use of nanomaterials has great potential is in the use of fertilizers, owing to a characteristic of NPs, which is the slow release of the same ones, the NPs retaining more material from the plants because of the higher surface tension of NPs compared with conventional surfaces (Duhan et al. 2017).

1.8.3 Applications in Electronics

Interest in electronics using NPs has grown considerably in recent years, mainly caused by the use of printed electronics containing functional paints with nanoparticulate materials, such as organic electronic molecules and metallic NPs (Kosmala et al. 2011). The main characteristics of NPs that make them attractive for incorporation into electronic and electrical equipment are their reversible assembly and easy manipulation (O'Brien et al. 2001).

1.8.4 Food Industry Applications

Nanotechnology in food processing has been a technological innovation in the food industry through the use of nanocarriers as food delivery systems that do not alter the basic morphology of food (Ezhilarasi et al. 2013). The use of nanotechnology in the food industry offers advantages such as reduced food disposal due to microbial infestation and increases the shelf life of various types of food (Pradhan et al. 2015; Prasad et al. 2017c). The use of nanostructured materials in food ingredients has been studied to provide better consistency, flavor, and texture (Singh et al. 2017).

The nanotechnology employed in food packaging based on nano-packaging has advantages, from antimicrobial films to nano-sensing for pathogen detection and mechanical strength, alerting consumers to the safety status of food and barrier properties (Mihindukulasuriya and Lim 2014).

1.8.5 Applications in Environmental Cleanup

Nanoparticles have shown great potential for environmental cleanup with respect to the treatment of soil and sediment contaminated with heavy metals, organic and inorganic solutes, surface waters, wastewater, and groundwater (Caliman et al. 2010).

Ensuring good water quality is a major concern at the global research level; in this sense, nanotechnology has been enhanced in water purification processes such as bioactive NPs, nanostructured catalytic membranes, nanocatalysts, NP-enhanced filtration, and nanosorbents (Khan et al. 2014).

The use of nanotechnology in bioremediation processes has shown promise, as described by Salvadori et al. (2017); through the use of the dead biomass of fungi the concomitant process of waste removal and the synthesis of NPs can be carried out. The contamination of aquatic and terrestrial ecosystems by pesticides is a serious environmental problem and several processes involving nanotechnology are being successfully used to minimize environmental impacts, such as the use of AuNPs produced by the *Rhizopus oryzae* fungus, which is able to adsorb organophosphorus pesticides (Das et al. 2009).

1.8.6 Applications in Catalysis

Nanocatalysts are of great importance as industrial catalysts, because the metallic NPs with large surface-area-to-volume ratios are efficient catalysts that facilitate the catalytic industrial processes occurring on the surface of the metals. The nanocatalysts have various advantages such as cost effectiveness, higher activity, durability or recycling potential, greater stability, and efficient recovery characteristics (Glaser 2012).

1.9 Conclusions and Future Prospects

In this chapter, the great potential of biomatrices in NP processing was exposed as a promising pathway in nanotechnology whose application extends across several industrial and environmental sectors.

The use of the green approach in nanotechnology has been a great attraction in view of the wide range of advantages over chemical and physical processes, such as low operational costs, environmental safety, easy handling, and easy scale-up. The use of microorganisms in addition to plants in a nanofactory initiates the milestone of a new era in nanoscience.

The nanomaterials generated by the emerging nanobiotechnology using the biomass of bacteria, protozoans, viruses, yeasts, fungi, algae, and plants belong to the chemical families of oxides, metals, chalcogenides, and carbonates. The nanomaterials are produced from precursors, for example, metals, and the whole organism, which may be alive or dead, or its extract, adjusting the reactional parameters such as temperature, reaction time, pH, concentration of the reagent and the biomatrix, controlling the shape and the size of the nanomaterials according to the various fields to be used.

However, there is still a gap and at the same time an undisputed challenge for researchers, to stagger the process from bench to industrial level for some of these biomaterials to make bioprocesses effectively competitive in the trade.

To achieve all these goals with predictability and effectiveness, the researchers should always take into consideration combined screening for the correct biomatrices such as bacteria, protozoans, viruses, yeasts, fungi, algae, plants, and other biomolecules, the optimal reaction parameters, and proper methodologies for the utilization/extraction of the biomatrices.

The vast body of literature found in recent years, reporting the use of biomatrices in NP construction, in addition to the great global interest in this new technology, shows the accelerated and promising evolution of nanobiotechnology in the near future.

References

Abboud Y, Saffaj T, Chagraoui A, Bouari AE, Brouzi K, Tanane O et al (2014) Biosynthesis, characterization and antimicrobial activity of copper oxide nanoparticles (CONPs) produced using brown alga extract (*Bifurcaria bifurcata*). Appl Nanosci 4:571–576

Abdel-Aziz SM, Prasad R, Hamed AA, Abdelraof M (2018) Fungal nanoparticles: A novel tool for a green biotechnology? In: Fungal Nanobionics: Principles and Applications (eds. Prasad R, Kumar V, Kumar M and Wang S), Springer Singapore Pte Ltd. 61–87

Abdel-Raouf N, Al-Enazi NM, Ibraheem IBM (2013) Green biosynthesis of gold nanoparticles using *Galaxaura elongata* and characterization of their antibacterial activity. Arab J Chem 10:3029–3039

Abdel-Raouf N, Al-Enazi NM, Ibraheem IBM, Alharbi RM, Alkhulaifi MM (2018) Biosynthesis of silver nanoparticles by using of the marine brown alga *Padina pavonia* and their characterization. Saudi J Biol Sci:1–9. https://doi.org/10.1016/j.sjbs.2018.01.007

Afreen RV, Ranganath E (2011) Synthesis of monodispersed silver nanoparticles by *Rhizopus stolonifer* and its antibacterial activity against MDR strains of *Pseudomonas aeruginosa* from burnt patients. Int J Environ Sci 1:1582–1592

Agnihotri M, Joshi S, Kumar AR, Zinjarde S, Kulkarni S (2009) Biosynthesis of gold nanoparticles by the tropical marine yeast *Yarrowia lipolytica* NCIM 3589. Mater Lett 63:1231–1234

Ahmad A, Senapati S, Khan MI, Sastry M (2003) Intracellular synthesis of gold nanoparticles by a novel alkalotolerant actinomycete, *Rhodococcus* species. Nanotechnology 14:824–828

Ahmad N, Sharma S, Singh VN, Shamsi SF, Fatma A, Mehta BR (2011) Biosynthesis of silver nanoparticles from *Desmodium triflorum*: a novel approach towards weed utilization. Biotechnol Res Int:1–8. https://doi.org/10.4061/2011/454090

Ahmad T, Wani IA, Manzoor N, Ahmed J, Asiri AM (2013) Biosynthesis, structural characterization and antimicrobial activity of gold and silver nanoparticles. Colloids Surf B Biointerfaces 107:227–234

Ahmed S, Ahmad M, Swami BL, Ikram S (2016) A review on plants extract mediated synthesis of silver nanoparticles for antimicrobial applications: a green expertise. J Adv Res 7:17–28

Alghuthaymi MA, Almoammar H, Rai M, Said-Galiev E, Abd-Elsalam KA (2015) Myconanoparticles: synthesis and their role in phytopathogens management. Biotechnol Equip 29:221–236

Aljabali AAA, Akkam Y, Al Zoubi MS, Al-Batayneh KM, Al Trad B, Alrob OA et al (2018) Synthesis of gold nanoparticles using leaf extract of *Ziziphus zizyphus* and their antimicrobial activity. Nanomaterials 8:1–17

Anand P, Isar J, Saran S, Saxena RK (2006) Bioaccumulation of copper by *Trichoderma viride*. Bioresour Technol 97:1018–1025

Anastas PT, Kirchhoff MM (2002) Origins, current status, and future challenges of green chemistry. Acc Chem Res 35:686–694

Anjum S, Abbasi BH, Shinwari ZK (2016) Plant-mediated green synthesis of silver nanoparticles for biomedical applications: challenges and opportunities. Pak J Bot 48:1731–1760

Ankamwar B, Chaudhary M, Sastry M (2005a) Gold nanotriangles biologically synthesized using tamarind leaf extract and potential application in vapor sensing. Synth React Inorg Metal-Org NanoMet Chem 35:19–26

Ankamwar B, Damle C, Ahmad A, Sastry M (2005b) Biosynthesis of gold and silver nanoparticles using *Emblica officinalis* fruit extract, their phase transfer and transmetallation in an organic solution. J Nanosci Nanotechnol 5:1665–1671

Attia YA, Farag YE, Mohamed YMA, Hussien AT, Youssef T (2016) Photo-extracellular synthesis of gold nanoparticles using Baker's yeast and their anticancer evaluation against Ehrlich ascites carcinoma cells. New J Chem 40:9395–9402

Azandehi PK, Moghaddam J (2015) Green synthesis, characterization and physiological stability of gold nanoparticles from *Stachys lavandulifolia* Vahl extract. Particuology 19:22–26

Aziz N, Fatma T, Varma A, Prasad R (2014) Biogenic synthesis of silver nanoparticles using *Scenedesmus abundans* and evaluation of their antibacterial activity. J Nanoparticle:689419. https://doi.org/10.1155/2014/689419

Aziz N, Faraz M, Pandey R, Sakir M, Fatma T, Varma A, Barman I, Prasad R (2015) Facile algae-derived route to biogenic silver nanoparticles: synthesis, antibacterial and photocatalytic properties. Langmuir 31:11605–11612. https://doi.org/10.1021/acs.langmuir.5b03081

Aziz N, Pandey R, Barman I, Prasad R (2016) Leveraging the attributes of *Mucor hiemalis*-derived silver nanoparticles for a synergistic broad-spectrum antimicrobial platform. Front Microbiol 7:1984. https://doi.org/10.3389/fmicb.2016.01984

Aziz N, Faraz M, Sherwani MA, Fatma T, Prasad R (2019) Illuminating the anticancerous efficacy of a new fungal chassis for silver nanoparticle synthesis. Front Chem 7:65. https://doi.org/10.3389/fchem.2019.00065

Baesman SM, Bullen TD, Dewald J, Zhang D, Curran S, Islam FS et al (2007) Formation of tellurium nanocrystals during anaerobic growth of bacteria that use Te oxyanions as respiratory electron acceptors. Appl Environ Microbiol 73:2135–2145

Balasooriya ER, Jayasinghe CD, Jayawardena UA, Ruwanthika RWD, Silva RM, Udagama PV (2017) Honey mediated green synthesis of nanoparticles: new era of safe nanotechnology. J Nanomater 1:1–10

Balci S, Bittner AM, Hahn K, Scheu C, Knez M, Kadri A et al (2006) Copper nanowires within the central channel of tobacco mosaic virus particles. Eletrochim Acta 51:6251–6357

Banerjee P, Satapathy M, Mukhopahayay A, Das P (2014) Leaf extract mediated green synthesis of silver nanoparticles from widely available Indian plants: synthesis, characterization, antimicrobial property and toxicity analysis. Bioresour Bioprocess 1:1–10

Bansal P, Duhan JS, Gahlawat SK (2014) Biogenesis of nanoparticles: a review. Afr J Biotechnol 13:2778–2785

Bar H, Bhui DK, Sahoo GP, Sarkar P, De SP, Misra A (2009) Green synthesis of silver nanoparticles using latex of *Jatropha curcas*. Colloids Surf A Physicochem Eng Asp 339:134–139

Barcikowski S, Devesa F, Moldenhauer K (2009) Impact and structure of literature on nanoparticle generation by laser ablation in liquids. J Nanopart Res 11:1883–1893

Baron S (1996) Medical microbiology. University of Texas Medical Branch at Galveston, Galveston, TX

Barud HS, Barrios C, Regiani T, Marques RFC, Verelst M, Dexpert-Ghys J et al (2008) Self-supported silver nanoparticles containing bacterial cellulose membrane. Mater Sci Eng C 28:515–518

Bell J, Chen Z, Olofinjana A (2001) Synthesis of amorphous carbon nitride using reactive ion beam sputtering deposition with grazing bombardment. Diam Relat Mater 10:2184–2189

Beveridge TJ, Murray RGE (1980) Sites of metal deposition in the cell wall of *Bacillus subtilis*. J Bacteriol 141:876–887

Bharde A, Wani A, Shouche Y, Joy PA, Prasad BLV, Sastry M (2005) Bacterial aerobic synthesis of nanocrystalline magnetite. J Am Chem Soc 127:9326–9327

Bhattacharya D, Singh S, Satnalika N (2009) Nanotechnology, big things from a tiny world: a review. Int J Sci Technol 2:29–38

Bhattacharyya A, Duraisamy P, Govindarajan M, Buhroo AA, Prasad R (2016) Nano-biofungicides: Emerging trend in insect pest control. In: Advances and Applications through Fungal Nanobiotechnology (ed. Prasad R), Springer International Publishing Switzerland 307–319

Bhuyan T, Mishra K, Khanuja M, Prasad R, Varma A (2015) Biosynthesis of zinc oxide nanoparticles from *Azadirachta indica* for antibacterial and photocatalytic applications. Mater Sci Semicond Process 32:55–61

Birla SS, Tiwari VV, Gade AK, Ingle AP, Yadav AP, Rai MK (2009) Fabrication of silver nanoparticles by *Phoma glomerata* and its combined effect against *Escherichia coli*, *Pseudomonas aeruginosa* and *Staphylococcus aureus*. Lett Appl Microbiol 48:173–179

Boroumand MA, Namvar F, Moniri M, Tahir MP, Azizi S, Mohamad R (2015) Nanoparticles biosynthesized by fungi and yeast: a review of their preparation, properties, and medical applications. Molecules 20:16540–16565

Bromley KM, Patil AJ, Perriman AW, Stubbs G, Mann S (2008) Preparation of high quality nanowires by tobacco mosaic virus templating of gold nanoparticles. J Mater Chem 18:4796–4801

Caliman FA, Robu BM, Smaranda C, Pavel VL, Gavrilescu M (2010) Soil and groundwater cleanup: benefits and limits of emerging technologies: a review. Clean Techn Environ Policy 13:241–268

Chung IM, Park I, Hyun KS, Thiruvengadam M, Rajakumar G (2016) Plant-mediated synthesis of silver nanoparticles: their characteristic properties and therapeutic applications. Nanoscale Res Lett 11:1–14

Cui YH, Lili L, Zhou NQ, Liu JH, Wang HJ, Tian J et al (2016) *In vivo* synthesis of nano-selenium by *Tetrahymena thermophila* SB210. Enzym Microb Technol 95:185–191

Dahoumane SA, Mechouet M, Alvarez FJ, Agathos SN, Jeffryes C (2016) Microalgae: an outstanding tool in nanotechnology. Bionatura 1:196–201

Dahoumane SA, Jeffryes C, Mechouet M, Agathos SN (2017a) Biosynthesis of inorganic nanoparticles: a fresh look at the control os shape, size and composition. Bioengineering 4:1–16

Dahoumane SA, Mechouet M, Wijesekera K, Filipe CDM, Sicard C, Bazylinski DA et al (2017b) Algae-mediated biosynthesis of inorganic nanomaterials as a promising route in nanobiotechnology – a review. Green Chem 19:552–587

Dameron CT, Reese RN, Mehra RK, Kortan AR, Carroll PJ, Steigerwald ML et al (1989) Biosynthesis of cadmium sulphide quantum semiconductor crystallites. Nature 338:596–597

Darbandi M, Thomann R, Nann T (2005) Single quantum dots in silica spheres by microemulsion synthesis. Chem Mater 17:5720–5725

Das SK, Das AR, Guha AK (2009) Gold nanoparticles: microbial synthesis and application in water hygiene management. Langmuir 25:8192–8199

Dhas TS, Kumar VG, Karthick V, Angel KJ, Govindaraju K (2014) Facile synthesis of silver chloride nanoparticles using marine alga and its antibacterial efficacy. Spectrochim Acta A 120:416–420

Douglas T, Young M (1998) Host-guest encapsulation of materials by assembled virus protein cages. Nature 393:152–155

Dubey M, Bhadauria S, Kushwah B (2009) Green synthesis of nanosilver particles from extract of *Eucalyptus hybrida* (safeda) leaf. Dig J Nanomater Biostruct 4:537–543

Duhan JS, Kumar R, Kumar N, Kaur P, Nehra K, Duhan S (2017) Nanotechnology: the new perspective in precision agriculture. Biotechnol Rep 15:11–23

Dujardin E, Peet C, Stubbs G, Culver JN, Mann S (2003) Organisation of metallic nanoparticles using Tobacco mosaic virus. Nano Lett 3:413–417

Dwivedi AD, Gopal K (2010) Biosynthesis of silver and gold nanoparticles using *Chenopodium album* leaf extract. Physicochem Eng Aspects 369:27–33

Elrahman SHA, Mostafa MAM (2015) Applications of nanotechnology in agriculture: an overview. Egypt J Soil Sci 55:1–19

Evans P, Matsunaga H, Kiguchi M (2008) Large-scale application of nanotechnology for wood protection. Nat Nanotechnol 3:577

Ezhilarasi PN, Karthik P, Chhanwal N, Anandharamakrishnan C (2013) Nanoencapsulation techniques for food bioactive components: a review. Food Bioprocess Technol 6:628–647

Fulekar MH, Pathak B (2017) Environmental nanotechnology. CRC Press, New York

Gade A, Bonde PP, Ingle AP, Marcato P, Duran N, Rai MK (2008) Exploitation of Aspergillus niger for synthesis of silver nanoparticles. J Biobased Mater Bioenergy 2:1–5

Gericke M, Pinches A (2006) Biological synthesis of metal nanoparticles. Hydrometallurgy 83:132–140

Ghosh S, Patil S, Ahire M, Kitture R, Jabgunde A, Kale S et al (2011) Synthesis of gold nanoanisotrops using *Dioscorea bulbifera* tuber extract. J Nanomater 8:1–8

Glaser JA (2012) Green chemistry with nanocatalysts. Clean Techn Environ Policy 14:513–520

Gopinath K, Karthika V, Sundaravadivelan C, Gowri S, Arumugam A (2015) Mycogenesis of cerium oxide nanoparticles using *Aspergillus niger* culture filtrate and their applications for antibacterial and larvicidal activities. J Nanostruct Chem 5:295–303

Govindaraju K, Kiruthiga V, Ganesh Kumar V, Singaravelu G (2009) Extracellular synthesis of silver nanoparticles by a marine alga, *Sargassum wightii* Grevilli and their antibacterial effects. J Nanosci Nanotechnol 9:5497–5501

Gow NAR, Gadd GM (1994) The growing fungus. Chapman & Hall, London

Guo Z, Liang X, Pereira T, Scaffaro R, Hahn HT (2007) CuO nanoparticle filled vinyl-ester resin nanocomposites: fabrication, characterization and property analysis. Compos Sci Technol 67:2036–2044

Harris AT, Bali R (2007) On the formation and extent of uptake of silver nanoparticles by live plants. J Nanopart Res 10:691–695

Haverkamp RG, Marshall AT, van Agterveld D (2006) Pick your carats: nanoparticles of gold–silver–copper alloy produced *in vivo*. J Nanopart Res 9:697–700

Hesgazy HS, Lamis D, Shabaan GH, Rabie GH, Diana SR (2015) Biosynthesis of silver nanoparticles using cell free callus exudates of *Medicago sativa* L. Pak J Bot 47:1825–1829

Hou L, Tong D, Jiang Y, Gao F (2014) Synthesis and organization of platinum nanoparticles and nanoshells on a native virus bioscaffold. Nano Brief Rep Rev 9:1–8

Husseiney MI, El-Aziz MA, Badr Y, Mahmoud MA (2007) Biosynthesis of gold nanoparticles using *Pseudomonas aeruginosa*. Spectrochim Acta A 67:1003–1006

Itohara D, Shinohara K, Yoshida T, Fujita Y (2016) p-channel and n-channel thin-film-transistor operation on sprayed ZnO nanoparticle layers. J Nanomater:1–6 https://doi.org/10.1155/2016/8219326

Jaidev LR, Narasimha G (2010) Fungal mediated biosynthesis of silver nanoparticles, characterization and antimicrobial activity. Colloids Surf B Biointerfaces 81:430–433

Jena J, Pradhan N, Dash BP, Sukla LB, Panda PK (2013) Biosynthesis and characterization of silver nanoparticles using microalga *Chlorococcum humicola* and its antibacterial activity. Int J Nanomater Bios 3:1–8

Jena J, Pradhan N, Dash BP, Panda PK, Mishra BK (2015) Pigment mediated biogenic synthesis of silver nanoparticles using diatom *Amphora* sp. and its antimicrobial activity. J Saudi Chem Soc 19:661–666

Jha AK, Prasad K, Prasad K (2009) A green low-cost biosynthesis of Sb_2O_3 nanoparticles. Biochem Eng J 43:303–306

Joshi N, Jain N, Pathak A, Singh J, Prasad R, Upadhyaya CP (2018) Biosynthesis of silver nanoparticles using *Carissa carandas* berries and its potential antibacterial activities. J Sol-Gel Sci Techn 86(3):682–689. https://doi.org/10.1007/s10971-018-4666-2

Juganson K, Mortimer M, Ivask A, Kasemets K, Kahru A (2013) Extracellular conversion of silver ions into silver nanoparticles by protozoan *Tetrahymena thermophila*. Environ Sci 15:244–250

Kaduková J, Velgosová O, Mrazíková A, Marcincákivá R (2014) The effect of culture age and initial silver concentration on biosynthesis of Ag nanoparticles. N Biotechnol Chim 13:28–37

Kalishwaralal K, Deepak V, Ram S, Pandian K, Muniasamy K, Kanth SBM et al (2010) Biosynthesis of silver and gold nanoparticles using *Brevibacterium casei*. Colloids Surf B Biointerfaces 77:257–262

Kar PK, Murmu S, Saha S, Tandon V, Acharya K (2014) Anthelmintic efficacy of gold nanoparticles derived from a phytopathogenic fungus *Nigrospora oryzae*. PLoS ONE 9(1):e84693. https://doi.org/10.1371/journal.pone.0084693

Kathiraven T, Sundaramanickam A, Shanmugan N, Balasubramanian T (2015) Green synthesis of silver nanoparticles using marine algae *Caulerpa racemosa* and their antibacterial activity against some human pathogens. Appl Nanosci 5:499–504

Kaviya SSJ, Viswanathan B (2011) Green synthesis of silver nanoparticles using *Polyalthia longifolia* leaf extract along with D-sorbitol. J Nanotech:1–5. https://doi.org/10.1155/2011/152970

Khalil KA, Fouad H, Elsarnagawy T, Almajhdi FN (2013) Preparation and characterization of electrospun PLGA/silver composite nanofibers for biomedical applications. Int J Electrochem Sci 8:3483–3493

Khan I, Farhan M, Pratichi S, Thiagarajan P (2014) Nanotechnology for environmental remediation. Res J Pharm Biol Chem Sci 5:1916–1927

Khatami M, Pourseyedi S, Khatami M, Hamidi H, Zaeifi M, Soltani L (2015) Synthesis of silver nanoparticles using seed exudates of *Sinapis arvensis* as a novel bioresource, and evaluation of their antifungal activity. Bioresour Bioprocess 2:19

Klaus T, Joerger R, Olsson E, Granqvist CG (1999) Silver-based crystalline nanoparticles, microbially fabricated. Proc Natl Acad Sci U S A 96:13611–13614

Klaus-Joerger T, Joerger R, Olsson E, Granqvist CG (2001) Bacteria as workers in the living factory: metal-accumulating bacteria and their potential for materials science. Trends Biotechnol 19:15–20

Knez M, Sumser M, Bittner AM, Wege C, Jeske H, Kooi S et al (2002) Electrochemical modification of individual nano-objects. J Electroanal Chem 522:70–74

Knez M, Bittner AM, Boes F, Wege C, Jeske H, Maiβ E et al (2003) Biotemplate synthesis of 3-nm nickel and cobalt nanowires. Nano Lett 3:1079–1082

Knez M, Sumser M, Bittner AM, Wege C, Jeske H, Martin TP et al (2004) Spatially selective nucleation of metal clusters on the tobacco mosaic virus. Adv Funct Mater 14:116–124

Knez M, Kadri A, Wege C, Gosele U, Jeske H, Nielsch K (2006) Atomic layer deposition on biological macromolecules: metal oxide coating of tobacco mosaic virus and ferritin. Nano Lett 6:1172–1177

Konishi Y, Tsukiyama T, Tachimi T, Saitoh N, Nomura T, Nagamine S (2007) Microbial deposition of gold nanoparticles by the metal-reducing bacterium *Shewanella algae*. Electrochim Acta 53:186–192

Kosmala A, Wright R, Zhang Q, Kirby P (2011) Synthesis of silver nano particles and fabrication of aqueous Ag inks for inkjet printing. Mater Chem Phys 129:1075–1080

Kowshik M, Deshmukh N, Vogel W, Urban J, Kulkarni SK, Paknikar KM (2002) Microbial synthesis of semiconductor CdS nanoparticles, their characterization, and their use in the fabrication of an ideal diode. Biotechnol Bioeng 78:583–588

Kowshik M, Ashtaputre S, Kulkani SK, Parknikar KMM (2003) Extracellular synthesis of silver nanoparticles by a silver-tolerant yeast strain MKY3. Nanotechnology 14:95–100

Krishnaraj C, Jagan EG, Rajasekar S (2010) Synthesis of silver nanoparticles using *Acalypha indica* leaf extracts and its antibacterial activity against water borne pathogens. Colloids Surf B Biointerfaces 76:50–56

Kumar M, Ando Y (2010) Chemical vapor deposition of carbon nanotubes: a review on growth mechanism and mass production. J Nanosci Nanotechnol 10:3739–3758

Kumar D, Karthik L, Kumar G, Roa KB (2011) Biosynthesis of silver nanoparticles from marine yeast and their antimicrobial activity against multidrug resistant pathogens. Pharmacologyonline 3:1100–1111

Lachance MA (2016) Paraphyly and (yeast) classification. Int J Syst Evol Microbiol 66:4924–4929

Lacour F, Guillois O, Portier X, Perez H, Herlin N, Reynaud C (2007) Laser pyrolysis synthesis and characterization of luminescent silicon nanocrystals. Physica E: Low Dimens Syst Nanostruct 38:11–15

Lee LA, Niu Z, Wang Q (2009) Viruses and virus-like protein assemblies chemically programmable nanoscale building blocks. Nano Res 2:349–364

Lengke M, Ravel B, Fleet ME, Wanger G, Gordon RA, Southam G (2006) Mechanism of gold bioaccumulation by filamentous cyanobacteria form gold (III) – chloride complex. Environ Sci Technol 40:6304–6309

Li YL, Kinloch IA, Windle AH (2004) Direct spinning of carbon nanotube fibers from chemical vapor deposition synthesis. Science 304:276–278

Li Y, Liang J, Tao Z, Chen J (2007) CuO particles and plates: synthesis and gas-sensor application. Mater Res Bull 43:2380–2385

Li X, Chen S, Hu W, Shi S, Shen W, Zhang X et al (2009) In situ synthesis of CDS nanoparticles on bacterial cellulose nanofibers. Carbohydr Polym 76:509–512

Lim HA, Mishra A, Yun SI (2011) Effect of pH on the extra cellular synthesis of gold and silver nanoparticles by *Saccharomyces cerevisae*. J Nanosci Nanotechnol 11:518–522

Lirdprapamongkol K, Warisnoicharoen W, Soisuwas S, Svati J (2014) Eco-friendly synthesis of fucoidan-stabilized gold nanoparticles. Am J Appl Sci 7:1038–1104

Lodish H, Berk A, Zipursky SL, Matsudaira P, Baltimore D, Darnell J (2000) Molecular cell biology. W. H. Freeman and Company, New York

Lohani A, Verma A, Joshi H, Yadav N, Karki N (2014) Nanotechnology-based cosmeceuticals. ISRN Dermatol:1–14. https://doi.org/10.1155/2014/843687

Love AJ, Makarov V, Yaminsky I, Kalinina NO, Taliansky ME (2014) The use of tobacco mosaic virus and cowpea mosaic virus for the production of novel metal nanomaterials. Virology 449:133–139

Mädler L, Kammler HK, Mueller R, Pratsinis SE (2002) Controlled synthesis of nanostructured particles by flame spray pyrolysis. J Aerosol Sci 33:369–389

Mann S, Frankel RB, Blakemore RP (1984) Structure, morphology, and crystal growth of bacterial magnetite. Nature 310:405–407

Mao C, Flynn CE, Hayhurst A, Sweeney R, Qi J, Georgiou G et al (2003) Viral assembly of oriented quantum dot nanowires. Proc Natl Acad Sci U S A 100:6946–6951

Mao C, Solis DJ, Reiss BD, Kottmann ST, Sweeney RY, Hayhurst A et al (2004) Virus-based toolkit for the directed synthesis of magnetic and semiconducting nanowires. Science 303:213–217

Marshall M, Beliaev A, Dohnalkova A, David W, Shi L, Wang Z (2007) C-type cytochrome-dependent formation of U(IV) nanoparticles by *Shewanella oneidensis*. PLoS Biol 4:1324–1333

Mewada A, Oza G, Pandey S, Sharon M (2012) Extracellular synthesis of gold using *Pseudomonas denitrificans* and comprehending its stability, vol 2, pp 493–499

Mihindukulasuriya SDF, Lim LT (2014) Nanotechnology development in food packaging: a review. Trends Food Sci Technol 40:149–167

Mishra A, Tripathy SK, Yun SI (2011) Bio-synthesis of gold and silver nanoparticles from *Candida guilliermondii* and their antimicrobial effect against pathogenic bacteria. J Nanosci Nanotechnol 11:243–248

Mishra PM, Sahoo SK, Naik GK, Parida N (2015) Biomimetic synthesis, characterization and mechanism of formation of stable silver nanoparticles using *Averrhoa carambola* L. leaf extract. Mater Lett 160:566–571

Mittal AK, Chisti Y, Banerjee UC (2013) Synthesis of metallic nanoparticles using plant extracts. Biotechnol Adv 31:346–356

Mukherjee P, Ahmad A, Mandal D, Senapati S, Sainkar SR, Khan MI et al (2001) Fungus-mediated synthesis of silver nanoparticles and their immobilization in the mycelial matrix: a novel biological approach to nanoparticle synthesis. Nano Lett 1:515–519

Murphy CJ (2002) Materials science: nanocubes and nanoboxes. Science 298:2139–2141

Nair B, Pradeep T (2002) Coalescence of nanoclusters and formation of submicron crystallites assisted by *Lactobacillus strains*. Cryst Growth Design 2:293–298

Namasivayam SKR, Avimanyu B (2011) Silver nanoparticle synthesis from *Lecanicillium lecanii* and evolutionary treatment on cotton fabrics by measuring their improved antibacterial activity with antibiotics against *Staphylococcus aureus* (ATCC 29213) and *E. coli* (ATCC 25922) strains. Int J Pharm Pharm Sci 3:190–195

Namba K, Stubbs G (1986) Structure of tobacco mosaic virus at 3.6 a resolution: implications for assembly. Science 231:1401–1406

Narayanan KB, Sakthivel N (2010) Biological synthesis of metal nanoparticles by microbes. Adv Colloid Interf Sci 156:1–13

Nayak D, Pradhan S, Ashe S, Rauta PR, Nayak B (2015) Biologically synthesized silver nanoparticles from three diverse family of plant extracts and their anticancer activity against epidermoid A431 carcinoma. J Colloid Interface Sci 457:329–338

Nazeruddin G, Prasad N, Waghmare S, Garadkar K, Mulla I (2014) Extracellular biosynthesis of silver nanoparticle using *Azadirachta indica* leaf extract and its anti-microbial activity. J Alloys Compd 583:272–277

Nealson KH, Scott J (2006) Ecophysiology of the genus *Shewanella* prokaryotes. Appl Environ Microbiol 6:1133–1151

Nezamdoost T, Bagherieh-Najjar MB, Aghdasi M (2014) Biogenic synthesis of stable bioactive silver chloride nanoparticles using *Onosma dichroantha* Boiss. Root extract. Mater Lett 137:225–228

Nissinen T, Ikonen T, Lama M, Riikonen J, Lehto VP (2016) Improved production efficiency of mesoporous silicon nanoparticles by pulsed electrochemical etching. Powder Technol 288:360–365

O'Brien S, Brus L, Murray CB (2001) Synthesis of monodisperse nanoparticles of barium titanite: toward a generalized strategy of oxide nanoparticle synthesis. J Am Chem Soc 123:12085–12086

Omajali JB, Mikheenko IP, Merroun ML, Wood J, Macaskie LE (2015) Characterization of intracellular palladium nanoparticles synthesized by *Desulfovibrio desulfuricans* and *Bacillus benzeovorans*. J Nanopart Res 264:1–17

Oscar L, Bakkiyaraj D, Nithya C, Thajuddin N (2014) Deciphering the diversity of microalgal bloom in wastewater – an attempt to construct potential consortia for bioremediation. JCPAM 3:92–96

Oscar FL, Vismaya S, Arunkumar M, Thajuddin N, Dhanasekaran D, Nithya C (2016) Algal nanoparticles: synthesis and biotechnological potentials. INTECH 7:157–182

Otari SV, Patil RM, Ghosh SJ, Thorat ND, Pawar SH (2015) Intracellular synthesis of silver nanoparticle by actinobacteria and its antimicrobial activity. Spectrochim Acta A 136:1175–1180

Pantidos N, Horsfall LE (2014) Biological synthesis of metallic nanoparticles by bacteria, fungi and plants. J Nanomed Nanotechnol 5:1–10

Parashar V, Parashar R, Sharma B, Pandey AC (2009) *Parthenium* leaf extract mediated synthesis of silver nanoparticles: a novel approach towards weed utilization. Dig J Nanomater Biostruct 4:45–50

Pei H, Zhu S, Yang M, Kong R, Zheng Y, Qu F (2015) Graphene oxide quantum dots@silver core-shell nanocrystals as turn-on fluorescent nanoprobe for ultrasensitive detection of prostate specific antigen. Biosens Bioelectron 74:909–914

Philipse AP, Maas D (2002) Magnetic colloids from magnetotactic bacteria: chain formation and colloidal stability. Langmuir 18:9977–9984

Pokorski JK, Steinmetz NF (2011) The art of engineering viral nanoparticles. Mol Pharm 8:29–43

Pradhan N, Singh S, Ojha N, Srivastava A, Barla A, Rai V et al (2015) Facets of nanotechnology as seen in food processing, packaging, and preservation industry. Biomed Res Int:1–17. https://doi.org/10.1155/2015/365672

Prakash A, Seema S, Ahmad N, Ghosh A, Sinha P (2010) Bacterial mediated extracellular synthesis of metallic nanoparticles. Int Res J Biotechnol 1:71–79

Prasad R (2014) Synthesis of silver nanoparticles in photosynthetic plants. J Nanoparticle:963961. https://doi.org/10.1155/2014/963961

Prasad R (2016) Advances and applications through fungal nanobiotechnology. Springer, Cham. ISBN:978-3-319-42989-2

Prasad R (2017) Fungal nanotechnology: applications in agriculture, industry, and medicine. Springer, Cham. ISBN:978-3-319-68423-9

Prasad K, Jha AK (2010) Biosynthesis of CdS nanoparticles: an improved green and rapid procedure. J Colloid Interface Sci 342:68–72

Prasad R, Swamy VS (2013) Antibacterial activity of silver nanoparticles synthesized by bark extract of *Syzygium cumini*. J Nanoparticle. https://doi.org/10.1155/2013/431218

Prasad K, Jha AK, Kulkarni AR (2007) *Lactobacillus* assisted synthesis of titanium nanoparticles. Nanoscale Res Lett 2:248–250

Prasad KS, Pathak D, Patel A, Dalwadi P, Prasad R, Patel P, Kaliaperumal SK (2011) Biogenic synthesis of silver nanoparticles using *Nicotiana tobaccum* leaf extract and study of their antibacterial effect. Afr J Biotechnol 9(54):8122–8130

Prasad R, Swamy VS, Varma A (2012) Biogenic synthesis of silver nanoparticles from the leaf extract of *Syzygium cumini* (L.) and its antibacterial activity. Int J Pharm Bio Sci 3(4):745–752

Prasad R, Kumar V, Prasad KS (2014) Nanotechnology in sustainable agriculture: present concerns and future aspects. Afr J Biotechnol 13(6):705–713

Prasad R, Pandey R, Barman I (2016) Engineering tailored nanoparticles with microbes: quo vadis. WIREs Nanomed Nanobiotechnol 8:316–330. https://doi.org/10.1002/wnan.1363

Prasad R, Bhattacharyya A, Nguyen QD (2017a) Nanotechnology in sustainable agriculture: Recent developments, challenges, and perspectives. Front Microbiol 8:1014. https://doi.org/10.3389/fmicb.2017.01014

Prasad R, Kumar M, Kumar V (2017b) Nanotechnology: An Agriculture paradigm. Springer Nature Singapore Pte Ltd. (ISBN: 978-981-10-4573-8)

Prasad R, Kumar V, Kumar M (2017c) Nanotechnology: Food and Environmental Paradigm. Springer Nature Singapore Pte Ltd. (ISBN 978-981-10-4678-0)

Prasad R, Jha A, Prasad K (2018a) Exploring the Realms of Nature for Nanosynthesis. Springer International Publishing (ISBN 978-3-319-99570-0). https://www.springer.com/978-3-319-99570-0

Prasad R, Kumar V, Kumar M, Shanquan W (2018b) Fungal nanobionics: principles and applications. Springer, Singapore. ISBN: 978-981-10-8666-3. https://www.springer.com/gb/book/9789811086656

Raheman F, Deshmukh S, Ingle A, Gade A, Rai M (2011) Silver nanoparticles: novel antimicrobial agent synthesized from an endophytic fungus *Pestalotia* sp. isolated from leaves of *Syzygium cumini* (L). Nano Biomed Eng 3:174–178

Rai M, Yadav A, Bridge P, Gade A (2009) Myconanotechnology: a new and emerging science. CAB, New York

Rajakumar G, Rahuman A, Roopan SM, Khanna VG, Elango G, Kamaraj C et al (2012) Fungus-mediated biosynthesis and characterization of TiO_2 nanoparticles and their activity against pathogenic bacteria. Spectrochim Acta A Mol Biomol Spectrosc 91:23–29

Rajaram K, Aiswarya DC, Sureshkumar P (2015) Green synthesis of silver nanoparticle using *Tephrosia tinctoria* and its antidiabetic activity. Mater Lett 138:251–254

Rajesh S, Raja DP, Rathi JM, Sahayaraj K (2012) Biosynthesis of silver nanoparticles using *Ulva fasciata* (Delile) ethyl acetate extract and its activity against *Xanthomonas campestris* pv. *malvacearum*. J Biopest 5:119–128

Rajeshkumar S, Malarkodi C, Paulkumar K, Vanaja M, Gnanajobitha G, Annadurai G (2013a) Intracellular and extracellular biosynthesis of silver nanoparticles by using marine bacteria *Vibrio alginolyticus*. IJNN 3:21–25

Rajeshkumar S, Malarkodi C, Vanaja M, Gnanajobitha G, Paulkumar K, Kannan C et al (2013b) Antibacterial activity of algae mediated synthesis of gold nanoparticles from *Turbinaria conoides*. Pharma Chem 5:224–229

Raliya R, Tarafdar JC (2013) ZnO nanoparticle biosynthesis and its effect on phosphorous mobilizing enzyme secretion and gum contents in clusterbean (*Cyamopsis tetragonoloba* L). Agric Res 2:48–57

Raliya R, Rathore I, Tarafdar JC (2013) Development of microbial nanofactory for zinc, magnesium, and titanium nanoparticles production using soil fungi. J Bionanosci 7:590–596

Ramezani F, Jebali A, Kazemi B (2012) A green approach for synthesis of gold and silver nanoparticles by *Leishmania sp*. Appl Biochem Biotechnol 168:1549–1555

Reddy NJ, Vali DN, Rani M, Rani SS (2014) Evaluation of antioxidant, antibacterial and cytotoxic effects of green synthesized silver nanoparticles by *Piper longum* fruit. Mater Sci Eng C 34:115–122

Renugadevi K, Aswini RV (2012) Microwave irradiation assisted synthesis of silver nanoparticles using *Azadirachta indica* leaf extract as a reducing agent and *in vitro* evaluation of its antibacterial and anticancer activity. Int J Nanomater Biostruct 2:5–10

Saha S, Sarkar J, Chattopadhyay D, Patra S, Chakraborty A, Acharya K (2010) Production of silver nanoparticles by a phytopathogenic fungus *Bipolaris nodulosa* and its antimicrobial activity. Dig J Nanomater Biostruct 5:887–895

Sahayaraj K, Rajesh S, Rathi JM (2012) Silver nanoparticles biosynthesis using marine alga *Padina pavonica* (Linn.) and its microbicidal activity. Dig J Nanomater Biostruct 7:1557–1567

Salvadori MR, Lepre LF, Ando RA, Nascimento CAO, Corrêa B (2013) Biosynthesis and uptake of copper nanoparticles by dead biomass of *Hypocrea lixii* isolated from the metal mine in the Brazilian Amazon region. PLoS One 8:1–8

Salvadori MR, Ando RA, Nascimento CAO, Corrêa B (2014a) Intracellular biosynthesis and removal of copper nanoparticles by dead biomass of yeast isolated from the wastewater of a mine in the Brazilian Amazonia. PLoS One 9:1–9

Salvadori MR, Ando RA, do Nascimento CAO, Corrêa B (2014b) Bioremediation from wastewater and extracellular synthesis of copper nanoparticles by the fungus *Trichoderma koningiopsis*. J Environ Sci Health A Tox Hazard Subst Environ Eng 49:1286–1295

Salvadori MR, Nascimento CAO, Corrêa B (2014c) Nickel oxide nanoparticles film produced by dead biomass of filamentous fungus. Sci Rep 4:1–6

Salvadori MR, Ando RA, Nascimento CAO, Corrêa B (2015) Extra and intracellular synthesis of nickel oxide nanoparticles mediated by dead fungal biomass. PLoS One 5:1–15

Salvadori MR, Ando RA, Muraca D, Knobel M, Nascimento CAO, Corrêa B (2016) Magnetic nanoparticles of Ni/NiO nanostructured in film form synthesized by dead organic matrix of yeast. RSC Adv 6:60683–60692

Salvadori MR, Ando RA, Nascimento CAO, Corrêa B (2017) Dead biomass of Amazon yeasts: a new insight into bioremediation and recovery of silver by intracellular synthesis of nanoparticles. J Environ Sci Health A Tox Hazard Subst Environ Eng 52:1112–1120

Sangeetha J, Thangadurai D, Hospet R, Harish ER, Purushotham P, Mujeeb MA, Shrinivas J, David M, Mundaragi AC, Thimmappa AC, Arakera SB, Prasad R (2017) Nanoagrotechnology for soil quality, crop performance and environmental management. In: Nanotechnology (eds. Prasad R, Kumar M, Kumar V), Springer Nature Singapore Pte Ltd. 73–97

Sarkar R, Kumbhakar P, Mitra AK (2010) Green synthesis of silver nanoparticles and its optical properties. Dig J Nanomater Biostruct 5:491–496

Sathishkumar M, Sneha K, Kwak IS, Mao J, Tripathy S, Yun YS (2009a) Phyto-crystallization of palladium through reduction process using *Cinnamom zeylanicum* bark extract. J Hazard Mater 171:400–404

Sathishkumar M, Sneha K, Won S, Cho CW, Kim S, Yun YS (2009b) *Cinnamon zeylanicum* bark extract and powder mediated green synthesis of nano-crystalline silver particles and its bactericidal activity. Colloids Surf B Biointerfaces 73:332–338

Saxena A, Tripathi RM, Singh RP (2010) Biological synthesis of silver nanoparticles by using onion (*Allium cepa*) extract and their antibacterial activity. Dig J Nanomater Biostruct 5:427–432

Saxena J, Sharma MM, Gupta S, Singha A (2014) Emerging role of fungi in nanoparticle synthesis and their applications. J Pharm Pharm Sci 3:1586–1613

Shahverdi AR, Minaeian S, Shahverdi HR, Jamalifar H, Nohi AA (2007) Rapid synthesis of silver nanoparticles using culture supernatants of *Enterobacteria*: a novel biological approach. Process Biochem 42:919–923

Shankar SS, Ahmad A, Pasricha R, Sastry M (2003) Bioreduction of chloroaurate ions by geranium leaves and its endophytic fungus yields gold nanoparticles of different shapes. J Mater Chem 13:1822–1826

Shankar SS, Rai A, Ahmad A, Sastry M (2005) Controlling the optical properties of lemongrass extract synthesized gold nanotriangles and potential application in infrared-absorbing optical coatings. Chem Mater 17:566–572

Sharma B, Purkayastha DD, Hazra S, Thajamanbi M, Bhat-tacharjee CR, Ghosh NN et al (2014) Biosynthesis of fluorescent gold nanoparticles using an edible freshwater red alga, *Lemanea fluviatilis* (L.) and antioxidant activity of biomatrix loaded nanoparticles. Bioprocess Biosyst Eng 37:2559–2565
Sharma A, Sharma S, Sharma K, Chetri SPK, Vashishtha A, Singh P et al (2015) Algae as crucial organisms in advancing nanotechnology: a systematic review. J Appl Phycol 28:1–16
Sharon M, Choudhary AK, Kumar R (2010) Nanotechnology in agricultural diseases and food safety. J Phytol 2:83–92
Shenton W, Douglas T, Young M, Stubbs G, Mann S (1999) Inorganic-organic nanotubes composites from template mineralization of tobacco mosaic virus. Adv Mater 11:253–256
Shetty P, Supraja N, Garud M, Prasad TNVKV (2014) Synthesis, characterization and antimicrobial activity of *Alstonia scholaris* bark-extract-mediated silver nanoparticles. J Nanostruct Chem 4:161–170
Shiying H, Zhirui G, Zhanga Y, Zhanga S, Wanga J, Ning G (2007) Biosynthesis of gold nanoparticles using the bacteria *Rhodopseudomonas capsulata*. Mater Lett 61:3984–3987
Shukla MK, Singh RP, Reddy CRK, Jha B (2012) Synthesis and characterization of agar-based silver nanoparticles and nanocomposite film with antibacterial applications. Bioresour Technol 107:295–300
Siddiqi KS, Husen A (2016) Fabrication of metal and metal oxide nanoparticles by algae and their toxic effects. Nanoscale Res Lett 363:1–11
Singh T, Shukla S, Kumar P, Wahla V, Bajpai VK, Rather IA (2017) Application of nanotechnology in food science: perception and overview. Front Microbiol 8:1–7
Skalickova S, Baron M, Sochor J (2017) Nanoparticles biosynthesized by yeast: a review of their application. Kvasny Prum 63:290–292
Song JY, Kwon EY, Kim BS (2010) Biological synthesis of platinum nanoparticles using *Diopyros kaki* leaf extract. Bioprocess Biosyst Eng 33:159–164
Sonkusre P, Nanduri R, Gupta P, Cameotra SS (2016) Improved extraction of intracellular biogenic selenium nanoparticles and their specificity for cancer chemoprevention. J Nanomed Nanotechnol 5:2–9
Sreejivungsa K, Suchaichit N, Moosophon P, Chompoosor A (2016) Light-regulated release of entrapped drugs from photoresponsive gold nanoparticles. J Nanomater:1–7. https://doi.org/10.1155/2016/4964693
Sreekanth TV, Ravikumar S, Eom IY (2014) Green synthesized silver nanoparticles using *Nelumbo nucifera* root extract for efficient protein binding, antioxidant and cytotoxicity activities. J Photochem Photobiol B 141:100–105
Srivastava S, Thakur IS (2006) Isolation and process parameter optimization of *Aspergillus* sp. for removal of chromium from tannery effluent. Bioresour Technol 97:1167–1173
Suganya KU, Govindaraju K, Kumar VG, Dhas TS, Karthick V, Singaravelu G et al (2015) Blue green alga mediated synthesis of gold nanoparticles and its antibacterial efficacy against Gram positive organisms. Mater Sci Eng C 47:351–356
Sundaram AP, Augustine R, Kannan M (2012) Extracellular biosynthesis of iron oxide nanoparticles by *Bacillus subtilis* strains isolated from rhizosphere soil. Biotechnol Bioprocess Eng 17:835–840
Swamy VS, Prasad R (2012) Green synthesis of silver nanoparticles from the leaf extract of *Santalum album* and its antimicrobial activity. J Optoelectron Biomed Mater 4(3):53–59
Swamy MK, Akhtar MS, Mohanty SK, Sinniah UR (2015) Synthesis and characterization of silver nanoparticles using fruit extract of *Momordica cymbalaria* and assessment of their in vitro antimicrobial, antioxidant and cytotoxicity activities. Spectrochim Acta A Mol Biomol Spectrosc 151:939–944
Sweeney RY, Mao C, Gao X, Burt JL, Belcher AM, Georgiou G et al (2004) Bacterial biosynthesis of cadmium sulfide nanocrystals. Chem Biol 11:1553–1559
Syed A, Ahmad A (2013) Extracellular biosynthesis of CdTe quantum dots by the fungus *F. oxysporum* and their anti-bacterial activity. Spectrochim Acta A 106:41–47

Tapasztó L, Dobrik G, Lambin P, Biró LP (2008) Tailoring the atomic structure of graphene nanoribbons by scanning tunnelling microscope lithography. Nat Nanotechnol 3:397–401
Thajuddin N, Subramanian G (2005) Cyanobacterial biodiversity and potential application in biotechnology. Curr Sci 89:47–57
Tian X, He W, Cui J, Zhang X, Zhou W, Yan S et al (2010) Mesoporous zirconium phosphate from yeast biotemplate. J Colloid Interface Sci 343:344–349
Varshney R, Bhadauria S, Gaur MS (2012) A review: biological synthesis of silver and copper nanoparticles. Nano Biomed Eng 4:99–106
Velusamy P, Kumar GV, Jeyanthi JD, Pachaiappan R (2016) Bio-inspired green nanoparticles: synthesis, mechanism, and antibacterial application. Toxicol Res 32:95–102
Vlachou E, Chipp E, Shale E, Wilson YT, Papini R, Moiemen NS (2007) The safety of nanocrystalline silver dressing on burns: a study of systemic silver absorption. Burns 33:979–985
Yan S, He W, Sun C, Zhang X, Zhao H, Li Z et al (2009) The biomimetic synthesis of zinc phosphate nanoparticles. Dyes Pigments 80:254–258
Yi G, Wu Z, Sayer M (1988) Preparation of Pb (Zr, Ti) O_3 thin films by sol gel processing: electrical, optical, and electro-optic properties. J Appl Phys 64:2717–2724
Yong P, Rowsen NA, Farr JPG, Harris IR, Macaskie LE (2002) Bioreduction and biocrystallization of palladium by *Desulfovibrio desulfuricans* NCIMB 8307. Biotechnol Bioeng 80:369–379
Yousefzadi M, Rahimi Z, Ghafori V (2014) The green synthesis, characterization and antimicrobial activities of silver nanoparticles synthesized from green alga *Enteromorpha flexuosa* (wulfen) J Agardh. Mater Lett 137:1–4
Yu J, Xu D, Guan HN, Wang C, Huang LK, Chi DF (2016) Facile one-step green synthesis of gold nanoparticles using *Citrus maxima* aqueous extracts and its catalytic activity. Mater Lett 166:110–112
Yuan J, Cen Y, Kong XJ, Shuang W, Liu CL, Yu RQ et al (2015) MnO_2-nanosheet-modified upconversion nanosystem for sensitive turn-on fluorescence detection of H_2O_2 and glucose in blood. ACS Appl Mater Interfaces 19:10548–10555
Zhang X, He X, Wang K, Wang Y, Li H, Tan W (2009) Biosynthesis of size-controlled gold nanoparticles using fungus *Penicillium sp.* J Nanosci Nanotechnol 9:5738–5744
Zhang YX, Zheng J, Gao G, Kong YF, Zhi X, Wang K et al (2011) Biosynthesis of gold nanoparticles using chloroplasts. Int J Nanomed 6:2899–2906
Zhang P, Wei R, Zeng J, Cai M, Xiao J, Yang D (2016) Thermal properties of silver nanoparticle sintering bonding paste for high-power led packaging. J Nanomater 2016:1–6
Zhao Y, Yeh Y, Liu R, You J, Qu F (2015a) Facile deposition of gold nanoparticles on core–shell Fe_3O_4@polydopamine as recyclable nanocatalyst. Solid State Sci 45:9–14
Zhao Y, Zheng Y, Zhao C, You J, Qu F (2015b) Hollow PDA-Au nanoparticles-enabled signal amplification for sensitive nonenzymatic colorimetric immunodetection of carbohydrate antigen 125. Biosens Bioelectron 71:200–206
Zheng D, Hu C, Gan T, Dang X, Hu S (2010) Preparation and application of a novel vanillin sensor based on biosynthesis of Au-Ag alloy nanoparticles. Sensors Actuators B Chem 148:247–252
Zhou W, He W, Zhang X, Yan S, Sun X, Han X (2009a) Biosynthesis of iron phosphate nanopowders. Powder Technol 194:106–108
Zhou W, He W, Zhong S, Wang Y, Zhao H, Li Z et al (2009b) Biosynthesis and magnetic properties of mesoporous Fe_3O_4 composites. J Magn Magn Mater 321:1025–1028
Zhou JC, Soto CM, Chen MS, Bruckman MA, Moore MH, Barry E et al (2012) Biotemplating rod-like viruses for the synthesis of copper nanorods and nanowires. J Nanobiotechnology 10:1–12
Zonooz NF, Salouti M (2011) Extracellular biosynthesis of silver nanoparticles using cell filtrate of *Streptomyces sp.* ERI-3. Sci Iran 18:1631–1635
Zuas O, Hamim N, Sampora Y (2014) Bio-synthesis of silver nanoparticles using water extract of *Myrmecodia pendan* (Sarang Semut plant). Mater Lett 123:156–159

Chapter 2
Green Synthesis and Biogenic Materials, Characterization, and Their Applications

Gamze Tan, Sedef İlk, Ezgi Emul, Mehmet Dogan Asik, Mesut Sam, Serap Altindag, Emre Birhanli, Elif Apohan, Ozfer Yesilada, Sandeep Kumar Verma, Ekrem Gurel, and Necdet Saglam

Contents

G. Tan · M. Sam
Aksaray University, Faculty of Science and Letters, Department of Biology, Aksaray, Turkey

S. İlk
Nigde Omer Halisdemir University, Faculty of Medicine, Department of Immunology, Nigde, Turkey

E. Emul · N. Saglam (✉)
Hacettepe University, Nanotechnology and Nanomedicine Division, Ankara, Turkey
e-mail: saglam@hacettepe.edu.tr

M. D. Asik
Ankara Yildirim Beyazit University, Musculoskeletal Regenerative Medicine, Ankara, Turkey

S. Altindag
Aksaray University, Graduate School of Science, Aksaray, Turkey

R. Prasad (ed.), *Microbial Nanobionics*, Nanotechnology in the Life Sciences,
https://doi.org/10.1007/978-3-030-16383-9_2

2.1 Introduction

Today, many organic and inorganic nanomaterials are synthesized by two main synthesis approaches used including top-down and bottom-up. The top-down approach (i.e., small pieces from large bulks) refers to the production of nanostructures from macrostructures using chemical and physical methods. Photolithography or electron-beam lithography, anodization, and ion and plasma corrosion are some of the well-known techniques of top-down approaches (Daraio and Jin 2012). The top-down approach is currently being extensively used for the production of semiconductor circuit elements and computer chips (Booker and Boysen 2011). The bottom-up approach (i.e., building larger structure by combining small pieces) refers to the construction of organic or inorganic structures starting from atoms, the basic unit of matter. The synthesis process adopting the bottom-up approach can be carried out using chemical, physical, and biological methods. One of the most emphasized points in nanomedicine is that nanomaterials synthesized by these production approaches and used on living organism exhibit high biocompatibility and, in other sense, cause low toxicity. Recently, the use of biological synthesis methods also called green method has increased in many areas because of the relatively high toxic effects of chemical methods on both the living organisms and the environment (Prasad 2016, 2017; Prasad et al. 2018a, b). The most important reason for this is the fact that the nanomaterials prepared with the green methods are more environmentally friendly, economical, and less toxic compared to chemical counterparts (Prasad 2014; Prasad et al. 2016; Abdel-Aziz et al. 2018). The basic principles of the green method are presented in Fig. 2.1 (Anastas and Warner 1998).

Plants and microorganisms like yeasts, fungi, and bacteria have been used for biological synthesis of nanoparticles. Microorganisms are mainly approved due to their appropriate ability like easy cultivation and fast growth rate (Fariq et al. 2017; Rai and Duran 2011; Prasad et al. 2016). They synthesize basically inorganic nanoparticles by way of mechanisms of reduction of metal ions within intracellular and/or extracellular routes (Li et al. 2011a, b) (Fig. 2.2).

Microbial nanoparticles are found to have strong antimicrobial properties thanks to their surface charge, size, geometry, and colloidal stability, which can be modified by biosynthesis parameters including incubation time, temperature, concentration, and pH (Patra and Baek 2014; Hosseini and Sarvi 2015). This antimicrobial efficiency is probably due to enhanced interaction between particles and microorganisms. Nanoparticles with small size and the high surface area could easily adhere to the cell membrane and enter into the plasma membrane (Sunkar and Nachiyar 2012). By adhering of nanoparticles to the cell membrane,

E. Birhanli · E. Apohan · O. Yesilada
Inonu University, Faculty of Science and Letters, Department of Biology, Malatya, Turkey

S. K. Verma · E. Gurel
Abant Izzet Baysal University, Faculty of Science and Literature, Department of Biology, Bolu, Turkey

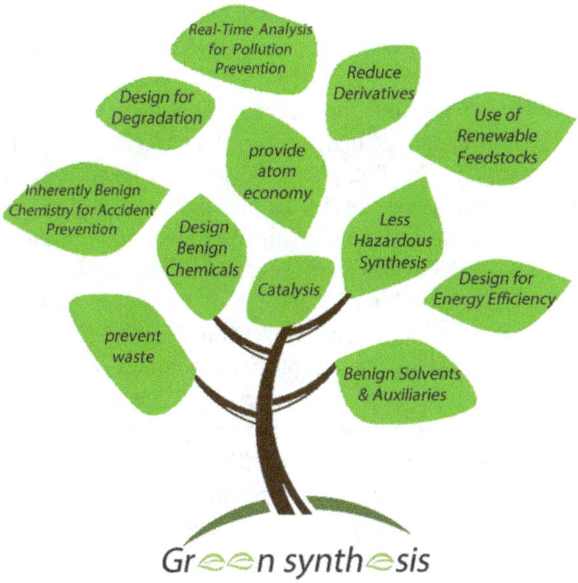

Fig. 2.1 The basic principles of the green synthesis

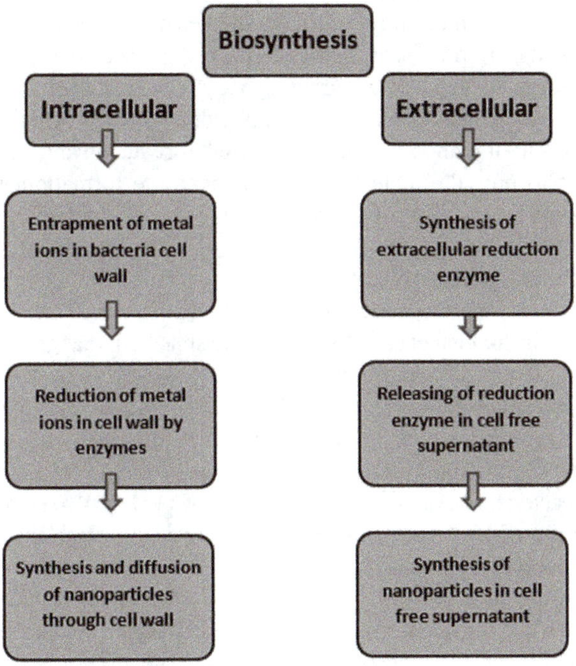

Fig. 2.2 Schematic flow diagram for intra- and extracellular biosynthesis of nanomaterials by microorganisms. (Fariq et al. 2017; Rai and Duran 2011)

they can cause structural damage to cellular membranes and components. By accessing of nanoparticles into the cell membrane, they can cause DNA damage via interfering the replication process (Fariq et al. 2017).

With green synthesis method, the nanomaterials can be synthesized based on microorganisms (bacteria, algae, fungi, etc.) or plants. As is known, microorganisms, bacteria in particular, have the ability to cope with stress when exposed to adverse environmental conditions. These special defense mechanisms can help them to suppress the toxic effects resulting from high metal exposure. For instance, the synthesis of gold nanoparticles (AuNPs) was carried out inside the cell by fungi *Verticillium* sp. (Mukherjee et al. 2001a, b) and outside the cell by *Fusarium oxysporum* and actinomycete *Thermomonospora* sp., respectively (Mukherjee et al. 2002; Ahmad et al. 2003). Sweeney et al. (2004) produced spherical cadmium sulfoxide nanocrystals in sizes 2–5 nm using *Escherichia coli*. In another study, the gold ions were reduced via *Pseudomonas aeruginosa* bacterium, and spherical gold nanoparticles were obtained at a size range of 15–30 nm (Husseiny et al. 2007). Using Gram-negative bacterium *Pseudomonas stutzeri*, AgNPs slightly larger than 200 nm were obtained in different shapes (Belliveau et al. 1987). Together with fungi and bacteria, hexagonal silver nanoparticles (AgNPs) were obtained at a size range of 2–5 nm using yeast strain MCF3 (Meenal et al. 2003). Similarly, using *Saccharomyces cerevisiae* yeast, the formation of spherical Sb_2O_3 particles in the 3–10 nm size range has been reported (Jha et al. 2009). These examples and areas of usage can be diversified. Plant-based synthesis applications have increased recently. The main reason for this is the fact that treatment with plants, namely, phytotherapy, has been used since ancient times, thanks to unique antioxidant (radical scavenging) contents of plants as well as their comparably less side effects. Today, there is a growing body of interest in plant-based synthesis methods that combine nanotechnology with phytotherapy and benefit from the great antioxidant content of plants. The plants are used as reducing and stabilizing agents for metal salts to provide nucleation and nanoparticle formation with important components such as polyphenols, sugars, alkaloids, phenolic acids, proteins, and terpenoids (Prasad 2014).

For instance, crystalline PtNPs in the size range 15–19 nm using *Diospyros kaki* extract (Song et al. 2009); spherical PdNPs with 10–15 nm in size using *Curcuma longa* extract (Sathishkumar et al. 2009); AuNPs in spherical, triangular, and hexagonal shapes in 20–100 nm size and spherical and cubic AgNPs in 10–100 nm using *Eucalyptus macrocarpa* (Poinern et al. 2013a, b); and AgNPs with the size of 31 nm using *Citrullus colocynthis* (Satyavani et al. 2011) were synthesized. Ali et al. (2015) synthesized AgNPs by microwave using *Eucalyptus globulus* leaf extracts and examined their antibacterial and antibiofilm activities. Roy et al. (2015) showed rapid colorimetric detection of Hg^{2+} ions with AgNPs synthesized using *Dahlia pinnata* leaf extract. Antimicrobial properties of AgNPs that were reduced using cranberry powder have been investigated and have been shown to contribute to wound healing in rats (Ashour et al. 2015). Characterization and contents of AgNPs prepared with biocompatibility using different amounts of garlic extract are shown (Von White et al. 2012).

Biosynthesized AgNPs have great antimicrobial activity against highly multiresistant strains of pathogenic microorganisms. Nowadays, researchers have made efforts to obtain eco-friendly and nontoxic process for the fabrication of AgNPs. It was shown that microorganisms could reduce Ag^+ ions to form spherical silver nanoparticles (Mukherjee et al. 2001a, b; Panáček et al. 2006; Fayaz et al. 2010). AgNPs were produced in the form of agglomerated or film structure by microorganisms: *Fusarium oxysporum*, *Verticillium*, or *Aspergillus flavus* (Vigneshwaran et al. 2007; Jain et al. 2011). AgNPs were also employed for enhanced antimicrobial properties in combination with commercial antibiotics against pathogenic microorganisms. The antimicrobial activities of antibiotics such as erythromycin, ampicillin, chloramphenicol, and kanamycin within biosynthesized AgNPs have increased antimicrobial effects against resistant bacteria (Fayaz et al. 2010) which ensure novel insight for the development of new antimicrobials. Although the antimicrobial activity of AgNPs synthesized by microorganism has been elucidated by many researchers, antimicrobial effects of other bio-fabricated nanoparticles and their action mechanisms are not yet to be clarified (Abdeen and Praseetha 2013; Malarkodi et al. 2013).

The antimicrobial properties of the AgNPs have been influenced by several factors including ionic strength, size, pH, and capping agent (Franci et al. 2015). The antimicrobial mechanism of AgNP is still controversial. The silver ion in production process by microorganisms must be in ionized form for positive charge and thus can be considered as a potential antimicrobial action, while these silver ions within positive charge can easily conjugate with cell membrane and nucleic acids of microbes that cause the deactivation of the enzyme in the cell (Wright et al. 1999; Klueh et al. 2000; Cao et al. 2001; Sharma et al. 2009).

Briefly, green synthesis of nanomaterials involves safe and non-polluting synthetic routes in production strategies and technologies compared to chemical strategies. The use of these alternative reaction conditions might enable to increase energy efficiency, to replace consumed resources with renewable and reusable ones, and to recycle by-products and processes. In this study, it is aimed to guide future research studies by drawing a perspective on production processes and application areas of nanomaterials synthesized by green chemistry method.

2.2 Nanomaterials and Their Characterization

The word nano is originated from a Latin word, which means dwarf. Ideal size range offered by nanotechnology refers to one thousand millionth of a particular unit; thus nanometer is one thousand millionth of a meter (i.e., 1 nm = 10^{-9} m). The branch of nanotechnology is the science that particularly deals with the processes that occur at molecular level and of nano length/scale/size. Nanotechnology has now become an allied science which is most commonly used in other fields of science like electronic, physics, and engineering for many decades. Recent exploration of nanotechnology in biomedical and pharmaceutical sciences results in successful improvement of conventional means of drug delivery system. This multidisciplinary science also covers several applications in other disciplines such as biophysics, molecular biology, and bioengineering, and nanotechnology has created potential impact in various fields like medicine including immunology, cardiology, endocrinology, ophthalmology, oncology, pulmonology, etc. In addition, it is highly utilized in specialized area like brain targeting, tumor targeting, and gene delivery. Nanotechnology also provides significant systems, devices, and materials for better pharmaceutical applications (Bhatia 2016).

The majority of the studies on nanoparticles voice the concern about their possible toxicity to plants and animals (Iavicoli et al. 2017). A relatively lesser number of studies have been conducted on NPs advocating their beneficial effects on plants. However, there is a recent thrust to develop novel applications involving nanotechnology, targeting specific delivery of chemicals, proteins, and nucleotides needed for genetic transformation of crops (Wang et al. 2016; Prasad et al. 2014, 2017). Nanotechnology holds the potential to provide new opportunities to the material scientists as well as plant science researchers to develop new tools for incorporation of nanoparticles into the plants, aiming for augmentation of existing functions and the addition of new ones, as well (Cossins 2014).

The nanomaterials can be categorized into four groups: (1) carbon-based nanoparticles (NPs) including fullerenes and carbon nanotubes such as single-walled carbon nanotubes (SW-CNTs), double-walled carbon nanotubes (DW-CNTs), and multi-walled carbon nanotubes (MW-CNTs); (2) inorganic NPs including metals (gold, silver, aluminum, and zerovalent iron), metal oxides (TiO_2, ZnO, Al_2O_3, Fe_3O_4, Fe_2O_4, NiO, CoO, CeO_2, etc.), and quantum dots (cadmium sulfide and cadmium selenide); (3) dendrimers, which are nanosized polymer networks built from branched molecules capable of being tailored to perform specific chemical functions; and (4) nanocomposites, which combine NPs with other NPs or with larger, bulk-like materials (Verma et al. 2018).

Characterization of NPs is based on the size, morphology, and surface charge, using various characterization techniques such as ultraviolet-visible spectroscopy (UV-Vis), transmission electron microscope (TEM), scanning electron microscope (SEM), high-resolution transmission electron microscope (HRTEM), field emission scanning electron microscope (FESEM), Fourier-transform infrared spectroscopy (FTIR), dynamic light scattering (DLS), thermogravimetric analysis (TGA), atomic force microscope (AFM), X-ray powder diffraction (XRD), X-ray photoelectron spectroscopy (XPS), power spectral density (PSD), Raman spectroscopy (RS), Inductively coupled plasma mass spectrometry (ICP-MS), chemical vapor deposition (CVD), energy dispersive X-ray analysis (EDX), and gas chromatography-mass spectrometry (GC-MS) (Verma et al. 2018).

2.2.1 Metallic Nanoparticles (MNPs) and Characterization

One of the goals of nanotechnology is the production of NPs, in size between 1 and 100 nm, such as silver, gold, copper, iron, and palladium. Among other nanostructures, metallic nanoparticles (MNPs) are good candidates as antimicrobial, antiviral, and cytotoxic agents for various medical applications. Their diameters, shapes, and dispersibility are very important for application. Given the unique properties, MNPs have promising potential as an antimicrobial and anticancer agent (Aziz et al. 2019).

MNPs can be prepared by various methods such as chemical, physical, and biological methods. However, chemical and physical methods have various major disadvantages such as MNPs from chemical methods may not be suitable for medical applications. On the other hand, green synthesis of MNPs by biological methods uses mild reaction conditions and nontoxic substances/compounds. This is a simple and eco-friendly method. Therefore, biological methods may be an alternative and eco-friendly way for production of MNPs. In biological methods, enzymes, bacteria, fungi, algae, and plant extracts could be used as reducing and stabilizing agent (Gurunathan et al. 2013a, b; Khan et al. 2015; Ortega et al. 2015, Venkatesan et al. 2016; Prasad et al. 2016).

MNPs have wide application potential due to the differences in their properties. For example, size, shape, morphology, and stability are important parameters for their antibacterial and cytotoxic effects. Therefore, NPs must be characterized before application. The primary indication of NP formation is the color change due to the surface plasmon vibrations exhibited by NPs (Patil and Kim 2017). Different NPs show different colors due to surface plasmon resonance depending on the particle type, size, and shape (Dhand et al. 2016). Optical properties of a solution could be analyzed by UV-visible spectroscopy, and therefore, the absorption spectra of the solution are determined. Surface plasmon band characteristic to NPs is indicative for the NP formation. TEM, SEM, and AFM give the information about size, shape, morphology, and elemental composition of the NPs. TEM can also be used to determine the dispersion of NPs in the solution. XRD is used to detect the crystal property of MNPs, and elemental composition of NPs is analyzed with EDX. FTIR is used to determine the different functional groups responsible for capping and stabilization of MNPs. Zeta potential gives the information about the stability of NPs (Patil and Kim 2017).

2.3 Antimicrobial Nanoparticles

The investigation of the nanoparticles that have antimicrobial effect is a very popular issue. Metals and metal oxides, carbon-based nanomaterials, and surfactant-based nano-fluids can be said as the materials which have antimicrobial effect (Li et al. 2008). Resistance developed against the antibiotics both increases the costs and delays the fight with microorganisms. During the studies with antimicrobial nanoparticles, the developing resistance of microorganisms against nanoparticles was not seen (Mühling et al. 2009). Nanoparticles exhibit the antimicrobial effects by using mechanisms such as induction of ROS generation, destruction of the cell wall, inhibition of ATP production and DNA replication (Weir et al. 2008). Cadmium sulfide (CdS), gold (Au), titanium dioxide (TiO_2), zinc oxide (ZnO), and silver (Ag) can be counted as one of the most used nanoparticles in this area (Saravanan and Nanda 2010). Between the considered nanomaterials, silver has a special place. It has a wide use for commercial purposes and a large research area.

Silver is used successfully against the virus and other eukaryotic microorganisms. The most promising results between the antimicrobial nanoparticles were achieved with silver (Sharma et al. 2009). Silver mainly causes cell to die by affecting the respiratory chain and the mechanism of the cell division. When AgNPs are used in combination with antibiotics such as vancomycin, amoxicillin, and penicillin, a positive synergistic effect was obtained in Gram-positive and Gram-negative organisms (Fayaz et al. 2010).

ZnO comes to the fore with the low production cost, the toxic effects against bacteria, biocompatibility, and the UV-blocking feature. ZnO NPs are used as drug delivery, medical filling material, and UV-blocking agent in medical industry (Zhou et al. 2006). ZnO is believed to damage the integrity of the cell by breaking down the lipids and proteins in the bacterial cell membrane, and as a result, it is believed to drag the cell to death (Zhou et al. 2006; Bhuyan et al. 2015).

TiO_2 NPs are a powerful bactericidal agent used for nearly 20 years. It gives an effective result against bacteria in 10 and 1000 ppm concentrations. It has a photocatalytic effect under the UV (Muranyi et al. 2010). Antimicrobial nanoparticles are extensively used in health sector, medical devices, white goods (refrigerators, washing machines, etc.), and textile industry. Nanostructures based on carbon such as carbon nanotube have a very strong antimicrobial activity. But their high cytotoxicity limits their application areas (Arias and Yang 2009). Although nanomaterials have plenty of application areas, there is a significant lack of information about the effects of the materials on human health and environment. There are limited studies on the toxicity of NPs.

2.4 Silver Nanoparticles and Medicinal Applications

Many studies on NP production have mainly focused on MNP production with biological methods. Silver nanoparticles (AgNPs) are very important in nanotechnology, because of their chemical stability, conductivity, and catalytic and biological activities such as antimicrobial, cytotoxic, and antioxidant activities (Rao et al. 2016). AgNPs have promising potential as antimicrobial and cytotoxic agents. Thus, there are various studies on biogenic synthesis and potential applications of AgNPs, especially as medicinal agents. AgNPs can show its antibacterial effect on DNA, proteins, membrane, and cell wall of bacteria (Singh et al. 2014; Khan et al. 2015; Aziz et al. 2015, 2016; Joshi et al. 2018). It was reported that AgNPs induce cellular apoptosis via activation of p53, p-Erk1/2, and caspase-3 signaling and downregulation of Bcl-2 in MDA-MB-231 human breast cancer cells (Gurunathan et al. 2015). Singh et al. (2014) reported that AgNPs induce the reactive oxygen species (ROS) generation and DNA fragmentation of Dalton's lymphoma and human carcinoma colo205 cells. AgNPs also show dose-dependent cytotoxicity against MDA-MB-231 cells through activation of the lactate dehydrogenase (LDH), caspase-3, and ROS generation and thus induce apoptosis (Gurunathan et al. 2013a, b). Therefore, here, antimicrobial and cytotoxic effects of AgNPs produced with bacteria, fungi, and algae were reviewed.

2.5 Biologically Prepared AgNPs as Antimicrobial Agents

2.5.1 *Algae*

Microorganisms can be used to produce NPs, and production of NPs using algae is an eco-friendly way of synthesis (Aziz et al. 2014, 2015). There are some studies on AgNP production and antimicrobial activities of biogenic NPs (Table 2.1).

Sargassum longifolium, a brown alga, was tested for producing AgNPs in a study conducted by Rajeskumar et al. (2014). They treated 1 mM $AgNO_3$ with 10 mL *Sargassum longifolium* extract solution. Maximum NP synthesis was occurred at pH 8.4, and the NP production was completed at the end of the 64 h of incubation. The antimicrobial effect of these NPs on three pathogenic fungi (*Aspergillus fumigatus*, *Candida albicans*, and *Fusarium* sp.) was determined based on the measurement of inhibition zones caused by NPs at different concentrations (50, 100, and 150 μL). Antifungal activity was high and it increased due to the increase concentration of NPs. The order of inhibition zone diameters caused

Table 2.1 Antimicrobial activities of AgNPs biosynthesized by algae

Algae	Nanoparticle shape	Nanoparticle size (nm)	Antimicrobial effect on	References
Sargassum longifolium	Spherical	–	*Aspergillus fumigatus*, *Candida albicans*, *Fusarium* sp.	Rajeshkumar et al. (2014)
Ecklonia cava	Spherical	43	*Escherichia coli*, *Staphylococcus aureus*	Venkatesan et al. (2016)
Gelidiella sp.	Cubical	–	*Escherichia coli*, *Staphylococcus aureus*, *Klebsiella pneumoniae*, *Bacillus* sp.	Devi and Bhimba (2013)
Laurencia papillosa	Cubical	–	*Bacillus subtilis*, *Staphylococcus aureus*, *Streptococcus pneumoniae*, *Escherichia coli*, *Klebsiella pneumoniae*, *Pseudomonas aeruginosa*, *Aspergillus flavus*, *Aspergillus fumigatus*, *Aspergillus niger*	Omar et al. (2017)
Scenedesmus sp. (IMMTCC-25)	Spherical	15–20 (intracellular) 5–10 (extracellular)	*Streptococcus mutans*, *Escherichia coli*	Jena et al. (2014)
Spirogyra varians	–	35	*Staphylococcus aureus*, *Bacillus cereus*, *Listeria monocytogenes*, *Salmonella typhimurium*, *Escherichia coli*, *Pseudomonas aeruginosa*, *Klebsiella* sp.	Salari et al. (2016)

by the AgNPs was *Fusarium* sp. > *Candida albicans* > *Aspergillus fumigatus* (Rajeshkumar et al. 2014).

It was stated that edible brown alga *Ecklonia cava* contains phloroglucinol, eckol, fucodiphlorethol G, 7-phloroeckol, and dieckol acting as the reducer for AgNPs. Therefore, this alga was used for AgNPs' production. Its aqueous extract was mixed with 1 mM $AgNO_3$ solution, and after 72 h, nearly spherical AgNPs with 43 nm sizes were obtained. These NPs showed a strong antibacterial effect on both *Escherichia coli* ATCC 10536 and *Staphylococcus aureus* ATCC 6538 (Venkatesan et al. 2016).

It was reported that AgNPs synthesized by macroalga *Gelidiella* sp. showed strong antibacterial effect on *Staphylococcus aureus* and *Bacillus* sp. On the other hand, they exhibited lower antibacterial activity on *Klebsiella pneumoniae* and *Escherichia coli* (Devi and Bhimba 2013). In a study performed by Omar et al. (2017), *Laurencia papillosa* (red alga) extract was used for the green synthesis of AgNPs. SEM showed many NPs in cubic forms on the surface of the cell. The antimicrobial activity studies on Gram-positive bacteria (*Bacillus subtilis*, *Staphylococcus aureus*, *Streptococcus pneumoniae*), Gram-negative bacteria (*Escherichia coli*, *Klebsiella pneumoniae*, *Pseudomonas aeruginosa*), and fungi (*Aspergillus flavus*, *Aspergillus fumigatus*, *Aspergillus niger*) showed that AgNPs had the highest antimicrobial activity on *B. subtilis* and *A. flavus*. The minimum inhibitory concentrations (MICs) determined were between 10 and 16 μg/mL for bacteria and 17 and 20 μg/mL for fungi (Omar et al. 2017).

Jena et al. (2014) investigated the intracellular and extracellular AgNP production potential of unicellular green microalga *Scenedesmus* sp. (IMMTCC-25). Fast intracellular accumulation and reduction of Ag ions were occurred depending on the treatment of algal biomass with 5 mM of $AgNO_3$. Intracellular silver accumulation, which was 25% within the first hour of incubation, reached 80% in 24 h and 93% in 72 h. AgNPs were well separated, spherical, and highly crystalline in form. Raw and boiled algal extracts were treated with 5 mM $AgNO_3$ solution for the synthesis of extracellular AgNPs, and the formation of the NPs was detected by UV-Vis spectrophotometer. The raw algal extract was also treated with $AgNO_3$ solution (5 mM), and many extracellular AgNPs in polydispersed form and 30–150 nm size were obtained after 3 days of incubation. Due to the increase of the incubation period, the particle size also increased, and the maximum particle size reached 50–60 nm. When the $AgNO_3$ solution was incubated with boiled algal extract, more stable AgNPs were obtained. These AgNPs with 5–10 nm size and uniform distribution had high stability of over 3 months at room temperature. The synthesized AgNPs showed significant antibacterial effects on *Escherichia coli* and *Streptococcus mutans* (Jena et al. 2014).

Washed, dried, and powdered *Spirogyra varians* incubated with 1 mM $AgNO_3$ produced AgNPs with an average size of 35 nm. Inhibition zone measurements with these NPs showed higher antibacterial activity on *Bacillus cereus*, *Pseudomonas aeruginosa*, and *Klebsiella* sp. compared to the antibiotic cephalothin and showed a significant antibacterial activity on *Staphylococcus aureus*, *Listeria monocytogenes*, and *E. coli*. While the MIC against *B. cereus* and *E. coli* was 0.25 mg/mL, it was 0.5 mg/

mL on the other bacteria tested. Minimum bactericidal concentration (MBC) values against *S. aureus*, *B. cereus*, *L. monocytogenes*, *Salmonella typhimurium*, *E. coli*, *P. aeruginosa*, and *Klebsiella* sp. were 1, 0.5, 1, 1, 0.25, 0.5, and 1 mg/mL, respectively. *E. coli* was more sensitive to these NPs than the other pathogenic bacteria tested (Salari et al. 2016).

2.5.2 Bacteria

Bacteria have a great potential for AgNPs' production. Various bacteria are known to produce AgNPs with antimicrobial activity (Table 2.2). The cell extract of cyanobacterium *Anabaena doliolum* was used for green synthesis of AgNPs by Singh et al. (2014). It was stated that the AgNPs synthesized by *Anabaena doliolum* were well dispersed, spherical, and 10–50 nm in size. Antibacterial effects of these biogenic AgNPs were investigated on two Gram-negative bacteria (*Klebsiella pneumoniae* DF12SA (HQ114261) and *Escherichia coli* DF39TA (HQ163793)) and a Gram-positive bacterium (*Staphylococcus aureus* DF8TA (JN642261)) by Kirby-Bauer disc diffusion method. All bacterial species tested were incubated with discs containing AgNP at concentrations of 5–500 μg/mL, and it was observed that the inhibition zones gradually increased due to the increased silver concentration. Moreover, the maximum inhibition zone diameters caused by 500 μg/mL AgNPs for *Klebsiella pneumoniae*, *Escherichia coli*, and *Staphylococcus aureus* were 36, 33, and 34 mm, respectively (Singh et al. 2014).

Deinococcus radiodurans, a radiation- and desiccation-resistant bacterium, was tested for green synthesis of AgNPs in a study performed by Kulkarni et al. (2015). The highest AgNP was obtained in the medium containing 2.5 mM $AgNO_3$ at pH 6.8 after 24 h of incubation at 32 °C. It was reported that the synthesized AgNPs were extracellular and spherical and had an average size of 16.82 nm. Three Gram-negative bacteria (*Escherichia coli* NCIM 2739, *Proteus vulgaris* NCIM 2027, and *Pseudomonas aeruginosa* NCIM 2948) and two Gram-positive bacteria (*Staphylococcus aureus* and *Bacillus subtilis*) were used for monitoring the antibacterial effects of these biosynthesized AgNPs by agar well diffusion method. In addition, *E. coli* and *S. aureus* were selected as the standard Gram-negative and Gram-positive organisms, and the antimicrobial activities of AgNPs were determined by standard dilution micromethod. According to the results of the agar well diffusion method, the largest inhibition zone was observed against *S. aureus*, a Gram-positive bacterium, while the largest inhibition zone observed against Gram-negative bacteria occurred for *E. coli*. Standard dilution micromethod showed that while the growth inhibition for *E. coli* was 91% when treated with 150 μg/mL AgNP for 3 h, it was 46% for *S. aureus* under the same condition (Kulkarni et al. 2015).

Wei et al. (2012) tested the cell-free extract of *Bacillus amyloliquefaciens* LSSE-62 (CGMCC No. 4157) for synthesizing the AgNPs by using solar radiation. The researchers incubated $AgNO_3$ with *B. amyloliquefaciens* cell-free extract in the

Table 2.2 Antimicrobial activities of AgNPs biosynthesized by bacteria

Bacteria	Nanoparticle shape	Nanoparticle size (nm)	Antimicrobial effect on	References
Anabaena doliolum	Spherical	10–50	*Klebsiella pneumonia* DF12SA (HQ114261), *Escherichia coli* DF39TA (HQ163793), *Staphylococcus aureus* DF8TA (JN642261)	Singh et al. (2014)
Deinococcus radiodurans	Spherical	16.82	*Escherichia coli* NCIM 2739, *Proteus vulgaris* NCIM 2027, *Pseudomonas aeruginosa* NCIM 2948, *Staphylococcus aureus*, *Bacillus subtilis*	Kulkarni et al. (2015)
Bacillus amyloliquefaciens LSSE-62 (CGMCC no. 4157}	Mostly circular, triangular	4.8–23.7	*Bacillus subtilis*, *Escherichia coli*	Wei et al. (2012)
Bacillus methylotrophicus DC3	Spherical	10–30	*Candida albicans* KACC 30062, *Salmonella enterica* ATCC 13076, *Escherichia coli* ATCC 10798, *Vibrio parahaemolyticus* ATCC 33844	Wang et al. (2016)
Brevibacterium frigoritolerans DC2	Spherical	10–30	*Vibrio parahaemolyticus* ATCC 33844, *Bacillus anthracis* NCTC 10340, *Salmonella enterica* ATCC 13076, *Bacillus cereus* ATCC 14579, *Escherichia coli* ATCC 10798, *Candida albicans* KACC 30062	Singh et al. (2015)
Exiguobacterium sp. KNU1	Spherical	5–50	*Salmonella typhimurium*, *Pseudomonas aeruginosa*, *Escherichia coli*, *Staphylococcus aureus*	Tamboli and Lee (2013)
Lactobacillus rhamnosus GG ATCC 53103	Mostly spherical, triangular, rod, hexagonal	2–15	*Escherichia coli* ATCC 35218, *Bacillus cereus* ATCC 10987, *Listeria monocytogenes* ATCC 15313, *Pseudomonas aeruginosa* ATCC 15442, *Klebsiella pneumoniae* ATCC 27736, *Aspergillus* spp., *Penicillium* spp.	Kanmani and Lim (2013)
Kinneretia THG-SQI4	Spherical	15–25	*Candida albicans* KACC 30062, *Candida tropicalis* KCTC 7909, *Bacillus cereus* ATCC 14579, *Bacillus subtilis* KACC 14741, *Staphylococcus aureus* ATCC 6538, *Pseudomonas aeruginosa* ATCC 6538, *Escherichia coli* ATCC 10798, *Vibrio parahaemolyticus* ATCC 33844	Singh et al. (2017)

Lactobacillus casei LPW2 *Lactobacillus fermentum* LPF6	Spherical Various in shapes (partially aggregated particles)	0.7–10.0 1.4–10.0	*Bacillus* sp., *Streptococcus pyogenes*, *Staphylococcus aureus*, *Klebsiella* sp., *Pseudomonas aeruginosa*	Adebayo-Tayo et al. (2017)
Nocardiopsis sp. MBRC-1	Spherical	30–90	*Escherichia coli* ATCC 10536, *Bacillus subtilis* ATCC 6633, *Enterococcus hirae* ATCC 10541, *Pseudomonas aeruginosa* ATCC 27853, *Shigella flexneri* ATCC 12022, *Staphylococcus aureus* ATCC 6538, *Aspergillus niger* ATCC 1015, *Aspergillus brasiliensis* ATCC 16404, *Aspergillus fumigates* ATCC 1022, *Candida albicans* ATCC 10231	Manivasagan et al. (2013)
Nostoc sp. strain HKAR-2	Spherical	51–100	*Ralstonia solanacearum*, *Xanthomonas campestris*, *Aspergillus niger*, *Trichoderma harzianum*	Sonker et al. (2017)
Novosphingobium sp. THG-C3	Spherical	8–25	*Staphylococcus aureus*, *Bacillus subtilis*, *Bacillus cereus*, *Pseudomonas aeruginosa*, *Escherichia coli*, *Vibrio parahaemolyticus*, *Salmonella enterica*, *Candida tropicalis*, *Candida albicans*	Singh et al. (2017)
Ochrobactrum sp.	Spherical	38–85	*Salmonella typhi*, *Salmonella paratyphi*, *Vibrio cholerae*, *Staphylococcus aureus*	Thomas et al. (2014)
Streptomyces kasugaensis strain NH28	Spherical	4.2–65	*Staphylococcus aureus* ATCC 6338, *Klebsiella pneumoniae* ATCC 700603, *Proteus mirabilis*, *Escherichia coli* ATCC 8739, *Bacillus subtilis* ATCC 6633, *Salmonella infantis*, *Pseudomonas aeruginosa* ATCC 10145	Skladanowski et al. (2016)
Rhodococcus NCIM 2891	Spherical	5–50	*Staphylococcus aureus*, *Klebsiella pneumoniae*, *Proteus vulgaris*, *Enterococcus faecalis*, *Pseudomonas aeruginosa*, *Escherichia coli*	Otari et al. (2015)

presence of sunlight for the synthesis of NPs. Furthermore, different solar intensities (30,000, 40,000, 50,000, 70,000 lx), various amounts of cell-free extracts (1, 2, 3, 4 mg/mL), and the supplementation of different concentrations of NaCl (0.5, 1, 2, and 3 mM) were investigated for optimizing the production of AgNPs. According to the results of optimization studies, high ratios of the parameters tested generally affected the production of AgNPs positively, so a part of the work was conducted with 70,000 lx light intensity, 3 mg/mL extract, and 2 mM NaCl supplement. It was determined that most of the synthesized AgNPs were circular and some were triangular, the sizes of NPs varied between 4.8 and 23.7 nm, and the mean NP size was 14.6 nm. *E. coli* DH5α, a Gram-negative bacterium, and *Bacillus subtilis* LSSE52, a Gram-positive bacterium, were used to determine the antibacterial effect of the AgNPs. Two different methods were used to determine the antibacterial activity. According to the first method, the tested bacteria were incubated in liquid media containing AgNP at different concentrations, followed by spectrophotometric measurements were done. *B. subtilis* growth was almost inhibited completely (96.72%) in the medium containing 9 μg/mL AgNPs, while *E. coli* was inhibited by 71.90%. In well diffusion method, the largest inhibition zone diameters against both bacteria were detected around the well containing the highest concentration (120 μg/mL) of AgNPs. *B. subtilis* was more sensitive to these NPs than *E. coli* (Wei et al. 2012).

Bacillus methylotrophicus DC3, isolated from the soil of Korean ginseng, was tested as the AgNP producer organism (Wang et al. 2016). The supernatant of the liquid bacterial cultures synthesized spherical AgNPs with the size between 10 nm and 30 nm. Antimicrobial activities of these AgNPs against four different pathogenic microorganisms (*Candida albicans* (KACC 30062), *Salmonella enterica* (ATCC 13076), *Escherichia coli* (ATCC 10798), and *Vibrio parahaemolyticus* (ATCC 33844)) were investigated by disc diffusion method. Significant inhibition zones occurred around the discs containing AgNPs for all the microorganisms tested (Wang et al. 2016).

Brevibacterium frigoritolerans DC2 was used to produce AgNPs extracellularly, and the antimicrobial activities of the synthesized AgNPs on various pathogenic microorganisms were investigated (Singh et al. 2015). Spherical AgNPs with 10–30 nm were obtained. Antimicrobial activity of the AgNPs was studied by using well diffusion method. In order to investigate the synergistic effect of AgNPs with the antibiotics, the commercial antibiotics named as lincomycin, oleandomycin, novobiocin, vancomycin, penicillin G, and rifampicin were separately tested on bacteria, while cycloheximide was tested on yeast. Antimicrobial effects of $AgNO_3$ and AgNPs and the combinations of commercial antibiotics with AgNPs against pathogenic microorganisms such as *Vibrio parahaemolyticus* ATCC 33844, *Bacillus anthracis* NCTC 10340, *Salmonella enterica* ATCC 13076, *Bacillus cereus* ATCC 14579, *Escherichia coli* ATCC 10798, and *Candida albicans* KACC 30062 were determined by measuring the inhibition zones. AgNP showed a stronger antimicrobial effect in comparison with $AgNO_3$ against all tested pathogenic microorganisms. Furthermore, antimicrobial activities of the commercial antibiotics on microorganisms were enhanced by addition of the synthesized AgNPs (Singh et al. 2015).

Well dispersed, spherical, and 5–50 nm AgNPs could be obtained using the extracellular extract of a newly isolated *Exiguobacterium* sp. KNU1. Three different Gram-negative bacteria (*Salmonella typhimurium*, *Pseudomonas aeruginosa*, and *Escherichia coli*) and Gram-positive bacterium (*Staphylococcus aureus*) were used for antimicrobial studies. Microdilution studies showed that these NPs were the most effective on *Escherichia coli* and the least effective on *Staphylococcus aureus*. It was concluded that this difference is probably from different cell walls of the Gram-negative bacterium *Escherichia coli*, which have a thin and permeable cell wall compared to *Staphylococcus aureus*, a Gram-positive bacterium. Accordingly, the AgNPs passing easily through *Escherichia coli* cell wall damaged the cell membrane and subsequently caused cell death by causing breaks in the double-stranded DNA (Tamboli and Lee 2013).

Kanmani and Lim (2013) used the bacterial exopolysaccharide as both the reducing and stabilizing agents. AgNPs in various sizes (2–15 nm) and shapes (predominantly spherical, triangular, rod, and hexagonal) were obtained. The antibacterial activities of different concentrations of AgNPs were determined with the agar well diffusion method by using the pathogen *Escherichia coli* ATCC 35218, *Listeria monocytogenes* ATCC 15313, a food-borne pathogen bacterium, and also the multidrug-resistant pathogens *Klebsiella pneumoniae* ATCC 27736 and *Pseudomonas aeruginosa* ATCC 15442. Similarly, the antifungal effects of AgNPs tested at different concentrations were performed with the agar well diffusion method by using *Aspergillus* spp. and *Penicillium* spp. as the pathogen fungi. Furthermore, antibiofilm activities of the AgNPs against the bacterial pathogens like *Escherichia coli* ATCC 35218, *Bacillus cereus* ATCC 10987, *Listeria monocytogenes* ATCC 15313, and *Pseudomonas aeruginosa* ATCC 15442 were also investigated. The results from the antibacterial studies indicated that all the tested bacteria were inhibited at high rates due to the increased AgNP concentration. The order of antibacterial effect caused by AgNPs on the tested bacteria was *Pseudomonas aeruginosa* > *Escherichia coli* > *Klebsiella pneumoniae* > *Listeria monocytogenes*. Both fungi were inhibited due to the increase in AgNP concentration, but *Aspergillus* spp. were more inhibited by AgNPs compared to *Penicillium* spp. While the AgNPs at a concentration of 56 μg/mL caused a biofilm inhibition of 100% in *Escherichia coli* and *Pseudomonas aeruginosa*, the biofilm inhibition in *Listeria monocytogenes* was detected as approximately 32% (Kanmani and Lim 2013).

Singh et al. (2017) synthesized AgNPs by using *Kinneretia* THG-SQI4 and investigated the antimicrobial activities of these AgNPs. The researchers used cell-free culture supernatant and synthesized spherical AgNPs with 15–25 nm. Antimicrobial activity was determined by disc diffusion method using two yeast species (*Candida albicans* KACC 30062, *Candida tropicalis* KCTC 7909), three Gram-positive bacteria (*Bacillus cereus* ATCC 14579, *Bacillus subtilis* KACC 14741, *Staphylococcus aureus* ATCC 6538), and also three Gram-negative bacteria such as *Pseudomonas aeruginosa* ATCC 6538, *Escherichia coli* ATCC 10798, and *Vibrio parahaemolyticus* ATCC 33844. The greatest inhibition zone (20.5 mm) was detected for *Candida albicans*, while the lowest inhibition zone (11.5 mm) was determined for *Bacillus cereus* and *Bacillus subtilis*. It was stated that the differences in antimicrobial activities might be due to differences in the cell structures of the tested microorganisms, their physiologies, metabolisms, and interactions with the charged AgNPs. In addition, the synergistic effect of AgNP with various antibiotics was also tested. *Salmonella enterica*, *Escherichia coli*, and *Pseudomonas aeruginosa* were resistant to all investigated antibiotics (erythromycin, lincomycin, novobiocin, penicillin G, vancomycin, and oleandomycin), but it was observed that *Vibrio parahaemolyticus* was sensitive to these antibiotics. According to the results of the antibacterial studies performed to determine the synergistic effects of AgNPs, the inhibition zones were observed in multidrug-resistant bacteria cultures when AgNPs were added to the antibiotics. However, the addition of AgNPs to the antibiotics resulted in a significant increase in the inhibition zones of *Vibrio parahaemolyticus* cultures, which were susceptible to the antibiotics tested (Singh et al. 2017).

Adebayo-Tayo et al. (2017) obtained AgNPs using the culture-free supernatants of two lactic acid bacteria, *Lactobacillus casei* LPW2 and *Lactobacillus fermentum* LPF6, and investigated various characteristic properties such as shape, size, and antibacterial activity of the obtained AgNPs (Adebayo-Tayo et al. 2017). SEM micrographs indicated that AgNPs from *Lactobacillus casei* supernatant were spherical in shape and their sizes varied between 0.7 nm and 10 nm, and AgNPs synthesized by the supernatant of *Lactobacillus fermentum* were in various forms (partially aggregated particles), and their sizes ranged between 1.4 nm and 10 nm. The antibacterial activity of the AgNPs was investigated using agar well diffusion method on some selected pathogenic indicator bacteria (*Bacillus* sp., *Streptococcus pyogenes*, *Staphylococcus aureus*, *Klebsiella* sp., and *Pseudomonas aeruginosa*). Among the tested bacteria, *Bacillus* sp. was determined as the most sensitive bacterium. On the other hand, *Pseudomonas aeruginosa* was detected as the most resistant bacterium. The antibacterial studies showed that the tested Gram-positive bacteria were more sensitive to AgNPs than Gram-negative ones and AgNPs obtained from the supernatant of *Lactobacillus casei* had a stronger antibacterial effect against the indicator bacteria used (Adebayo-Tayo et al. 2017).

Spherical AgNPs (30–90 nm) could be obtained using the culture supernatant of *Nocardiopsis* sp. MBRC-1 (Manivasagan et al. 2013). The researchers used various bacteria (*Escherichia coli* ATCC 10536, *Bacillus subtilis* ATCC 6633, *Enterococcus hirae* ATCC 10541, *Pseudomonas aeruginosa* ATCC 27853, *Shigella flexneri* ATCC 12022, *Staphylococcus aureus* ATCC 6538) and fungi (*Aspergillus niger* ATCC 1015, *Aspergillus brasiliensis* ATCC 16404, *Aspergillus fumigates* ATCC 1022, *Candida albicans* ATCC 10231) to detect the antimicrobial effects and MIC values of the AgNPs. Antibacterial and antifungal activities of AgNPs were determined by well diffusion method. Concentration-dependent antimicrobial activity was detected. The results of the study showed that the antimicrobial effects of these NPs at the concentrations of 40 and 50 μg/mL were stronger than amoxicillin and nystatin antibiotics. The maximum antimicrobial activity was against *Bacillus subtilis*, *Pseudomonas aeruginosa*, and *Candida albicans*. Moreover, MIC values for *B. subtilis*, *P. aeruginosa*, and *C. albicans* were 7, 10, and 10 μg/mL, respectively (Manivasagan et al. 2013).

The cell extract of the cyanobacterium *Nostoc* sp. strain HKAR-2 was used for green synthesis of AgNPs (Sonker et al. 2017). SEM and TEM images showed that the synthesized AgNPs were spherical in shape and their dimensions were between 51 nm and 100 nm. The antimicrobial activities of AgNPs at different concentrations (5, 10, and 15 μg/mL) were investigated against two plant pathogenic bacteria (*Ralstonia solanacearum* and *Xanthomonas campestris*) and two fungi (*Aspergillus niger* and *Trichoderma harzianum*) by using well diffusion method. While the inhibition zone diameters caused by AgNPs at 5, 10, and 15 μg/mL concentrations were detected as 15, 25, and 25 mm for *R. solanacearum*, respectively, the inhibition zone diameters at the same concentrations for *X. campestris* were determined as 18, 23, and 23 mm, respectively. The inhibition zones caused by pure $AgNO_3$ were determined as 11 and 13 mm for *R. solanacearum* and *X. campestris*, respectively. AgNPs at 5, 10, and 15 μg/mL concentrations caused the inhibition zone of 0.3, 0.3, and 0.5 mm for *A. niger* and 0.4, 0.4, and 0.5 mm inhibition zones for *T. harzianum*, respectively (Sonker et al. 2017).

Supernatant of *Novosphingobium* sp. THG-C3, a bacterial strain isolated from soil, was used for AgNP biosynthesis (Du et al. 2017). Most of AgNPs synthesized were spherical in shape and 8–25 nm in size. The antimicrobial activities of AgNPs (500 mg/L) against Gram-positive (*Staphylococcus aureus*, *Bacillus subtilis*, and *Bacillus cereus*) and Gram-negative (*Pseudomonas aeruginosa*, *Escherichia coli*, *Vibrio parahaemolyticus*, and *Salmonella enterica*) bacteria and also two fungi (*Candida tropicalis* and *Candida albicans*) were determined by disc diffusion method. The order of the inhibition zone diameters of the tested microorganisms was detected as *Candida albicans* (19.7 mm) > *Candida tropicalis* (18.7 mm) > *Vibrio parahaemolyticus* (16.3 mm) > *Pseudomonas aeruginosa* (16.0 mm) > *Staphylococcus aureus* (15.3 mm) > *Bacillus cereus* (13.7 mm) > *Bacillus subtilis* (13.0 mm) ≥*Escherichia coli* (13.0 mm) > *Salmonella enterica* (12.7 mm). In addition, the combinations of the synthesized AgNPs with commercial antibiotics (erythromycin, novobiocin, oleandomycin, lincomycin, penicillin G, and vancomycin) were also treated with some bacteria (*Pseudomonas aeruginosa*, *Salmonella enterica*, *Escherichia coli*, and *Vibrio parahaemolyticus*), and the combinations of AgNPs with the commercial antibiotics showed stronger antibacterial activity than antibiotics alone (Du et al. 2017).

In a study conducted by Thomas et al. (2014), the AgNP production ability of *Ochrobactrum* sp. isolated from the seawater was investigated. AgNPs synthesized by the bacterial supernatant were spherical in shape and 38–85 nm in size. The researchers used four indicator bacteria including *Salmonella typhi*, *Salmonella paratyphi*, *Vibrio cholerae*, and *Staphylococcus aureus* for finding the antibacterial capacities of the obtained AgNPs. $AgNO_3$ did not cause any inhibition on the bacteria. On the other hand, AgNPs caused the inhibition zones of 14, 15, 16, and 15 mm for *S. typhi*, *S. paratyphi*, *V. cholerae*, and *S. aureus*, respectively (Thomas et al. 2014).

Skladanowski et al. (2016) investigated the AgNP synthesis capacity of *Streptomyces kasugaensis* strain NH28 and also the antibacterial properties of the synthesized AgNPs. TEM analysis showed that the AgNPs were polydispersed, spherical in shape, and 4.2–65 nm in size. The order of the antibacterial effect of

AgNPs on the indicator bacteria tested was *Staphylococcus aureus* ≥ *Klebsiella pneumoniae* ≥ *Proteus mirabilis* ≥ *Escherichia coli* > *Bacillus subtilis* > *Salmonella infantis* > *Pseudomonas aeruginosa*. The lowest MIC value (1.25 μg/mL) of the synthesized AgNPs inhibited the growth of *S. aureus*, *K. pneumoniae*, *P. mirabilis*, and *E. co*li at 48, 29, 23, and 14% rates, respectively. While the MIC value of the obtained AgNPs against *B. subtilis* was found to be 2.5 μg/mL, it was detected as 10 μg/mL against *P. aeruginosa* and *S. infantis* (Skladanowski et al. 2016).

Otari et al. (2015) used *Rhodococcus* NCIM 2891 for the intracellular synthesis of AgNPs and also investigated the antibacterial effects of the produced AgNPs on pathogenic Gram-positive (*Staphylococcus aureus*) and Gram-negative (*Klebsiella pneumoniae*, *Proteus vulgaris*, *Enterococcus faecalis*, *Pseudomonas aeruginosa*, and *Escherichia coli*) bacteria. The NPs obtained were spherical in shape and ranged from 5 to 50 nm in size. The bactericidal and bacteriostatic effects of the obtained AgNPs were investigated by monitoring the different growth phases of the indicator pathogenic bacteria treated with AgNPs (10, 30, and 50 μg/mL of medium) at various concentrations. The lowest concentration of AgNPs (10 μg/mL) was determined as the most effective inhibitory concentration for *Klebsiella pneumoniae*. This low AgNP concentration (10 μg/mL) caused only the log phase of other organisms to be delayed. AgNPs at 30 μg/mL concentration indicated a partially inhibitory effect on the growth of *Escherichia coli*, *Staphylococcus aureus*, and *Pseudomonas aeruginosa*. The highest concentration of AgNP tested, 50 μg/mL, completely inhibited the growth of all bacteria (Otari et al. 2015).

2.5.3 Fungi

Several fungi can synthesize AgNPs. There are various studies on antimicrobial activity of the AgNPs synthesized by fungal sources (Table 2.3). Jaidev and Narasimha (2010) used *Aspergillus niger* for fungal mediated biosynthesis of AgNPs and investigated the antimicrobial activity of the synthesized AgNPs. The AgNPs formed were in polydispersed form and spherical shape with dimensions of 3–30 nm. *Aspergillus niger*, *Staphylococcus* sp., *Bacillus* sp., and also *Escherichia coli* were used as the test organisms for detection of the antimicrobial activity of AgNP synthesized extracellularly by *Aspergillus niger*. AgNPs showed antimicrobial activity against *Aspergillus niger* as well as against Gram-positive (*Staphylococcus* sp. and *Bacillus* sp.) and Gram-negative (*Escherichia coli*) bacteria. The inhibition zone diameters caused by AgNPs against *Aspergillus niger*, *Staphylococcus* sp., *Bacillus* sp., and *Escherichia coli* were measured as 1.2, 0.9, 0.8, and 0.8 cm, respectively (Jaidev and Narasimha 2010).

The boiled and filtered biomass filtrate of *Aspergillus terreus* was also able to produce AgNPs. Antibacterial effects of these AgNPs against nine reference bacteria (*Pseudomonas aeruginosa* ATCC 27853, *Serratia marcescens* ATCC 27137, *Shigella flexneri* ATCC 12022, *Salmonella typhi* ATCC 13311, *Escherichia coli* ATCC 25922, *Proteus mirabilis* ATCC 43071, *Klebsiella pneumoniae* ATCC

Table 2.3 Antimicrobial activities of AgNPs biosynthesized by fungi

Fungi	Nanoparticle shape	Nanoparticle size (nm)	Antimicrobial effect on	References
Aspergillus niger	Spherical	3–30	*Aspergillus niger*, *Staphylococcus* sp., *Bacillus* sp., *Escherichia coli*	Jaidev and Narasimha (2010)
Aspergillus terreus	Spherical, oval	16.54	*Pseudomonas aeruginosa* ATCC 27853, *Serratia marcescens* ATCC 27137, *Shigella flexneri* ATCC 12022, *Salmonella typhi* ATCC 13311, *Escherichia coli* ATCC 25922, *Proteus mirabilis* ATCC 43071, *Klebsiella pneumoniae* ATCC 700603, *Enterococcus faecalis* ATCC 29212, *Staphylococcus aureus* ATCC 259323, *Escherichia coli* MDREC1, *Klebsiella pneumoniae* MDRKP2, *Pseudomonas aeruginosa* MDRPA3	Rani et al. (2017)
Aspergillus versicolor	Spherical	5–30	*Sclerotinia sclerotiorum*, *Botrytis cinerea*	Elgorban et al. (2016)
Colletotrichum sp. ALF2–6	Spherical, triangular, hexagonal	5–60	*Escherichia coli* MTCC 7410, *Salmonella typhi* MTCC 733, *Bacillus subtilis* MTCC 121, *Staphylococcus aureus* MTCC 7443	Azmath et al. (2016)
Cryphonectria sp. (NCBI accession no. HQ432805)	–	30–70	*Staphylococcus aureus* ATCC 25923, *Escherichia coli* ATCC 39403, *Salmonella typhi* ATCC 51812, *Candida albicans* accession no. NCIM-3100	Dar et al. (2013)
Saccharomyces cerevisiae PTCC 5052	Spherical	5–20	Fluconazole-susceptible and fluconazole-resistant strains of *Candida albicans*	Niknejad et al. (2015)

700603, *Enterococcus faecalis* ATCC 29212, and *Staphylococcus aureus* ATCC 259323) and also three multidrug-resistant bacteria (*Escherichia coli* MDREC1, *Klebsiella pneumoniae* MDRKP2, and *Pseudomonas aeruginosa* MDRPA3) were screened using agar well diffusion method. The MIC value of the AgNPs was also determined in the antimicrobial studies. Antibacterial tests indicated that the antibacterial activities of AgNP increase due to the increase in AgNP concentration, whereas the fungal filtrate and $AgNO_3$ used as the controls had no any antibacterial effect on the bacteria. Among the tested bacterial strains, *Salmonella typhi* ATCC 13311 was detected as the most susceptible bacterium with the highest zone of inhibition (16.67 mm). The lowest inhibition zone was observed in the multidrug-resistant bacterium *Klebsiella pneumoniae* MDRKP2 (13.33 mm). On the other hand, *Klebsiella pneumoniae* MDRKP2 was found the least susceptible bacterium. Furthermore, the MIC values observed for *Salmonella typhi* ATCC 13311 and *Klebsiella pneumoniae* MDRKP2 were 11.43 and 308 μg/mL, respectively (Rani et al. 2017).

AgNPs were also synthesized using biomass filtrate of *Aspergillus versicolor* (Elgorban et al. 2016). The TEM results revealed that the synthesized AgNPs were spherical in shape and varied in size from 5 to 30 nm. Pathogenic white mold (*Sclerotinia sclerotiorum*) and gray mold (*Botrytis cinerea*) isolated from strawberry were used as the test organisms to observe the antifungal effects of biosynthesized AgNPs. The highest inhibitory effect on the fungal growth was observed at 150 mg/L for both of the fungi. Accordingly, the highest inhibition rates were 80.38 and 74.39% for *Sclerotinia sclerotiorum* and *Botrytis cinerea* treated with 150 mg/L AgNPs, respectively (Elgorban et al. 2016).

Azmath et al. (2016) utilized the cell-free extract of the endophytic fungus *Colletotrichum* sp. ALF2–6 for mycosynthesis of AgNPs. The obtained NPs were evaluated for their bactericidal effect against *Escherichia coli* MTCC 7410, *Salmonella typhi* MTCC 733, *Bacillus subtilis* MTCC 121, and *Staphylococcus aureus* MTCC 7443. The NPs obtained were polydispersed form with spherical, triangular, and hexagonal shapes and 5–60 nm in size. The antibacterial studies showed that the number of the colony gradually reduced when the concentration of NPs increased from 0 to 100 μg/mL. Among the pathogenic bacteria tested, *Staphylococcus aureus* was more sensitive to AgNPs at 50 μg/mL concentration, while the other tested bacteria are susceptible to AgNPs at 100 μg/mL concentration. Similarly, MIC value of AgNPs was measured as 12.5 μg/mL for *Staphylococcus aureus*, *Salmonella typhi*, and *Bacillus subtilis*, while this value was determined as 25 μg/mL for *Escherichia coli* (Azmath et al. 2016).

Dar et al. (2013) studied the AgNPs' synthesis activity of the biomass filtrate of the newly identified pathogenic bacterial isolate *Cryphonectria* sp. (NCBI accession no. HQ432805). The synthesized AgNPs displayed a little monodispersity in the range of 30–70 nm in size. The antimicrobial effects of AgNPs were detected by disc diffusion method. In order to investigate the synergistic effect of the AgNPs with the standard antibiotics, the combinations of AgNPs with streptomycin and amphotericin were also tested. According to the inhibition zone measurements, the synthesized AgNPs exhibited more antimicrobial activity on the tested indicator microorganisms compared to $AgNO_3$ and the standard antibiotics. In addition, the combinations of the synthesized AgNPs with the utilized standard antibiotics caused the significant increases in the antimicrobial activities (Dar et al. 2013).

Saccharomyces cerevisiae PTCC 5052 was also tested to synthesize AgNPs and investigate the antifungal activity of the obtained AgNPs against the strains of *Candida albicans* ATCC 10261. The obtained AgNPs were in spherical shape and predominantly smaller than 50 nm in size (5–20 nm). Studies on antimicrobial activity demonstrated that the MIC values (μg/mL) of AgNPs for both fluconazole-susceptible and fluconazole-resistant *Candida albicans* strains were similar and ranged from 2 to 4 μg/mL (Niknejad et al. 2015).

2.6 Biologically Prepared AgNPs as Cytotoxic Agents

2.6.1 Algae

Various algal sources could be used to produce AgNPs with cytotoxic activity (Table 2.4). The macroalga *Gracilaria edulis* extracts were used for the production of biogenic AgNPs with anticancer activity. Anticancer activity of these AgNPs was tested against human prostate cancer cell lines (PC3), and normal African monkey kidney cell line was used as healthy cells. While IC_{50} value against PC3 cells was 39.60 μg/mL, it was 68.49 μg/mL against Vero cells after 48 h incubation (Priyadharshini et al. 2014).

Biogenic AgNPs with 39.5 μg/mL IC_{50} value against HT29 cell line were synthesized using the macroalga *Gelidiella* sp. (Devi and Bhimba 2013). The spherical AgNPs with an average size about 43 nm, biosynthesized with *Ecklonia cava* extracts, had IC_{50} value of 59 μg/mL against human cervical cancer cells (Venkatesan et al. 2016). AgNPs biosynthesized from *Turbinaria turbinata* marine alga also showed anticancer effect against Ehrlich cell carcinoma (ECC) in mice. It was reported that AgNPs induced dose-dependent reduction in tumor size and also apoptosis via caspase-3 activation (El-Sonbaty 2013).

2.6.2 Bacteria

Bacteria are effective microbial sources for NP synthesis. The AgNPs obtained from bacterial sources have cytotoxic activity against several cancer cell lines (Table 2.5). AgNPs synthesized by *Bacillus tequilensis* and *Calocybe indica* showed cytotoxic effect on breast cancer cell line with the IC_{50} values of 10 μg/mL and 2 μg/mL, respectively. Leakage of LDH, activation of ROS, and terminal deoxynucleotidyl transferase dUTP nick-end labeling (TUNEL)-positive cells in MDA-MB-231 breast cancer cells confirmed the dose-dependent toxic effect of AgNPs. Meanwhile, Western blot analyses showed that these NPs induce cellular apoptosis via activation of p53, p-Erk1/2, and caspase-3 signaling and downregulation of Bcl-2 (Gurunathan et al. 2015). Furthermore, same research team also obtained stable AgNPs (about 20 nm) using *Bacillus funiculus* and tested their cytotoxic effects on

Table 2.4 Cytotoxic effect of AgNPs biosynthesized by algae

Alga	Shape	Size	Cytotoxicity on	References
Gracilaria edulis	Spherical	55–99	PC3	Priyadharshini et al. (2014)
Gelidiella sp.	Cubical		HT29	Devi and Bhimba (2013)
Ecklonia cava	Nearly spherical	43	HeLa	Venkatesan et al. (2016)
Turbinaria turbinata	–	8–16 nm	Ehrlich cell carcinoma	El-bialy et al. (2017))

Table 2.5 Cytotoxic effect of AgNPs biosynthesized by bacteria

Bacteria	Shape	Size (nm)	Cytotoxicity on	References
Bacillus tequilensis	Spherical	20	MDA-MB-21	Gurunathan et al. (2015)
Bacillus funiculus	Spherical	10–20	MDA-MB-21	Gurunathan et al. (2013a, b)
Anabaena doliolum	Spherical	10–50	Dalton's lymphoma, colo205	Singh et al. (2014)
Nocardiopsis valliformis	Spherical	5–50	HeLa	Rathod et al. (2016)
Nostoc sp. HKAR-2	Spherical	51–100	MCF-7	Sonker et al. (2017)
Pseudomonas putida	Spherical	6–10	HEp-2	Gopinath et al. (2017)
Streptomyces sp. NH28	Spherical	4.2–65	L929	Skladanowski et al. (2016)
Streptomyces xinghaiensis	Spherical	64 (±49)	HeLa, mouse fibroblast (normal cell line}	Wypij et al. (2018)

MDA-MB-231 cell line. AgNPs showed dose-dependent cytotoxicity on breast cancer cell line through activation of lactate dehydrogenase (LDH), caspase 3, and ROS generation, eventually leading to induction of apoptosis. They reported that these AgNPs inhibit the growth of the tumor cells and offer a new method to develop molecule for cancer therapy (Gurunathan et al. 2013a, b).

AgNPs with 10–50 nm in size were biosynthesized by cell extracts of the cyanobacterium *Anabaena doliolum*, and their cytotoxic effect on Dalton's lymphoma and human carcinoma colo205 was tested. These AgNPs affected the survival of both cell types through induction of ROS generation and DNA fragmentation. It was concluded that DNA fragmentation is probably due to the activation of intracellular caspase enzyme and oxidative stress. It was also reported that AgNP treatment inhibits the growth of tumor and cancer cells and induces apoptosis (Singh et al. 2014).

Spherical and polydispersed AgNPs about 5–50 nm in size were biosynthesized by the cell-free filtrate of alkaliphilic actinobacterium *Nocardiopsis valliformis* OT1 strain, and their IC_{50} value against human cervical cancer cell line (HeLa) was found to be 100 μg/mL. These AgNPs showed high cytotoxic effect on these cells (Rathod et al. 2016).

The aqueous extract of cyanobacterium *Nostoc* sp. strain HKAR-2 was also able to produce AgNPs. These AgNPs (51–100 nm) from cyanobacterium showed a dose-dependent cytotoxic activity against MCF-7 cells, and the IC_{50} value on human breast cancer MCF-7 cells was 27.5 μg/mL (Sonker et al. 2017). While the AgNPs synthesized using culture supernatant of *Pseudomonas putida* MVP2 showed no cytotoxicity under 25 μg/mL, it was toxic above 50 μg/mL on human epidermoid larynx carcinoma (Hep-2) cells (Gopinath et al. 2017).

AgNPs produced by *Streptomyces xinghaiensis* OF1 strain showed dose dependent in vitro cytotoxic activity against mouse fibroblasts (3 T3) and HeLa cell lines. The IC_{50} values on 3 T3 and HeLa cell line were 4 and 3.8 μg ml/L, respectively (Wypij et al. 2018). Kulkarni et al. (2015) also reported the dose-dependent decrease in percent viability of the cells treated with AgNPs synthesized using *D. radiodurans*.

The LD_{50} of AgNPs was 7–8 μg/mL against MCF-7 human breast cancer cell line. AgNPs produced using the cell filtrate of *Streptomyces* sp. NH28 were also tested for the cytotoxicity against L929 mouse fibroblast cell line. The IC_{50} value detected was 64.5 μg/mL on mouse fibroblast cell line. Therefore, it was reported that these biogenic AgNPs exhibit low cytotoxicity and thus, AgNPs could be a promising and safe antibacterial agents (Skladanowski et al. 2016). AgNPs (93 nm) produced by cell-free culture supernatant of *Stenotrophomonas maltophilia* showed no toxicity to liver function, RBCs, splenocytes, and HeLa cell. These biogenic AgNPs were reported as safe to use in drug formulations (Oves et al. 2013).

2.6.3 Fungi

Fungal sources have a great efficiency to produce various AgNPs with cytotoxic activity (Table 2.6). The biosynthesis of AgNPs with in vitro cytotoxic activity on HCT-116, MCF-7, and HepG2 cell lines, using cell filtrates of *Aspergillus fumigatus*

Table 2.6 Cytotoxic effect of AgNPs biosynthesized by fungi

Fungal species	Shape	Size (nm)	Cytotoxicity on	References
Aspergillus fumigatus, Candida albicans, Penicillium italicum, Syncephalastrum racemosum, Fusarium oxysporum, Aspergillus ochraceus	Spherical	13.88 ± 4.11	HCT-116, MCF-7, HepG2	Magdi et al. (2014)
Agaricus bisporus	Spherical	8–20	MCF-7	El-sonbaty (2013)
Fusarium oxysporum	Spherical	5–13	MCF-7	Husseiny et al. (2015)
Fusarium oxysporum	Spherical	40 ± 5	MCF-7	Salaheldin et al. (2016)
Ganoderma neo-japonicum Imazeki	Spherical	5	MDA-MB-231	Gurunathan et al. (2013b)
Guignardia mangiferae	Spherical	5–30	HeLa, MCF-7, Vero (normal cell line}	Balakumaran et al. (2015)
Humicola sp.	Spherical	5–25	NIH3T3, MDA-MB-231	Syed et al. (2013)
Cryptococcus laurentii (BNM 0525}	–	35 ± 10	MCF7, T47D, MCF10-A	Ortega et al. (2015)
Penicillium aurantiogriseum (IMI 89372}	Spherical	12.7	MCF-7, MCT	Elshawy et al. (2016)
Pleurotus ostreatus	Spherical	4–15	MCF-7	Yehia and Al-Sheikh (2014)
Pestalotiopsis microspora VJ1/VS1	Spherical	2–10	B16F10, SKOV3, A549, PC3	Netala et al. (2016)

(RCMB 02568), *Candida albicans* (RCMB 05031), *Penicillium italicum* (RCMB 03924), *Syncephalastrum racemosum* (RCMB 05922), *Fusarium oxysporum* (RCMB 08213), and *Aspergillus ochraceus* (RCMB 036254), was reported. AgNPs had a potent cytotoxic activity with IC_{50} values of 1.2 μg/mL, 1.4 μg/mL, and 2.1 μg/mL against human hepatocellular carcinoma (HepG2), human colon carcinoma (HCT-116), and human breast cancer (MCF-7) cells, respectively. On the other hand, 39.6 μg/mL induced 50% of normal Vero cell mortality (Magdi et al. 2014).

Spherical AgNPs with 8–20 nm in size were also synthesized using aqueous extract of *Agaricus bisporus*. These AgNPs showed dose-dependent cytotoxicity against MCF-7 breast cancer cells with IC_{50} value of 50 μg/mL (El-Sonbaty 2013). In another study on cytotoxicity of biogenic AgNPs on MCF-7 cells, it was reported that AgNPs synthesized with *Fusarium oxysporum* also showed cytotoxic activity on MCF-7 with the IC_{50} of 121.23 μg/mL (Husseiny et al. 2015). Salaheldin et al. (2016) described the cytotoxic effect of AgNPs synthesized using aqueous filtrate of the biomass and also aqueous mycelial-free filtrate of *Fusarium oxysporum*, against a human breast carcinoma cell line (MCF-7) and the normal WISH cell line. The IC_{50} was 14 μg/mL for MCF-7 cells and 42 μg/mL for WISH cells (Salaheldin et al. 2016).

The cytotoxic effect of biologically synthesized AgNPs using hot aqueous extracts of the mycelia of *Ganoderma neo-japonicum* Imazeki on human breast cancer cells (MDA-MB-231) was also examined. These AgNPs were found to be significantly toxic to the cells at concentrations of 6 μg/mL and higher after 24 h of incubation. The authors suggested that AgNPs possess cytotoxic effects with apoptotic features and proposed that the ROS generated by AgNPs have a significant role in apoptosis (Gurunathan et al. 2013a, b).

AgNPs produced by the mycelial-free filtrate of *Guignardia mangiferae* showed cytotoxic activity against African monkey kidney (Vero), HeLa (cervical), and MCF-7 (breast) cells. The IC_{50} values were 63.37, 27.54, and 23.84 μg/mL against Vero, HeLa, and MCF-7 cells, respectively, after 24-h incubation. These NPs had higher rate of cytotoxicity against MCF-7 cells followed by HeLa cells (Balakumaran et al. 2015).

Thermophilic fungus *Humicola* sp. on NIH3T3 was also able to produce AgNPs. The cell viability of the mouse embryonic fibroblast cell line and MDA-MB-231 human breast carcinoma cell line was reduced especially in high AgNP concentrations, and these NPs were nontoxic to cancer and normal cells up to concentrations of 50 μg/mL and, thus, could be used in targeted drug delivery systems (Syed et al. 2013).

The cytotoxic effect of biosynthesized AgNPs by culture supernatant of *Cryptococcus laurentii* (BNM 0525) against MCF7 and T47D human breast cancer cells and MCF10-A normal breast cells was described. AgNPs showed higher inhibition efficacy in tumor lines than in normal lines of breast cells. It was reported that cell viability of breast tumor lines MCF7 and T47D was considerably decreased above 2.5 μg/mL concentration. This effect was probably due to the endocytosis activity. The apoptotic assays demonstrated that cell viability was induced by apoptotic mechanisms generated through the intrinsic pathway of caspase activation (Ortega et al. 2015)

Yehia and Al-Sheikh (2014) showed that it is possible to synthesize AgNPs using culture supernatant from *Pleurotus ostreatus* and it was determined that the AgNPs had effective cytotoxic activity against MCF-7 cell line. An increase in AgNPs' concentration induced the greater inhibition of cell proliferation (Yehia and Al-Sheikh 2014).

It was shown that AgNPs synthesized using aqueous culture filtrate of *Pestalotiopsis microspora* had a significant cytotoxic effect on different cell lines. The authors reported that these AgNPs selectively inhibit cell proliferation in all of cell line, but no significant effect was observed on normal cells (Netala et al. 2016).

2.7 Biological Response to Nanomaterials

Systematic researches of biological systems are not as simple as physical and chemical researches. Several works have been done to investigate the effects of NPs on microorganisms. Since microorganisms have different physiological properties, NPs show different activity on different microorganisms. It is clear that different species of microorganisms can exhibit potentially toxic converse susceptibility to nanomaterials. For example, the studies on comparative toxicity of processed silver nanocrystals, Gram-negative and Gram-positive bacteria, Gram-negative *E. coli*, and *S. oneidensis* have been found to be more resistant than Gram-positive *B. subtilis*. It has also been found that Gram-positive and Gram-negative organisms have sensitivity differences to various forms of nanoparticles (Marambio-Jones and Hoek 2010).

Typically, Gram-positive organisms have been found to be more susceptible to potentially toxic nanomaterials, and this increased susceptibility is likely due to differences in bacterial cell membrane and cell wall structures. Lipopolysaccharides of the outer membrane of Gram-negative bacteria constitute resistance bases against nanoparticles. In a study on toxic effects of copper and silver nanoparticles on Gram-negative (*E. coli*) and Gram-positive (*B. subtilis*) microorganisms have indicated that Gram-positive organisms have been affected more strongly by metal nanoparticles than gram-negative organisms (Yoon et al. 2007).

Despite the release of heavy metal ions (Cd^{2+}), toxicity is linked to the formation of the main hydroxyl radicals. It was also found that the toxic effects of titanium and Ag nanoparticles on Gram-positive microorganisms were higher than those of lipopolysaccharide layer deficiency (Li et al. 2011a, b).

Most of the studies evaluating nanoparticle-dependent microbial toxicity have used relatively well-characterized model microorganisms such as *E. coli*, *P. aeruginosa*, *S. aureus*, and *B. subtilis*. However, the diversity of microorganisms found in natural environments is astonishing and often gives similar responses to nanoparticles. Numerous studies have recently reported on the long-term effects of nanoparticles on soil, water, wastewater, and landfill ecosystems. Microorganisms that support plant growth, including rhizobacteria (*P. aeruginosa*, *P. putida*, *P. fluorescens*, *B. subtilis*) and nitrification and denitrification bacteria, have shown varying degrees of inhibition in both pure culture conditions and aqueous nanoparticle suspensions (Das et al. 2012). In a study with copper metal oxide nanoparticles

(80–160 nm) with different size distributions, it was found that antibacterial activity against all bacterial strains was demonstrated when evaluated in terms of antibacterial activity against *K. pneumoniae*, *P. aeruginosa*, *S. paratyphi*, and *Shigella*. It has been suggested that NPs provide this effect because they can cross cell membranes without encountering any barrier due to their size. The authors also suggested that NPs could form complexes with enzymes and proteins within the cell, thereby destroying cellular function that leads to cell death. In a study of the bacterial effects of TiO_2 and ZnO nanoparticles in the soil flora, it has been found that ZnNPs are more toxic than titanium NPs while inhibiting bacterial growth of the nanoparticles severely, resulting in a decrease in total bacterial mass (Ge et al. 2011). Transformation of polyhydroxyalkanoates and removal of glycogen and biological phosphorus are not affected, while TiO_2 NPs cause significant inhibition of biological nitrogen removal after prolonged exposure due to degradation of ammonium oxidation. In a study examining the effects of commercial AgNPs in natural river waters and sediments, it has been suggested that the particles reduce respiration relative to dissolved silver concentrations and lead to a ten times increase in phosphate concentration in stream water (Zheng et al. 2011).

The thing what is meant here is that not all transformed nanoparticles are poisoned. Conversely, recent research has shown that sulfur-rich, highly toxic silver ions in the wastewater treatment process can convert to stable silver sulfide (Ag_2S) form. A better understanding of environmental transformations will be possible by a better understanding of nanomaterial interactions with microbial systems (Colman et al. 2012).

The main problem in natural ecosystems such as soil and water is the occurrence of ROS compounds known to be toxic to the cells due to metal ions released from the nanomaterials. For this reason, the increase in the ability of the nanoparticles to produce ROS and the increase in their toxicity are directly proportional. Moreover, this concept of toxicity is more important because NPs have a much higher surface area than microparticles.

2.8 Conclusion

Nanotechnology is an atomic and/or molecular level engineering process and focuses on understanding, controlling, and producing physical, chemical, and biological systems in nanometer size. In other words, nanotechnology is defined as the functionalizing, creating, and manipulating of materials, tools, and systems in molecular level. The target of this discipline is improving the obtained novel properties of these materials in every size and every field.

Different physical and chemical methods have been widely used for synthesis of MNPs, but they are expensive and harmful to the environment and living organisms. Therefore, many biological systems have become very popular to produce those materials. They can transform inorganic metal ions into MNPs via reduction processes. MNPs such as gold and silver are the most common ones due to their unique advantages in nano-world. Their antimicrobial and cytotoxic properties still have been evaluated, and many researches have been presented in literature as summarized in this chapter.

Briefly, green synthesis has significant potential advantages in comparison with those produced by eco-friendly and cost-effective approaches. Therefore, these opportunities of using living organisms for synthesis of nanomaterials will continue to interest researchers to investigate the new mechanisms of biosynthesis in the future.

References

Abdeen S, Praseetha P (2013) Diagnostics and treatment of metastatic cancers with magnetic nanoparticles. J Nanomedine Biotherapeutic Discov 3(2):115

Abdel-Aziz SM, Prasad R, Hamed AA, Abdelraof M (2018) Fungal nanoparticles: a novel tool for a green biotechnology. In: Prasad R, Kumar V, Kumar M, Shanquan W (eds) Fungal nanobionics. Springer Nature, Singapore, pp 61–87

Adebayo-Tayo BC, Popoola AO, Ajunwa OM (2017) Bacterial synthesis of silver nanoparticles by culture free supernatant of lactic acid bacteria isolated from fermented food samples. Biotechnol J Int 19(1):1–13

Ahmad A, Senapati S, Khan MI, Kumar R, Sastry M (2003) Extracellular biosynthesis of monodisperse gold nanoparticles by a novel extremophilic actinomycete, *Thermomonospora* sp. Langmuir 19(8):3550–3553

Ali K, Ahmed B, Dwivedi S, Saquib Q, Al-Khedhairy AA, Musarrat J (2015) Microwave accelerated green synthesis of stable silver nanoparticles with eucalyptus globulus leaf extract and their antibacterial and antibiofilm activity on clinical isolates. PLoS One 10(7):e0131178

Anastas PT, Warner JC (1998) Green chemistry: theory and practice. Oxford University Press, New York

Arias LR, Yang L (2009) Inactivation of bacterial pathogens by carbon nanotubes in suspensions. Langmuir 25(5):3003–3012

Ashour AA, Raafat D, El-Gowelli HM, El-Kamel AH (2015) Green synthesis of silver nanoparticles using cranberry powder aqueous extract: characterization and antimicrobial properties. Int J Nanomedicine 10:7207–7221

Aziz N, Fatma T, Varma A, Prasad R (2014) Biogenic synthesis of silver nanoparticles using *Scenedesmus abundans* and evaluation of their antibacterial activity. J Nanopart. https://doi.org/10.1155/2014/689419

Aziz N, Faraz M, Pandey R, Sakir M, Fatma T, Varma A, Barman I, Prasad R (2015) Facile algae-derived route to biogenic silver nanoparticles: synthesis, antibacterial and photocatalytic properties. Langmuir 31(42):11605–11612

Aziz N, Pandey R, Barman I, Prasad R (2016) Leveraging the attributes of *Mucor hiemalis*-derived silver nanoparticles for a synergistic broad-spectrum antimicrobial platform. Front Microbiol 7:1984

Aziz N, Faraz M, Sherwani MA, Fatma T, Prasad R (2019) Illuminating the anticancerous efficacy of a new fungal chassis for silver nanoparticle synthesis. Front Chem 7:65. https://doi.org/10.3389/fchem.2019.00065

Azmath P, Baker S, Rakshith D, Satish S (2016) Mycosynthesis of silver nanoparticles bearing antibacterial activity. Saudi Pharm J 24(2):140–146

Balakumaran M, Ramachandran R, Kalaichelvan P (2015) Exploitation of endophytic fungus, *Guignardia mangiferae* for extracellular synthesis of silver nanoparticles and their in vitro biological activities. Microbiol Res 178:9–17

Belliveau BH, Starodub ME, Cotter C, Trevors JT (1987) Metal resistance and accumulation in bacteria. Biotechnol Adv 5(1):101–127

Bhatia S (2016) Nanoparticles types, classification, characterization, fabrication methods and drug delivery applications. Natural polymer drug delivery systems. Springer, Cham

Bhuyan T, Mishra K, Khanuja M, Prasad R, Varma A (2015) Biosynthesis of zinc oxide nanoparticles from *Azadirachta indica* for antibacterial and photocatalytic applications. Mater Sci Semicond Process 32:55–61

Booker RD, Boysen E (2011) Nanotechnology for dummies. John Wiley & Sons, Hoboken

Cao Y, Jin R, Mirkin CA (2001) DNA-modified core-shell Ag/Au nanoparticles. J Am Chem Soc 123(32):7961–7962

Colman BP, Wang SY, Auffan M, Wiesner MR, Bernhardt ES (2012) Antimicrobial effects of commercial silver nanoparticles are attenuated in natural streamwater and sediment. Ecotoxicology 21(7):1867–1877

Cossins D (2014) Next generation: nanoparticles augment plant functions. The scientist, exploring life, inspiring innovation, March 16. https://www.the-scientist.com/daily-news/next-generation-nanoparticles-augment-plant-functions-37804

Dar MA, Ingle A, Rai M (2013) Enhanced antimicrobial activity of silver nanoparticles synthesized by *Cryphonectria* sp. evaluated singly and in combination with antibiotics. Nanomedicine-Nanotechnol 9(1):105–110

Daraio C, Jin S (2012) Synthesis and patterning methods for nanostructures useful for biological applications. Nanotechnology for biology and medicine: at the building block level. Springer, New York

Das P, Xenopoulos MA, Williams CJ, Hoque ME, Metcalfe CD (2012) Effects of silver nanoparticles on bacterial activity in natural waters. Environ Toxicol Chem 31(1):122–130

Devi JS, Bhimba BV (2013) Silver nanoparticles: anti-bacterial and in vitro cytotoxic activity. Int J Biol Ecol Environ Sci 2(2):25–27

Dhand V, Soumya L, Bharadwaj S, Chakra S, Bhatt D, Sreedhar B (2016) Green synthesis of silver nanoparticles using *Coffea arabica* seed extract and its antibacterial activity. Mater Sci Eng 58:36–43

Du J, Singh H, Yi TH (2017) Biosynthesis of silver nanoparticles by *Novosphingobium* sp. THG-C3 and their antimicrobial potential. Artif Cells Nanomed Biotechnol 45(2):211–217

El-bialy B, Hamouda R, Khalifa KS, Hamza HA (2017) Cytotoxic effect of biosynthesized silver nanoparticles on Ehrlich ascites tumor cells in mice. Int J Pharmacol 13(2):134–144

Elgorban AM, Aref SM, Seham SM, Elhindi KM, Bahkali AH, Sayed SR, Manal MA (2016) Extracellular synthesis of silver nanoparticles using *Aspergillus versicolor* and evaluation of their activity on plant pathogenic fungi. Mycosphere 7(6):844–852

Elshawy OE, Helmy EA, Rashed LA (2016) Preparation, characterization and in vitro evaluation of the antitumor activity of the biologically synthesized silver nanoparticles. Adv Nanoparticles 5:149–156

El-Sonbaty S (2013) Fungus-mediated synthesis of silver nanoparticles and evaluation of antitumor activity. Cancer Nanotechnol 4(4–5):73–79

Fariq A, Khan T, Yasmin A (2017) Microbial synthesis of nanoparticles and their potential applications in biomedicine. J Appl Biomed 15(4):241–248

Fayaz AM, Balaji K, Girilal M, Yadav R, Kalaichelvan PT, Venketesan R (2010) Biogenic synthesis of silver nanoparticles and their synergistic effect with antibiotics: a study against gram-positive and gram-negative bacteria. Nanomed-Nanotechnol 6(1):103–109

Franci G, Falanga A, Galdiero S, Palomba L, Rai M, Morelli G, Galdiero M (2015) Silver nanoparticles as potential antibacterial agents. Molecules 20(5):8856–8874

Ge Y, Schimel JP, Holden PA (2011) Evidence for negative effects of TiO_2 and ZnO nanoparticles on soil bacterial communities. Environ Sci Technol 45(4):1659–1664

Gopinath V, Priyadarshini S, Loke MF, Arunkumar J, Marsili E, MubarakAli D, Velusamy P, Vadivelu J (2017) Biogenic synthesis, characterization of antibacterial silver nanoparticles and its cell cytotoxicity. Arab J Chem 10(8):1107–1117

Gurunathan S, Han JW, Eppakayala V, Jeyaraj M, Kim JH (2013a) Cytotoxicity of biologically synthesized silver nanoparticles in MDA-MB-231 human breast cancer cells. Biomed Res Int 2013(535796):1–10

Gurunathan S, Raman J, Malek NA, John PA, Vikineswary S (2013b) Green synthesis of silver nanoparticles using *Ganoderma neo-japonicum* Imazeki: a potential cytotoxic agent against breast cancer cells. Int J Nanomedicine 8:4399–4413

Gurunathan S, Park JH, Han JW, Kim JH (2015) Comparative assessment of the apoptotic potential of silver nanoparticles synthesized by *Bacillus tequilensis* and *Calocybe indica* in MDA-MB-231 human breast cancer cells: targeting p53 for anticancer therapy. Int J Nanomedicine 10:4203–4222

Hosseini MR, Sarvi MN (2015) Recent achievements in the microbial synthesis of semiconductor metal sulfide nanoparticles. Mater Sci Semicond Process 40:293–301

Husseiny MI, El-Aziz MA, Badr Y, Mahmoud MA (2007) Biosynthesis of gold nanoparticles using *Pseudomonas aeruginosa*. Spectrochim Acta A 67(3):1003–1006

Husseiny SM, Salah TA, Anter HA (2015) Biosynthesis of size controlled silver nanoparticles by *Fusarium oxysporum*, their antibacterial and antitumor activities. Beni-Suef Univ J Basic Appl Sci 4(3):225–231

Iavicoli I, Leso V, Beezhold DH, Shvedova AA (2017) Nanotechnology in agriculture: opportunities, toxicological implications, and occupational risks. Toxicol Appl Pharmacol 329:96–111

Jaidev L, Narasimha G (2010) Fungal mediated biosynthesis of silver nanoparticles, characterization and antimicrobial activity. Colloid Surface B 81(2):430–433

Jain N, Bhargava A, Majumdar S, Tarafdar J, Panwar J (2011) Extracellular biosynthesis and characterization of silver nanoparticles using *Aspergillus flavus* NJP08: a mechanism perspective. Nanoscale 3(2):635–641

Jena J, Pradhan N, Nayak RR, Dash BP, Sukla LB, Panda PK, Mishra BK (2014) *Microalga Scenedesmus* sp.: a potential low-cost green machine for silver nanoparticle synthesis. J Microbiol Biotechnol 24(4):522–533

Jha AK, Prasad K, Prasad K, Kulkarni AR (2009) Plant system: nature's nanofactory. Colloid Surface B 73(2):219–223

Joshi N, Jain N, Pathak A, Singh J, Prasad R, Upadhyaya CP (2018) Biosynthesis of silver nanoparticles using *Carissa carandas* berries and its potential antibacterial activities. J Sol-Gel Sci Technol 86(3):682–689

Kanmani P, Lim ST (2013) Synthesis and structural characterization of silver nanoparticles using bacterial exopolysaccharide and its antimicrobial activity against food and multidrug resistant pathogens. Process Biochem 48(7):1099–1106

Khan T, Khan MA, Nadhman A (2015) Synthesis in plants and plant extracts of silver nanoparticles with potent antimicrobial properties: current status and future prospects. Appl Microbiol Biotechnol 99(23):9923–9934

Klueh U, Wagner V, Kelly S, Johnson A, Bryers J (2000) Efficacy of silver-coated fabric to prevent bacterial colonization and subsequent device-based biofilm formation. J Biomed Mater Res 53(6):621–631

Kulkarni RR, Shaiwale NS, Deobagkar DN, Deobagkar DD (2015) Synthesis and extracellular accumulation of silver nanoparticles by employing radiation-resistant *Deinococcus radiodurans*, their characterization, and determination of bioactivity. Int J Nanomedicine 10:963–974

Li QL, Mahendra S, Lyon DY, Brunet L, Liga MV, Li D, Alvarez PJJ (2008) Antimicrobial nanomaterials for water disinfection and microbial control: potential applications and implications. Water Res 42(18):4591–4602

Li M, Noriega-Trevino ME, Nino-Martinez N, Marambio-Jones C, Wang J, Damoiseaux R, Ruiz F, Hoek EM (2011a) Synergistic bactericidal activity of Ag-TiO2 nanoparticles in both light and dark conditions. Environ Sci Technol 45(20):8989–8995

Li X, Xu H, Chen ZS, Chen G (2011b) Biosynthesis of nanoparticles by microorganisms and their applications. J Nanomater 2011:270974

Magdi HM, Mourad MH, El-Aziz MMA (2014) Biosynthesis of silver nanoparticles using fungi and biological evaluation of mycosynthesized silver nanoparticles. Egypt J Exp Biol (Bot) 10(1):1–12

Malarkodi C, Chitra K, Rajeshkumar S, Gnanajobitha G, Paulkumar K, Vanaja M, Annadurai G (2013) Novel eco-friendly synthesis of titanium oxide nanoparticles by using *Planomicrobium* sp. and its antimicrobial evaluation. Der Pharmacia Sinica 4(3):59–66

Manivasagan P, Venkatesan J, Senthilkumar K, Sivakumar K, Kim SK (2013) Biosynthesis, antimicrobial and cytotoxic effect of silver nanoparticles using a novel Nocardiopsis sp. MBRC-1. Biomed Res Int 2013:287638

Marambio-Jones C, Hoek EM (2010) A review of the antibacterial effects of silver nanomaterials and potential implications for human health and the environment. J Nanopart Res 12(5):1531–1551

Meenal K, Shriwas A, Sharmin K, Vogel W, Urban J, Kulkarni SK, Paknikar KM (2003) Extracellular synthesis of silver nanoparticles by a silver-tolerant yeast strain MKY3. Nanotechnology 14(1):95–100

Mühling M, Bradford A, Readman JW, Somerfield PJ, Handy RD (2009) An investigation into the effects of silver nanoparticles on antibiotic resistance of naturally occurring bacteria in an estuarine sediment. Mar Environ Res 68(5):278–283

Mukherjee P, Ahmad A, Mandal D, Senapati S, Sainkar SR, Khan MI, Parishcha R, Ajaykumar P, Alam M, Kumar R (2001a) Fungus-mediated synthesis of silver nanoparticles and their immobilization in the mycelial matrix: a novel biological approach to nanoparticle synthesis. Nano Lett 1(10):515–519

Mukherjee P, Ahmad A, Mandal D, Senapati S, Sainkar SR, Khan MI, Ramani R, Parischa R, Ajayakumar P, Alam M (2001b) Bioreduction of $AuCl_4^-$ ions by the fungus, *Verticillium* sp. and surface trapping of the gold nanoparticles formed. Angew Chem Int Ed 40(19):3585–3588

Mukherjee P, Senapati S, Mandal D, Ahmad A, Khan MI, Kumar R, Sastry M (2002) Extracellular synthesis of gold nanoparticles by the fungus Fusarium oxysporum. Chembiochem 3(5):461–463

Muranyi P, Schraml C, Wunderlich J (2010) Antimicrobial efficiency of titanium dioxide-coated surfaces. J Appl Microbiol 108(6):1966–1973

Netala VR, Bethu MS, Pushpalatha B, Baki VB, Aishwarya S, Rao JV, Tartte V (2016) Biogenesis of silver nanoparticles using endophytic fungus *Pestalotiopsis microspora* and evaluation of their antioxidant and anticancer activities. Int J Nanomedicine 11:5683–5696

Niknejad F, Nabili M, Ghazvini RD, Moazeni M (2015) Green synthesis of silver nanoparticles: advantages of the yeast *Saccharomyces cerevisiae* model. Curr Med Mycol 1(3):17–24

Omar HH, Bahabri FS, El-Gendy AM (2017) Biopotential application of synthesis nanoparticles as antimicrobial agents by using *Laurencia papillosa*. Int J Pharmacol 13(3):303–312

Ortega FG, Fernández-Baldo MA, Fernández JG, Serrano MJ, Sanz MI, Diaz-Mochón JJ, Lorente JA, Raba J (2015) Study of antitumor activity in breast cell lines using silver nanoparticles produced by yeast. Int J Nanomedicine 10:2021–2031

Otari S, Patil R, Ghosh S, Thorat N, Pawar S (2015) Intracellular synthesis of silver nanoparticle by Actinobacteria and its antimicrobial activity. Spectrochim Acta A-M 136:1175–1180

Oves M, Khan MS, Zaidi A, Ahmed AS, Ahmed F, Ahmad E, Sherwani A, Owais M, Azam A (2013) Antibacterial and cytotoxic efficacy of extracellular silver nanoparticles biofabricated from chromium reducing novel OS4 strain of Stenotrophomonas maltophilia. PLoS One 8(3):e59140

Panáček A, Kvitek L, Prucek R, Kolář M, Večeřová R, Pizúrová N, Sharma VK, Nevěčná TJ, Zbořil R (2006) Silver colloid nanoparticles: synthesis, characterization, and their antibacterial activity. J Phys Chem B 110(33):16248–16253

Patil MP, Kim GD (2017) Eco-friendly approach for nanoparticles synthesis and mechanism behind antibacterial activity of silver and anticancer activity of gold nanoparticles. Appl Microbiol Biotechnol 101(1):79–92

Patra JK, Baek KH (2014) Green nanobiotechnology: factors affecting synthesis and characterization techniques. J Nanomater 2014:417305

Poinern GEJ, Chapman P, Le X, Fawcett D (2013a) Green biosynthesis of gold nanometre scale plates using the leaf extracts from an indigenous Australian plant *Eucalyptus macrocarpa*. Gold Bull 46(3):165–173

Poinern GEJ, Chapman P, Shah M, Fawcett D (2013b) Green biosynthesis of silver nanocubes using the leaf extracts from *Eucalyptus macrocarpa*. Nano Bulletin 2(1):1–7

Prasad R (2014) Synthesis of silver nanoparticles in photosynthetic plants. J Nanoparticles 2014:963961

Prasad R (2016) Advances and applications through fungal nanobiotechnology. Springer, Cham

Prasad R (2017) Fungal nanotechnology: applications in agriculture, industry, and medicine. Springer International Publishing, Cham

Prasad R, Kumar V, Prasad KS (2014) Nanotechnology in sustainable agriculture: present concerns and future aspects. Afr J Biotechnol 13(6):705–713

Prasad R, Pandey R, Barman I (2016) Engineering tailored nanoparticles with microbes: quo vadis. WIREs Nanomed Nanobiotechnol 8:316–330

Prasad R, Bhattacharyya A, Nguyen QD (2017) Nanotechnology in sustainable agriculture: recent developments, challenges, and perspectives. Front Microbiol 8:1014

Prasad R, Kumar V, Kumar M, Shanquan W (2018a) Fungal nanobionics: principles and applications. Springer, Singapore

Prasad R, Jha A, Prasad K (eds) (2018b) Exploring the realms of nature for nanosynthesis. Nanotechnology in the life sciences, 1st ed. Springer International Publishing, Cham, Switzerland. http://dx.doi.org/10.1007/978-3-319-99570-0

Priyadharshini RI, Prasannaraj G, Geetha N, Venkatachalam P (2014) Microwave-mediated extracellular synthesis of metallic silver and zinc oxide nanoparticles using macro-algae (*Gracilaria edulis*) extracts and its anticancer activity against human PC3 cell lines. Appl Biochem Biotechnol 174(8):2777–2790

Rai M, Duran N (2011) Metal nanoparticles in microbiology. Springer Science & Business Media, Berlin

Rajeshkumar S, Malarkodi C, Paulkumar K, Vanaja M, Gnanajobitha G, Annadurai G (2014) Algae mediated green fabrication of silver nanoparticles and examination of its antifungal activity against clinical pathogens. Int J Metals 2014:1–8. http://dx.doi.org/10.1155/2014/692643

Rani R, Dimple S, Jena N, Kundu A, De Sarkar A, Hazra KS (2017) Controlled formation of nanostructures on MoS2 layers by focused laser irradiation. Appl Phys Lett 110(083101):1–5

Rao PV, Nallappan D, Madhavi K, Rahman S, Jun Wei L, Gan SH (2016) Phytochemicals and biogenic metallic nanoparticles as anticancer agents. Oxidative Med Cell Longev 2016(3685671):1–15

Rathod D, Golinska P, Wypij M, Dahm H, Rai M (2016) A new report of *Nocardiopsis valliformis* strain OT1 from alkaline Lonar crater of India and its use in synthesis of silver nanoparticles with special reference to evaluation of antibacterial activity and cytotoxicity. Medical Microbiol Immun 205(5):435–447

Roy K, Sarkar Chandan K, Ghosh Chandan K (2015) Rapid colorimetric detection of Hg2+ ion by green silver nanoparticles synthesized using Dahlia pinnata leaf extract. Green Process Synth 4(6):455–461

Salaheldin TA, Husseiny SM, Al-Enizi AM, Elzatahry A, Cowley AH (2016) Evaluation of the cytotoxic behavior of fungal extracellular synthesized Ag nanoparticles using confocal laser scanning microscope. Int J Mol Sci 17(329):1–11

Salari Z, Danafar F, Dabaghi S, Ataei SA (2016) Sustainable synthesis of silver nanoparticles using macroalgae *Spirogyra varians* and analysis of their antibacterial activity. J Saudi Chem Soc 20(4):459–464

Saravanan M, Nanda A (2010) Extracellular synthesis of silver bionanoparticles from *Aspergillus clavatus* and its antimicrobial activity against MRSA and MRSE. Colloid Surface B 77(2):214–218

Sathishkumar M, Sneha K, Yun Y (2009) Palladium nanocrystal synthesis using *Curcuma longa* tuber extract. Int J Mater Sci 4(1):11–17

Satyavani K, Ramanathan T, Gurudeeban S (2011) Green synthesis of silver nanoparticles by using stem derived callus extract of bitter apple (Citrullus colocynthis). Dig J Nanomater Biostruct 6(3):1019–1024

Sharma VK, Yngard RA, Lin Y (2009) Silver nanoparticles: green synthesis and their antimicrobial activities. Adv Colloid Interfac 145(1–2):83–96

Singh G, Babele PK, Shahi SK, Sinha RP, Tyagi MB, Kumar A (2014) Green synthesis of silver nanoparticles using cell extracts of *Anabaena doliolum* and screening of its antibacterial and antitumor activity. J Microbiol Biotechnol 24(10):1354–1367

Singh P, Kim YJ, Singh H, Wang C, Hwang KH, Farh MEA, Yang DC (2015) Biosynthesis, characterization, and antimicrobial applications of silver nanoparticles. Int J Nanomedicine 10:2567–2577

Singh H, Du J, Yi TH (2017) Kinneretia THG-SQI4 mediated biosynthesis of silver nanoparticles and its antimicrobial efficacy. Artificial Cell Nanomed B 45(3):602–608

Skladanowski M, Golinska P, Rudnicka K, Dahm H, Rai M (2016) Evaluation of cytotoxicity, immune compatibility and antibacterial activity of biogenic silver nanoparticles. Med Microbiol Immunol 205(6):603–613

Song JY, Kwon EY, Kim BS (2009) Biological synthesis of platinum nanoparticles using *Diopyros kaki* leaf extract. Bioprocess Biosyst Eng 33(1):159–164

Sonker AS, Richa JP, Rajneesh VK (2017) Characterization and in vitro antitumor, antibacterial and antifungal activities of green synthesized silver nanoparticles using cell extract of Nostoc sp. strain HKAR-2. Can J Biotech 1(1):26–37

Sunkar S, Nachiyar CV (2012) Biogenesis of antibacterial silver nanoparticles using the endophytic bacterium *Bacillus cereus* isolated from *Garcinia xanthochymus*. Asian Pac J Trop Biomed 2(12):953–959

Sweeney RY, Mao C, Gao X, Burt JL, Belcher AM, Georgiou G, Iverson BL (2004) Bacterial biosynthesis of cadmium sulfide nanocrystals. Chem Biol 11(11):1553–1559

Syed A, Saraswati S, Kundu GC, Ahmad A (2013) Biological synthesis of silver nanoparticles using the fungus *Humicola* sp. and evaluation of their cytotoxicity using normal and cancer cell lines. Spectrochim Acta A 114:144–147

Tamboli DP, Lee DS (2013) Mechanistic antimicrobial approach of extracellularly synthesized silver nanoparticles against Gram positive and Gram negative bacteria. J Hazard Mater 260:878–884

Thomas R, Janardhanan A, Varghese RT, Soniya EV, Mathew J, Radhakrishnan EK (2014) Antibacterial properties of silver nanoparticles synthesized by marine *Ochrobactrum* sp. Braz J Microbiol 45(4):1221–1227

Venkatesan J, Kim SK, Shim MS (2016) Antimicrobial, antioxidant, and anticancer activities of biosynthesized silver nanoparticles using marine algae *Ecklonia cava*. Nanomaterials (eBasel) 6(12):1–18

Verma SK, Das AK, Patel MK, Shah A, Kumar V, Gantait S (2018) Engineered nanomaterials for plant growth and development: a perspective analysis. Sci Total Environ 630:1413–1435

Vigneshwaran N, Ashtaputre N, Varadarajan P, Nachane R, Paralikar K, Balasubramanya R (2007) Biological synthesis of silver nanoparticles using the fungus *Aspergillus flavus*. Mater Lett 61(6):1413–1418

Von White G, Kerscher P, Brown RM, Morella JD, McAllister W, Dean D, Kitchens CL (2012) Green synthesis of robust, biocompatible silver nanoparticles using garlic extract. J Nanomater 2012:1–12

Wang C, Kim YJ, Singh P, Mathiyalagan R, Jin Y, Yang DC (2016) Green synthesis of silver nanoparticles by *Bacillus methylotrophicus*, and their antimicrobial activity. Artif Cells Nanomed Biotechnol 44(4):1127–1132

Wei X, Luo M, Li W, Yang L, Liang X, Xu L, Kong P, Liu H (2012) Synthesis of silver nanoparticles by solar irradiation of cell-free *Bacillus amyloliquefaciens* extracts and AgNO3. Bioresour Technol 103(1):273–278

Weir E, Lawlor A, Whelan A, Regan F (2008) The use of nanoparticles in anti-microbial materials and their characterization. Analyst 133(7):835–845

Wright JB, Lam K, Hansen D, Burrell RE (1999) Efficacy of topical silver against fungal burn wound pathogens. Am J Infect Control 27(4):344–350

Wypij M, Czarnecka J, Swiecimska M, Dahm H, Rai M, Golinska P (2018) Synthesis, characterization and evaluation of antimicrobial and cytotoxic activities of biogenic silver nanoparticles synthesized from Streptomyces xinghaiensis OF1 strain. World J Microbiol Biotechnol 34(23):1–13

Yehia RS, Al-Sheikh H (2014) Biosynthesis and characterization of silver nanoparticles produced by *Pleurotus ostreatus* and their anticandidal and anticancer activities. World J Microbiol Biotechnol 30(11):2797–2803

Yoon KY, Byeon JH, Park JH, Hwang J (2007) Susceptibility constants of *Escherichia coli* and *Bacillus subtilis* to silver and copper nanoparticles. Sci Total Environ 373(2–3):572–575

Zheng X, Chen Y, Wu R (2011) Long-term effects of titanium dioxide nanoparticles on nitrogen and phosphorus removal from wastewater and bacterial community shift in activated sludge. Environ Sci Technol 45(17):7284–7290

Zhou J, Xu NS, Wang ZL (2006) Dissolving behavior and stability of ZnO wires in biofluids: a study on biodegradability and biocompatibility of ZnO nanostructures. Adv Mater 18(18):2432–2435

Chapter 3
Biological Synthesis of Nanoparticles by Different Groups of Bacteria

Nariman Marooufpour, Mehrdad Alizadeh, Mehrnaz Hatami, and Behnam Asgari Lajayer

Contents

N. Marooufpour
Department of Plant Protection, Faculty of Agriculture, University of Tabriz, Tabriz, Iran

M. Alizadeh
Department of Plant Pathology, College of Agriculture, Tarbiat Modares University, Tehran, Iran

M. Hatami (✉)
Department of Medicinal Plants, Faculty of Agriculture and Natural Resources, Arak University, Arak, Iran

Institute of Nanoscience and Nanotechnology, Arak University, Arak, Iran
e-mail: m-hatami@araku.ac.ir

B. Asgari Lajayer
Department of Soil Science, Faculty of Agriculture, University of Tabriz, Tabriz, Iran

R. Prasad (ed.), *Microbial Nanobionics*, Nanotechnology in the Life Sciences,
https://doi.org/10.1007/978-3-030-16383-9_3

3.1 Introduction

Nanoscience is a rapidly developing field that covers a wide range of application in a large variety of areas of science and technology. The Greek prefix "nano" used in nanoscience, nanomaterials, or nanoparticles means "1 billionth," while 1 nanometer (1 nm) is 1/109 m. An accepted definition of nanoscale materials is materials that are in the 1–100 nm size range in at least one dimension. The dimension factor is important because it allows materials such as carbon nanotubes, which are several micrometers long by few nanometers wide, to be included in the definition (Dobias et al. 2011).

The increasing demand for nanomaterials should be accompanied by "green" synthesis methods in an effort to reduce generated hazardous waste from this industry. Green chemistry would help minimize the use of unsafe products and maximize the efficiency of chemical processes (Sharma et al. 2009). An advantage of biogenic synthesis, over conventional chemical synthesis, is the safer and easier handling of microbial cultures and the simpler downstream processing of biomass as compared to synthetic methods (Rai et al. 2011). Hence, biogenic NP synthesis represents a very interesting greener and more environmentally friendly manufacturing alternative, due to the use of chemicals of lower toxicity and to the use of lower ambient temperatures and lower pressures in the synthesis (Dobias et al. 2011; Prasad 2014, Prasad et al. 2016; Abdel-Aziz et al. 2018). Microbial synthesis is one of such processes, a green chemistry approach that interlinks nanotechnology and microbial biotechnology (Li et al. 2011).

In order to overcome the limitations posed by these conventional methods, there has been a growing demand to develop eco-friendly and rapid synthesis of nanomaterials with the desired size and shape. Consequently, researchers have developed biogenic principles to synthesize nanomaterials by using biological resources such as plants and microorganisms or their products (Schröfel et al. 2011; Prasad 2016, 2017, Prasad et al. 2018).

Microbial synthesis of metal nanoparticles can take place either intracellularly or extracellularly (Jain et al. 2011). Intracellular synthesis of nanoparticles requires additional steps such as ultrasound treatment or reactions with suitable detergents to release the synthesized nanoparticles (Kalimuthu et al. 2008). At the same time extracellular biosynthesis is cheap, and it requires simpler downstream processing. This favors large-scale production of silver nanoparticles to explore its potential applications. Because of this, many studies were focused on extracellular methods for the synthesis of metal nanoparticles (Durán et al. 2005; Prasad et al. 2016).

In natural environment also, microbes produce nanomaterials as part of their metabolism and, hence, can be utilized for various applications discussed in this chapter. The microbes reproduce fast; therefore this characteristic can be well exploited for their use in various aspects. Their use in various applications is well known to everyone in the field of biological sciences. Biotechnology has joined hands and has emerged as an initiative for the study of microbes and its various characteristics in the form of "microbiology." Microorganisms are of size 10–6 nm, and they are referred to as nanofactories, meaning generators of nanoparticles. Since they are present in nature, they are also called as biofactories (Deepak et al. 2011).

Biosynthesis of nanoparticles by microorganisms is a green and eco-friendly technology. Diverse microorganisms, both prokaryotes and eukaryotes, are used for synthesis of metallic nanoparticles, viz., silver, gold, platinum, zirconium, palladium, iron, cadmium, and metal oxides such as titanium oxide, zinc oxide, etc. (Hasan 2015). These microorganisms include bacteria, actinomycetes, fungi, and algae. The synthesis of nanoparticles may be intracellular or extracellular according to the location of nanoparticles (Mann 2001; Hulkoti and Taranath 2014; Prasad et al. 2016).

Biosynthesis of metal nanoparticles by bacteria is due to their defense mechanism (resistance mechanism), the resistance caused by the bacterial cell on metal ions in the environment is responsible for its nanoparticle synthesis (Saklani et al. 2012), and the cell wall being negatively charged interacts electrostatically with the positively charged metal ions. The enzymes present within the cell wall bioreduce the metal ions to nanoparticles, and finally the smaller-sized nanoparticles get diffused of through the cell wall, and the nanoparticles are produced (Mukherjee et al. 2001).

3.2 Nanoparticles and Their Applications

Nanotechnology has become one of the most important technologies in allareas of science. It relies on the synthesis and modulation of nanoparticles, which requires significant modifications of the properties of metals (Visweswara Rao and Hua Gan 2015). Nanomaterials have in fact been used unknowingly for thousands of years; for example, gold nanoparticles that were used to stain drinking glasses also cured certain diseases. Scientists have been progressively able to observe the shape- and size-dependent physiochemical properties of nanoparticles by using advanced techniques. Recently, the diverse applications of metal nanoparticles have been explored in biomedical, agricultural, environmental, and physiochemical areas (Visweswara Rao and Hua Gan 2015; Rai et al. 2016; Abbasi et al. 2016; Giljohann et al. 2010; Pereira et al. 2015; Prasad et al. 2014, 2017). For instance, gold nanoparticles have been applied for the specific delivery of drugs, such as paclitaxel, methotrexate, and doxorubicin (Rai et al. 2016). Gold nanoparticles have been also used for tumor detection, angiogenesis, genetic disease and genetic disorder diagnosis, photo imaging, and photo thermal therapy. Iron oxide nanoparticles have been applied for cancer therapy, hyperthermia, drug delivery, tissue repair, cell labeling, targeting and immunoassays, detoxification of biological fluids, magnetic resonance imaging, and magnetically responsive drug delivery therapy (Khlebtsov and Dykman 2011; Huang et al. 2007; Iv et al. 2015). Silver nanoparticles have been used for many antimicrobial purposes, as well as in anticancer, anti-inflammatory, and wound treatment applications (Ahamed et al. 2010). Due to their biocompatible, nontoxic, self-cleansing, skin-compatible, antimicrobial, and dermatological behaviors, zinc and titanium nanoparticles have been used in biomedical, cosmetic, ultraviolet (UV)-blocking agents, and various cutting-edge processing applications (Ambika and Sundrarajan 2015;

Zahir et al. 2015; Bhuyan et al. 2015). Copper and palladium nanoparticles have been applied in batteries, polymers, plastic plasmonic wave guides, and optical limiting devices (Momeni and Nabipour 2015; Nasrollahzadeh and Sajadi 2015). Moreover, they were found to be antimicrobial in nature against many pathogenic microorganisms. Additionally, metal nanoparticles have been used in the spatial analysis of various biomolecules, including several metabolites, peptides, nucleic acids, lipids, fatty acids, glycosphingolipids, and drug molecules, to visualize these molecules with higher sensitivity and spatial resolution (Waki et al. 2015). In addition, the unique properties of nanoparticles make them well suited for designing electrochemical sensors and biosensors (Peng and Miller 2011). For example, nanosensors have been developed for the detection of algal toxins, mycobacteria, and mercury present in drinking water (Selid et al. 2009). Researchers also developed nanosensors by utilizing nanomaterials for hormonal regulation and for detecting crop pests, viruses, soil nutrient levels, and stress factors. For instance, nanosensors for sensing auxin and oxygen distribution have been developed (Koren et al. 2015).

3.3 Factors Affecting Synthesis of Nanoparticles

Shape and size of nanoparticles depend on the physical and chemical factors. The optimum metal ion concentration, pH, and temperature of reaction mixture play key role in nanoparticle synthesis.

3.3.1 Concentration of Metal Ion

Increasing the concentration of silver ions 1–5 mM in reaction mixture revealed that in 1 mM concentration, the nanoparticle synthesis and size reduction started quickly due to more availability of functional groups in the extract. While increasing the substrate concentration, the large size and aggregation of nanoparticles occurred due to the occurrence of silver ions and functional group (Vanaja et al. 2013).

3.3.2 pH

pH plays an important role in the nanoparticle synthesis of extract with silver ions. Alkaline pH 8.2 showed a sharp peak at 460 nm with maximum production of silver nanoparticles. The sharp peak indicated formation of spherical shape of silver nanoparticles, thus indicating alkaline pH is more suitable for synthesis of nanoparticles. pH plays a role in shape and size control in nanoparticle synthesis. Another report suggests increase in absorption was seen with a decrease in pH and also indicated the production of bigger particles with decrease in pH (Prakash and Soni 2011).

3.3.3 Temperature

Temperature is one of the important physical parameters for synthesis of nanoparticles. Synthesis of nanoparticles while increasing the reaction temperature. The higher rate of reduction occurs at higher temperature due to the consumption of metal ions in the formation of nuclei, whereas the secondary reduction stops on the surface preformed nuclei. The broadening peak obtained at low temperature shows the formation of large-sized nanoparticles, and the narrow peak obtained at high temperature indicates the nanoparticles synthesized are smaller in size. It can be stabilized that higher temperature is optimum for nanoparticle synthesis (Vanaja et al. 2013).

3.3.4 Time

In a study, synthesis of nanoparticles at various time intervals was studied after reaction for 1 h, the AgNPs obtained showed a UV-visible spectroscopy absorption peak, and the intensity of the peak increases as the reaction time is increased, which indicated the continued reduction of the silver ions. The increase of the absorbance with the reaction time indicates that the concentration of AgNPs increases. When the reaction time reached 3 hours, the absorbance increased, and the wavelength value was slightly shifted. This phenomenon continued for reaction times of 6–24 h, indicating that the size of particles was decreased. At the end of the reaction, i.e., 48 h, the absorbance was considerably increased, and there was no significant change in wavelength (430 nm), compared with the 24 h reaction time. The transmission electron microscopy (TEM) results indicate that the samples obtained over a longer time period retained a narrower particle size distribution; the average size of all prepared AgNPs was 20 nm (Darroudi et al. 2011).

3.4 Common Methodologies for Synthesis of Metal Nanoparticles Using Microbes

3.4.1 Extracellular Mechanism

The test strain (culture) is grown in suitable media and incubated on orbital shaker at 150 rpm at 37 °C. After incubation the broth is centrifuged, and the supernatant is used for synthesis of nanoparticles. The supernatant is added to separate reaction vessels containing the metal ions in suitable concentrations and incubated for a period of 72 h. The color change of the reaction mixture suggests the presence of nanoparticles in the solution, and bioreduction of silver ions in the solution is monitored by sampling the aqueous solution and measuring the absorption spectrum using a UV-visible spectrophotometer. The morphology and uniformity of silver nanoparticles are investigated by X-ray diffraction (XRD) and scanning electron microscopy (SEM), while the interaction between protein and AgNPs is analyzed using Fourier transform infrared spectroscopy (FTIR) (Jeevan et al. 2012).

3.4.2 Intracellular Mechanism

The culture is grown in suitable liquid media incubated on shaker at optimal temperature. After incubation the flask is kept at static condition to allow the biomass to settle following which the supernatant is discarded and sterile distilled water is added for washing the cells. The flask iskept steady for 30 min to settle the biomass post which the supernatant is again discarded. This step is repeated for three times. The biomass is then separated from the sterile distilled water by centrifugation for 10 min. The wet biomass is exposed to 50 ml of sterilized aqueous solution of metals at various dilutions and incubated on shaker at suitable temperature till visual color change is observed. The change in color from pale yellow to brownish color indicates the formation of silver nanoparticles, pale yellow to pinkish color indicates the formation of gold nanoparticles, and the formation of whitish yellow to yellow color indicates the formation of manganese and zinc nanoparticles (Waghmare et al. 2011).

In summary, the extracellular synthesis of nanoparticles involves trapping the metal ions on the surface of the cells and reducing them in the presence of enzymes, while in intracellular synthesis ions are transported into the microbial cell to form nanoparticles in the presence of enzymes (Kalabegishvili et al. 2012). The biosynthesized nanoparticles have been used in a variety of applications including drug carriers for targeted delivery, cancer treatment, gene therapy and DNA analysis, antibacterial agents, biosensors, separation science, and magnetic resonance imaging (Li et al. 2011).

3.5 Nanoparticle Synthesis by Bacteria

The most important challenge in nanotechnology today is to cost-effectively tailor the optical, electric, and electronic property of NPs by controlling the configuration as well as monodispersity. This goal could be achieved using bacterial organisms in an organized manner (Gericke and Pinches 2006). In the last few years, fabrication of AgNPs has increased extensively owing to its immense applications (Morones et al. 2005). Bacteria possess remarkable ability to reduce heavy metal ions and are one of the best candidates for nanoparticle synthesis. For instance, some bacterial species have developed the ability to resort to specify defense mechanisms to quell stresses like toxicity of heavy metal ions or metals. It was observed that some of them could survive and grow even at high metal ion concentrations (e.g., *Pseudomonas stutzeri* and *P. aeruginosa*) (Bridges et al. 1979; Haefeli et al. 1984).

Among the milieu of natural resources, prokaryotic bacteria have been most extensively researched for synthesis of metallic nanoparticles. One of the reasons for "bacterial preference" for nanoparticle synthesis is their relative ease of manipulation (Slawson et al. 1992).

In nature, bacteria are frequently exposed to diverse and sometimes extreme environmental situations. Survival in these harsh conditions ultimately depends on their ability to resist the effects of environmental stresses. Natural defense mechanisms exist in bacteria to deal with a variety of stresses such as toxicity arising from high concentrations of metallic ions in the environment. The major bacterial species used for the synthesis of metallic nanoparticles include *Acinetobacter* sp., *Escherichia coli*, *Klebsiella pneumoniae*, *Lactobacillus* spp., *Bacillus cereus*, *Corynebacterium* sp., and *Pseudomonas* sp. (Mohanpuria et al. 2008; Iravani 2014; NVKV Prasad et al. 2011). Bacteria are known to synthesize metallic nanoparticles by either intracellular or extracellular mechanisms.

The first synthesis of Ag nanoparticles by bacteria was reported in 2000. Joerger et al. (2000) used *Pseudomonas stutzeri* AG259 to synthesize Ag nanoparticles with size less than 200 nm. Bacteria were grown on Lennox L (LB) agar substrate, containing 50 mmol/L $AgNO_3$, at 30 °C for 48 h in the dark (Lengke et al. 2006). In 2008, biosynthesis of silver nanocrystals by *B. licheniformis* was studied. Aqueous silver ions were reduced to silver nanoparticles when added to the biomass of *B. licheniformis*. This was indicated by the change in color from whitish yellow to brown. The probable mechanism for the formation of silver nanoparticles involves the enzyme nitrate reductase (Kalimuthu et al. 2008).

In 2008, silver nanoparticles in the range of 50 nm were synthesized by the supernatant of *B. licheniformis* when silver nitrate was added to it. The synthesized silver nanoparticles were highly stable. Also, the time required for reaction completion was 24 h (Kalimuthu et al. 2008). Biosynthesis of silver nanoparticles using microorganisms is rather slow. However, finding microorganisms to synthesize Ag nanoparticles is an important aspect. Shahverdi et al. (2007b) reported on the rapid synthesis of metallic nanoparticles of silver using the reduction of aqueous Ag^+ ion using the culture supernatants of *Klebsiella pneumoniae*, *Escherichia coli* (Lee 1996), and *Enterobacter cloacae* (*Enterobacteriaceae*). The synthetic process was quite fast, and silver nanoparticles were formed within 5 min of the silver ion coming into contact with the cell filtrate (Shahverdi et al. 2007a). However, the culture supernatants of different bacteria from *Enterobacteriaceae* are potential candidates for the rapid synthesis of silver nanoparticles. In 2009, investigated was the effect of different visible-light irradiation on the formation of silver nanoparticles from silver nitrate using the culture supernatant of *Klebsiella pneumoniae*. In addition, the study experimentally investigated the liquid mixing process effect on silver nanoparticle synthesis by visible-light irradiation. That study successfully synthesized evenly dispersed silver nanoparticles of uniform size and shape in the range of 1–6 nm and average size of 3 nm (Mokhtari et al. 2009). Another report focused on the synthesis of metallic bio-nanoparticles of silver using a reduction of aqueous Ag^+ ion with the culture supernatants of *Staphylococcus aureus*. The observation indicated that the reduction of the Ag^+ ions took place extracellularly. Also, the reaction between this supernatant and Ag^+ ions was carried out in bright conditions for 5 min (Nanda and Saravanan 2009).

Moreover, Brock and Gustafson (1976) reported that *Thiobacillus ferrooxidans*, *T. thiooxidans*, and *Sulfolobus acidocaldarius* were able to reduce ferric ion to the ferrous state when growing on elemental sulfur as an energy source. *T. thiooxidans* was able to reduce ferric iron at low pH medium aerobically. The ferrous iron formed was stable to autoxidation and *T. thiooxidans* was unable to oxidize ferrous iron, but the bioreduction of ferric iron using *T. ferrooxidans* was not aerobic because of the rapid bacterial reoxidation of the ferrous iron in the presence of oxygen (Brock and Gustafson 1976). Other biomineralization phenomena, such as the formation of tellurium (Te) in *Escherichia coli* K12 (Taylor 1999), the direct enzymatic reduction of Tc (VII) by resting cells of *Shewanella* (previously *Alteromonas*) *putrefaciens* and *Geobacter metallireducens* (previously known as strain GS-15) (Lloyd et al. 1999), and the reduction of selenite to selenium by *Enterobacter cloacae*, *Desulfovibrio desulfuricans*, and *Rhodospirillum rubrum* (Kessi et al. 1999), have been reported, as well. Mullen et al. (1989) examined the ability of *Bacillus cereus*, *B. subtilis*, *E. coli*, and *P. aeruginosa* for removing Ag^+, Cd^{2+}, Cu^{2+}, and La^{3+} from solution. They found that bacterial cells were capable of binding large quantities of metallic cations. Moreover, some of these bacteria are able to synthesize inorganic materials like the magnetotactic bacteria, which synthesize intracellular magnetite NPs (Lovley et al. 1987).

Pseudomonas stutzeri AG259 has been reported to fabricate Ag particles (Joerger et al. 2000), which are accumulated within the periplasmic space of bacterial cell of 200 nm. *Lactobacillus*, a common bacterial strain present in the buttermilk, synthesizes both Au and Ag NPs under standard conditions (Nair and Pradeep 2002). Rapid synthesis of metallic NPs of Ag using the reduction of aqueous Ag^+ has been achieved in the cultural supernatants of *Klebsiella pneumoniae*, *Escherichia coli*, and *Enterobacter cloacae* (Shahverdi et al. 2007b). Recently detailed studies confirmed that synthesis of Ag can be triggered through the liquid mixing process developed in the visible-light spectrum by *Klebsiella pneumoniae* (Mokhtari et al. 2009). Extracellular biosynthesis of 40 nm Ag NPs by the culture supernatant of *Bacillus licheniformis* has been customized as an easy way to work out the process (Kalimuthu et al. 2008). Varshney et al. (2011) have reported a rapid biological synthesis technique for the synthesis of spherical Cu nanoparticles in the size range of 8–15 nm using nonpathogenic *Pseudomonas stutzeri*. Recently, an innovative approach has been used for the synthesis of copper nanoparticles where *P. stutzeri* bacterial strain was used for copper nanoparticle synthesis from electroplating wastewater. The bacterial strain was isolated from soil and found that it produced 50–150 nm-sized cubical copper nanoparticles (Varshney et al. 2011). Prakash et al. (2011) reported extracellular synthesis of silver nanoparticles by bacteria *Bacillus cereus* collected from the riverine belt of Gangetic Plain of India. Synthesized nanoparticles were spherical in shape and in the range of 10–30 nm in size. Antibacterial effect of the synthesized AgNPs was tested with gram-negative and gram-positive bacteria *E. coli* and *Streptococcus* in varying strength of nanoparticles, and it was observed that the lowest concentration up to 50 ppm was sufficient to inhibit bacterial growth. *Bacillus* species has depicted to synthesize metal nanoparticles; researchers showed the ability of bacteria to decrease silver and fabrication of extracellularly, consistently circulated nanoparticles, ranging from 10 to 20 nm in size (Sunkar and Nachiyar 2012). The silver-producing bacteria isolated from the silver mines exhibit the silver nanoparticles accumulated in the periplasmic space of *Pseudomonas stutzeri* AG259 (Slawson et al. 1994). Bacteria are also used to synthesize gold nanoparticles. Sharma et al. (2012) reported that whole cells of a novel strain of *Marinobacter pelagius are* applicable for stable, monodisperse gold nanoparticle formation. Prasad et al. (2007) had reported the use of *Lactobacillus* strains to synthesize the titanium nanoparticles. The understanding of natural processes will apparently help in the discovery of entirely new and unexplored methodology of metal nanoparticle synthesis.

In one of the earliest studies in this technology, Slawson et al. (1992) found that a silver-resistant bacterial strain isolated from silver mines, *Pseudomonas stutzeri* AG259, accumulated AgNPs within the periplasmic space. Of note, the particle size ranged from 35 to 46 nm (Slawson et al. 1992). Interestingly, Klaus and group observed that when this bacterium was placed in concentrated aqueous solution (50 mM), particles of larger size (200 nm) were formed (Klaus et al. 1999). Klaus et al. (1999) attributed the difference in particle size (in comparison with the report of Slawson et al. 1992) to the differences in cell growth and metal incubation conditions. An important application of such a bacterium would be in industrial Ag recovery. Intriguingly, the exact mechanism(s) of AgNPs synthesis by this bacterium is still unclear. However, we have investigated the molecular basis of biochemical synthesis of AgNPs from *Morganella* sp. RP-42, an insect midgut isolate (Parikh et al. 2008). We observed that *Morganella* sp. RP-42 when exposed to silver nitrate ($AgNO_3$) produced extracellular crystalline AgNPs of size 20 ± 5 nm. Three gene homologues (silE, silP, and silS) were identified in silver-resistant *Morganella* sp. The homologue of silE from *Morganella* sp. showed 99% nucleotide sequence similarity with the previously reported gene, silE, which encodes a periplasmic silver-binding protein (Parikh et al. 2008). This is the only report that elucidates the molecular evidence of silver resistance in bacteria, which could be linked to synthesis mechanism. In an elegant study by Nair and Pradeep (2002), common *Lactobacillus strains* present in buttermilk were exposed to large concentrations of metal ions to produce microscopic gold, silver, and gold-silver alloy crystals of well-defined morphology. The bacteria produced these intracellularly, and, remarkably, the cells preserved their viability even after crystal growth (Nair and Pradeep 2002). Notably, even cyanobacteria have been observed to produce AgNPs. For example, the biosynthesis of AgNPs has been successfully conducted using *Plectonema boryanum* UTEX 485, a filamentous cyanobacterium (Lengke et al. 2006). The authors posit that the mechanisms of AgNPs production via cyanobacteria could involve metabolic processes from the use of nitrate at 25 °C and/or organics released from the dead cyanobacteria at 25 to 100 °C (Lengke et al. 2006). Among the first reports of intracellular semiconductor nanoparticle synthesis, Sweeney et al. (2004) demonstrated that *E. coli*, when incubated with cadmium chloride ($CdCl_2$) and sodium sulfide (Na_2S), spontaneously formed cadmium sulfide

(CdS) semiconductor nanocrystals. They showed that the formation of nanocrystals was markedly affected by physiologic parameters. Indeed, the entry into stationary phase increased the yield by 20-fold (Sweeney et al. 2004). In line with these observations, Cunningham and Lundie (1993) found that *Clostridium thermoaceticum* precipitates CdS at the cell surface as well as in the medium when exposed to $CdCl_2$ in the presence of cysteine hydrochloride as a source of sulfide in the growth medium. In a separate report of bacterial synthesis of nanoparticles, Watson et al. (1999) demonstrated that sulfate-reducing bacteria synthesize strongly magnetic iron sulfide (FeS) nanoparticles on their surfaces. The magnetic nanoparticles (about 20 nm in size) were separated from the solution by a high gradient field of 1 Tesla. Of note, bacterially produced FeS is an adsorbent for a wide range of heavy metals and some anions. Furthermore, magnetite is a common product of bacterial iron reduction and could be a potential physical indicator of biological activity in geological settings (Watson et al. 1999). Interestingly, Bharde et al. (2005) have demonstrated magnetite nanoparticle synthesis by Acinetobacter, a nonmagnetotactic bacteria. In prior studies, biosynthesis of magnetite was found to be extremely slow (often requiring 1 week) under strictly anaerobic conditions. However, this study reported that *Acinetobacter* spp. were capable of magnetite synthesis by reaction with suitable aqueous iron precursors under fully aerobic conditions (Bharde et al. 2005). Importantly, the extracellular magnetite nanoparticles showed excellent magnetic properties (Bharde et al. 2005). Bacteria have also been used to synthesize gold nanoparticles. For example, microbial synthesis of gold nanoparticles was achieved by Konishi et al. (2004) using the mesophilic bacterium *Shewanella algae* with H_2 as the electron donor. The authors used varying pH conditions in their study. When the solution pH was 7, gold nanoparticles of 10–20 nm were synthesized in the periplasmic space of *S. algae* cells. Interestingly, when the solution pH was decreased to 1, larger-sized gold nanoparticles (50–500 nm) were precipitated extracellularly. In an analogous study, He et al. (2007) showed that the bacteria *Rhodopseudomonas capsulata* produces gold nanoparticles of different sizes and shapes. He et al. (2007) incubated *R. capsulata* biomass and aqueous chloroauric acid (HAuCl4) solution at pH values ranging from 7 to 4. They found that at pH 7, spherical gold nanoparticles in the range of 10–20 nm were formed. In contrast, at pH 4, a number of nanoplates were produced. In both these studies, the solution pH was an important factor in controlling the morphology of biogenic gold particles and location of gold deposition. These observations are in line with the findings of Klaus et al. (1999) who observed that variations in incubation conditions lead to variations in particle size. Of note, gold nanoparticles can be used for a variety of applications (e.g., direct electrochemistry of proteins) (Du et al. 2007). Ahmad et al. (2003c) have performed a series of studies on bacterial synthesis of gold nanoparticles. In one such study, they used extremophilic actinomycetes like *Thermomonospora* sp. to efficiently synthesize monodisperse gold nanoparticles. By comparing this with their earlier work on gold nanoparticle synthesis from a fungus, *Fusarium oxysporum*, they postulated that reduction of metal ions and stabilization of the gold nanoparticles occur by an enzymatic process. Furthermore, they attributed the synthesis of monodisperse gold nanoparticles by

Thermomonospora sp. to extreme biological conditions (i.e., alkaline and slightly elevated temperature conditions) (Ahmad et al. 2003a, c). In a separate study, these authors used alkalotolerant *Rhodococcus* sp. (Ahmad et al. 2003b) for intracellular synthesis of good-quality monodisperse gold nanoparticles. Notably, the metal ions were not toxic to the cells as evidenced by the continued growth of the cells even after the biosynthesis of gold nanoparticles.

Bacteria among microorganisms and prokaryotes have received the most attention in the area of AuNPs synthesis (Whiteley et al. 2011). For the first time microbial synthesis of AuNPs was reported in *Bacillus subtilis* 168 which revealed the presence of 5–25 nm octahedral NPs inside the cell wall (Beveridge and Murray 1980). In *Rhodopseudomonas capsulata*, spherical AuNPs with 10–20 nm range have been observed (He et al. 2007) at lower concentration and nanowires with network at higher concentration (He et al. 2008). Six cyanobacteria have been reported for production of AuNPs. *Plectonema* sp. (Lengke et al. 2006; Brayner et al. 2007), *Anabaena* sp., *Calothrix* sp., and *Leptolyngbya* sp. have been exploited for the AuNPs synthesis (Lengke et al. 2007). Single-cell protein of *Spirulina platensis* was also shown to produce AuNPs and Au core-Ag shell NPs (Govindaraju et al. 2008). An overview on bacterial synthesis of AuNPs is given in Table 3.1. If one tries to group the AuNPs-producing bacteria according to the 9th edition of *Bergey's Manual of Systematic Bacteriology* (2005), the members belonging to groups glidobacteria and *Beta-*, *Epsilon-*, and *Zetaproteobacteria* have not been reported so far (Shedbalkar et al. 2014).

For example, Ag nanoparticles have been synthesized using *Pseudomonas stutzeri* AG259 bacterium via a mechanism involving the NADH-dependent reductase enzyme that donates an electron and oxidizes to NAD. The electron transfer results in the biological reduction of Ag ions to Ag nanoparticles (Ahmad et al. 2003a). In a similar study, Husseiny et al. (2007) were able to reduce Au ions using *Pseudomonas aeruginosa* that resulted in the extracellular synthesis of Au nanoparticles. However, some other researchers have also shown the noninvolvement of biological enzymes. For example, Liu et al. (2009) were able to produce Au nanoparticles from dried cells of *Bacillus megaterium*. A similar study by Sneha et al. (2010) using a *Corynebacterium* sp. also revealed that a nonenzymatic reduction mechanism was involved in nanoparticle formation. The reduction of nanoparticles is believed to be the result of a combination of several factors. The first factor is the presence of some organic functional groups at the cell wall that induce reduction, and the second depends on the appropriate environmental parameters such as pH and temperature being present (Lin et al. 2001). For example, the dried biomass of *Lactobacillus* sp. A09 and *Bacillus megaterium* D01 can reduce Ag ions via the interaction of functional groups present on the cell wall to produce silver nanoparticles (Jin-Zhou et al. 2000). Size, shape, and composition of a nanoparticle can be significantly influenced by pH and temperature (Hulkoti and Taranath 2014). For example, particle size is an important factor since novel and unique physicochemical properties are more pronounced at smaller sizes. Therefore, there is a need to optimize synthesis parameters during nanoparticle formation to enhance the overall particle properties. In particular, selecting the appropriate culture media for a specific

Table 3.1 The major bacterial species that have been used to synthesize a variety of nanoparticles along with extracellular or intracellular methods

Bacteria	Nanoparticles	Synthesis method	References
Bacillus flexus	Ag	Extracellular	Priyadarshini et al. (2013)
Lactobacillus spp.	Ag	Extracellular	Ranganath et al. (2012)
Klebsiella pneumoniae	Se	Intracellular	Fesharaki et al. (2010)
Streptomyces sp.	Mn and Zn	Intracellular	Waghmare et al. (2011)
Streptomyces sp.	Ag	Extracellular	Chauhan et al. (2013)
Pseudomonas aeruginosa	Ag	Extracellular	Shivakrishna et al. (2013)
Thermomonospora sp.	Gold	Extracellular	Ahmad et al. (2003c)
Rhodococcus sp.	Gold	Intracellular	Ahmad et al. (2003b)
Pseudomonas aeruginosa	Gold	Extracellular	Husseiny et al. (2007)
Pseudomonas fluorescens	Cu	Extracellular	Shantkriti and Rani (2014)
Bacillus subtilis	Au	Intracellular	Castro et al. (2014)
Pseudomonas stutzeri	Ag triangles and hexagons	Intracellular	Castro et al. (2014)
Lactobacillus sp.	Au, Ag, Au-Ag alloys	Intracellular	Castro et al. (2014)
Desulfovibrio desulfuricans	Pd	Intracellular	Castro et al. (2014)
S. oneidensis	Pd	Intracellular	Castro et al. (2014)
Rhodopseudomonas capsulata	Au	Extracellular	Castro et al. (2014)
Cupriavidus necator	Pd	Intracellular	Castro et al. (2014)
Pseudomonas putida	Pd	Intracellular	Castro et al. (2014)
Paracoccus denitrificans	Pd	Intracellular	Castro et al. (2014)
Pseudomonas aeruginosa	Au, Ag, Pd, Fe, Rh, Ni, Ru, Pt	Extracellular	Castro et al. (2014)
Pyrobaculum islandicum	Au	Extracellular	Castro et al. (2014)
G. sulfurreducens	Au	Extracellular	Castro et al. (2014)
Pyrococcus furiosus	Au	Extracellular	Castro et al. (2014)
Morganella sp.	Ag	Intracellular	Castro et al. (2014)
Bacillus licheniformis	Ag	Intracellular	Castro et al. (2014)
Pseudomonas deceptionensis	Ag	Extracellular	Singh et al. (2016)
Weissella oryzae	Ag	Intracellular	Singh et al. (2016)
Bacillus methylotrophicus	Ag	Extracellular	Singh et al. (2016)
Brevibacterium frigoritolerans	Ag	Extracellular	Singh et al. (2016)
Bhargavaea indica	Ag and Au	Extracellular	Singh et al. (2016)
Bacillus amyloliquefaciens	CdS	Extracellular	Singh et al. (2016)
Bacillus pumilus	Ag	Extracellular	Singh et al. (2016)
Bacillus persicus	Ag	Extracellular	Singh et al. (2016)
Bacillus licheniformis	Ag	Extracellular	Singh et al. (2016)
Lysinibacillus sphaericus	Ag	–	Gou et al. (2015)
Lactobacillus mindensis	Ag_2O	–	Dhoondia and Chakraborty (2012)

(continued)

Table 3.1 (continued)

Bacteria	Nanoparticles	Synthesis method	References
L. acidophilus	Ag	Extracellular	Mohseniazar et al. (2011)
L. casei	Ag	Extracellular	Mohseniazar et al. (2011)
L. reuteri	Ag	Extracellular	Mohseniazar et al. (2011)
Escherichia coli	Cd	Intracellular	Shah et al. (2015)
Pseudomonas aeruginosa	Au	Extracellular	Shah et al. (2015)
Pseudomonas stutzeri	Ag	Intracellular	Shah et al. (2015)
Streptomyces sp.	Ag	Extracellular	Zarina and Nanda (2014)
Bacillus cereus	Ag	Extracellular	Prakash et al. (2011)
Escherichia coli	Ag	–	Kushwaha et al. (2015)
Escherichia coli	Cu	Extracellular	Shobha et al. (2014)
Mycobacterium psychrotolerans	Cu	Extracellular	Shobha et al. (2014)
M. morganii	Cu	Extracellular	Shobha et al. (2014)
Pseudomonas sp.	Cu	Extracellular	Shobha et al. (2014)
Pseudomonas stutzeri	Cu	Extracellular	Shobha et al. (2014)
Streptomyces sp.	Cu	Extracellular	Shobha et al. (2014)
Planomicrobium sp.	TiO_2	Extracellular	Malarkodi et al. (2013)
Bacillus subtilis	Au	Intracellular	Southam and Beveridge (1996)
Bacillus subtilis	Co_3O_4	Extracellular	Shim et al. (2010)
Bacillus subtilis	TiO_2	–	Kirthi et al. (2011)
Bacillus licheniformis	Au	–	Kalishwaralal et al. (2009)
Bacillus licheniformis	Ag	Extracellular	Kalishwaralal et al. (2009)
Escherichia coli	CdS	Intracellular	Sweeney et al. (2004)
Escherichia coli	CdTe	Extracellular	Bao et al. (2010)
Escherichia coli	Au	Intracellular	Du et al. (2007)
Escherichia coli	Pt	–	Attard et al. (2012)
Acinetobacter sp.	Fe_3O_4	Extracellular	Bharde et al. (2005)
Acinetobacter sp.	Si/SiO_2	Extracellular	Singh et al. (2008)
Magnetospirillum gryphiswaldense	Fe_3O_4/Fe_3S_4	Intracellular	Schübbe et al. (2003)
Geobacter sulfurreducens	Pd	Extracellular	Yates et al. (2013)
Klebsiella pneumoniae	Ag	Extracellular	Mokhtari et al. (2009)
Klebsiella pneumoniae	Se	Intracellular	Fesharaki et al. (2010)
Lactobacillus sp.	TiO_2	Extracellular	Jha et al. (2009)
Morganella psychrotolerans	Ag	Extracellular	Ramanathan et al. (2010)
Clostridium thermoaceticum	CdS	Extracellular	Cunningham and Lundie (1993)
Desulfobacteraceae spp.	ZnS	–	Labrenz et al. (2000)
Shewanella oneidensis	UO_2	Extracellular	Marshall et al. (2006)
Shewanella algae	Au	Intracellular	Konishi et al. (2006)
Shewanella algae	Pt	–	Konishi et al. (2007)
Pseudomonas aeruginosa	Au	Extracellular	Husseiny et al. (2007)

(continued)

Table 3.1 (continued)

Bacteria	Nanoparticles	Synthesis method	References
Rhodopseudomonas capsulata	Au	Extracellular	He et al. (2008)
Rhodopseudomonas palustris	CdS	Intracellular	Bai et al. (2009)
Bacillus licheniformis	Ag	Extracellular	Kalimuthu et al. (2008)
Bacillus licheniformis	Ag	Extracellular	Kalishwaralal et al. (2009)
Klebsiella pneumoniae	Ag	–	Shahverdi et al. (2007b)
Escherichia coli	Ag	–	Shahverdi et al. (2007b)
Enterobacter cloacae	Ag	–	Shahverdi et al. (2007b)
Pseudomonas stutzeri	Ag	–	Joerger et al. (2000)
Klebsiella pneumoniae	Ag	–	Mokhtari et al. (2009)
Staphylococcus aureus	Ag	Extracellular	Nanda and Saravanan (2009)
Thermoanaerobacter sp.	Cu	Extracellular	Jang et al. (2015)
Thiobacillus thioparus	Fe_2O_3	–	Elcey et al. (2014)
Escherichia coli	CdS	Intracellular	Sweeney et al. (2004)
Azoarcus sp.	Se	Intracellular	Fernández-Llamosas et al. (2016)
Geobacillus sp.	Au	Intracellular	Correa-Llantén et al. (2013)
Marinobacter pelagius	Au	–	Sharma et al. (2012)
Myxococcus virescens	Ag	Extracellular	Wrótniak-Drzewiecka et al. (2014)
Streptomyces spp.	Ag	Extracellular	Tsibakhashvil et al. (2010)
Bacillus subtilis	Ag	Extracellular	Saifuddin et al. (2009)
Thermomonospora sp.	Au	Extracellular	Ahmad et al. (2003c)
Escherichia coli	Ag	Extracellular	El-Shanshoury et al. (2011)
Bacillus subtilis	Ag	Extracellular	El-Shanshoury et al. (2011)
Arthrobacter sp.	Au	Extracellular	Kalabegishvili et al. (2012)
Arthrobacter globiformis	Au	Extracellular	Kalabegishvili et al. (2012)
Rhodococcus sp.	Au	Intracellular	Ahmad et al. (2003b)
Bacillus sp.	Ag	Extracellular	Das et al. (2014)
Bacillus megaterium	Ag	Extracellular	Saravanan et al. (2011)
Pantoea agglomerans	Se	Intracellular	Torres et al. (2012)
Bacillus licheniformis	Ag	Extracellular	Shivaji et al. (2011)
Proteus mirabilis	Ag	Intra-/extracellular	Samadi et al. (2009)
Pseudomonas fluorescens	Au	Extracellular	Syed et al. (2016)
Pseudomonas stutzeri	Ag	Intracellular	Klaus et al. (1999)
Morganella sp.	Ag	Extracellular	Parikh et al. (2008)

(continued)

Table 3.1 (continued)

Bacteria	Nanoparticles	Synthesis method	References
Lactobacillus strains	Ag and Au	Intracellular	Nair and Pradeep (2002)
Plectonema boryanum	Ag	Intracellular	Lengke et al. (2007)
Escherichia coli	CdS	Intracellular	Sweeney et al. (2004)
Clostridium thermoaceticum	CdS	Intra-/extracellular	Cunningham and Lundie (1993)
Acinetobacter spp.	Fe_2O_3	Extracellular	Bharde et al. (2005)
Shewanella algae	Au	Intra-/extracellular	Thakkar et al. (2010)
Rhodopseudomonas capsulata	Au	Extracellular	He et al. 2007
Escherichia coli	Au	Intracellular	Du et al. (2007)
Thermomonospora sp.	Au	Extracellular	Ahmad et al. (2003c)
Rhodococcus sp.	Au	Intracellular	Ahmad et al. (2003b)
Klebsiella pneumoniae	Ag	Extracellular	Shahverdi et al. (2007b)
Pseudomonas aeruginosa	Au	Extracellular	Husseiny et al. (2007)
Shewanella oneidensis	U (IV)	Extracellular	Marshall et al. (2006)
Lactobacillus acidophilus	Se	–	Visha et al. (2015)
Pseudomonas alcaliphila	Se	–	Zhang et al. (2011)
Morganella sp.	Metallic Cu	Extracellular	Saif Hasan et al. (2008)
Pseudomonas sp.	Ag	Extracellular	Yadav et al. (2015)
Ochrobactrum sp.	Ag	Extracellular	Thomas et al. (2014)

bacteria and the particular metallic salt is important since these two parameters form the basis of nanoparticle synthesis and can influence particle yield (Roh et al. 2001; Nair and Pradeep 2002; Yong et al. 2002). Studies by He et al. (2008) using bacterium *Rhodopseudomonas capsulata* have shown that particle size and morphology can be influenced by both metallic salt concentration and medium pH. At pH 6, dilute concentrations of $AuCl_4$ tended to produce spherical Au nanoparticles ranging in size from 10 to 20 nm. Upon increasing the salt concentration, this reaction tended to produce Au nanowires at pH 6 (He et al. 2008). Also, when the pH was changed to 4, dilute salt concentrations tended to produce both spheres and triangular nanometer scale plates (Husseiny et al. 2007). The studies clearly indicated that controlling medium pH directly influenced nanoparticle morphology during formation. Table 3.1 summarizes the major bacterial species that have been used to synthesize a variety of nanoparticles along with extracellular or intracellular methods.

3.6 Conclusion

Bio-based approaches are still in the development stages, and stability and aggregation of the biosynthesized NPs, control of crystal growth, shape, size, and size distribution are the most important experienced problems. Furthermore, biologically synthesized NPs in comparison with chemically synthesized ones are more polydisperse. The properties of NPs can be controlled by optimization of important parameters which control the growth condition of organisms, cellular activities, and enzymatic processes (optimization of growth and reaction conditions).Mechanistic aspects have not been clearly and deeply described and discussed. Thus, more elaborated studies are needed to know the exact mechanisms of reaction and identify the enzymes and proteins which involve nanoparticle biosynthesis. The large-scale synthesis of NPs using bacteria is interesting because it does not need any hazardous, toxic, and expensive chemical materials for synthesis and stabilization processes. It seems that by optimizing their action conditions and selecting the best bacteria, these natural nanofactories can be used in the synthesis of stable NPs with well-defined sizes, morphologies, and compositions.

References

Abbasi E, Milani M, Fekri Aval S, Kouhi M, Akbarzadeh A, Tayefi Nasrabadi H, Nikasa P, Joo SW, Hanifehpour Y, Nejati-Koshki K (2016) Silver nanoparticles: synthesis methods, bio-applications and properties. Crit Rev Microbiol 42(2):173–180

Abdel-Aziz SM, Prasad R, Hamed AA, Abdelraof M (2018) Fungal nanoparticles: A novel tool for a green biotechnology? In: Fungal Nanobionics (eds. Prasad R, Kumar V, Kumar M, and Shanquan W), Springer Nature Singapore Pte Ltd. 61–87

Ahamed M, AlSalhi MS, Siddiqui M (2010) Silver nanoparticle applications and human health. Clin Chim Acta 411(23–24):1841–1848

Ahmad A, Mukherjee P, Senapati S, Mandal D, Khan MI, Kumar R, Sastry M (2003a) Extracellular biosynthesis of silver nanoparticles using the fungus *Fusarium oxysporum*. Colloids Surf B Biointerfaces 28(4):313–318

Ahmad A, Senapati S, Khan MI, Kumar R, Ramani R, Srinivas V, Sastry M (2003b) Intracellular synthesis of gold nanoparticles by a novel alkalotolerant actinomycete, *Rhodococcus* species. Nanotechnology 14(7):824–828

Ahmad A, Senapati S, Khan MI, Kumar R, Sastry M (2003c) Extracellular biosynthesis of monodisperse gold nanoparticles by a novel extremophilic actinomycete, *Thermomonospora* sp. Langmuir 19(8):3550–3553

Ambika S, Sundrarajan M (2015) Green biosynthesis of ZnO nanoparticles using *Vitex negundo* L. extract: spectroscopic investigation of interaction between ZnO nanoparticles and human serum albumin. J Photochem Photobiol B 149:143–148

Attard G, Casadesús M, Macaskie LE, Deplanche K (2012) Biosynthesis of platinum nanoparticles by Escherichia coli MC4100: can such nanoparticles exhibit intrinsic surface enantioselectivity? Langmuir 28(11):5267–5274

Bai H, Zhang Z, Guo Y, Yang G (2009) Biosynthesis of cadmium sulfide nanoparticles by photosynthetic bacteria *Rhodopseudomonas palustris*. Colloids Surf B Biointerfaces 70(1):142–146

Bao H, Lu Z, Cui X, Qiao Y, Guo J, Anderson JM, Li CM (2010) Extracellular microbial synthesis of biocompatible CdTe quantum dots. Acta Biomater 6(9):3534–3541

Beveridge T, Murray R (1980) Sites of metal deposition in the cell wall of *Bacillus subtilis*. J Bacteriol 141(2):876–887

Bharde A, Wani A, Shouche Y, Joy PA, Prasad BL, Sastry M (2005) Bacterial aerobic synthesis of nanocrystalline magnetite. J Am Chem Soc 127(26):9326–9327

Bhuyan T, Mishra K, Khanuja M, Prasad R, Varma A (2015) Biosynthesis of zinc oxide nanoparticles from *Azadirachta indica* for antibacterial and photocatalytic applications. Mater Sci Semicond Process 32:55–61

Brayner R, Barberousse H, Hemadi M, Djedjat C, Yéprémian C, Coradin T, Livage J, Fiévet F, Couté A (2007) Cyanobacteria as bioreactors for the synthesis of Au, Ag, Pd, and Pt nanoparticles via an enzyme-mediated route. J Nanosci Nanotechnol 7(8):2696–2708

Bridges K, Kidson A, Lowbury E, Wilkins M (1979) Gentamicin-and silver-resistant pseudomonas in a burns unit. Br Med J 1(6161):446–449

Brock TD, Gustafson J (1976) Ferric iron reduction by sulfur-and iron-oxidizing bacteria. Appl Environ Microbiol 32(4):567–571

Castro L, Blázquez ML, González FG, Ballester A (2014) Mechanism and applications of metal nanoparticles prepared by bio-mediated process. Rev Adv Sci Eng 3(3):199–216

Chauhan R, Kumar A, Abraham J (2013) A biological approach to the synthesis of silver nanoparticles with *Streptomyces* sp. JAR1 and its antimicrobial activity. Sci Pharm 81(2):607–624

Correa-Llantén DN, Muñoz-Ibacache SA, Castro ME, Muñoz PA, Blamey JM (2013) Gold nanoparticles synthesized by *Geobacillus* sp. strain ID17 a thermophilic bacterium isolated from Deception Island, Antarctica. Microb Cell Factories 12(1):75–80

Cunningham DP, Lundie L (1993) Precipitation of cadmium by *Clostridium thermoaceticum*. Appl Environ Microbiol 59(1):7–14

Darroudi M, Ahmad MB, Zamiri R, Zak A, Abdullah AH, Ibrahim NA (2011) Time-dependent effect in green synthesis of silver nanoparticles. Int J Nanomedicine 6:677–681

Das VL, Thomas R, Varghese RT, Soniya E, Mathew J, Radhakrishnan E (2014) Extracellular synthesis of silver nanoparticles by the Bacillus strain CS 11 isolated from industrialized area. 3. Biotech 4(2):121–126

Deepak V, Kalishwaralal K, Pandian SRK, Gurunathan S (2011) An insight into the bacterial biogenesis of silver nanoparticles, industrial production and scale-up. In: Metal nanoparticles in microbiology. Springer, Berlin, Heidelberg, pp 17–35

Dhoondia ZH, Chakraborty H (2012) Lactobacillus mediated synthesis of silver oxide nanoparticles. J Nanosci Nanotechnol 2:15–22

Dobias J, Suvorova EI, Bernier-Latmani R (2011) Role of proteins in controlling selenium nanoparticle size. Nanotechnology 22(19):195605

Du L, Jiang H, Liu X, Wang E (2007) Biosynthesis of gold nanoparticles assisted by *Escherichia coli* DH5α and its application on direct electrochemistry of hemoglobin. Electrochem Commun 9(5):1165–1170

Durán N, Marcato PD, Alves OL, De Souza GI, Esposito E (2005) Mechanistic aspects of biosynthesis of silver nanoparticles by several *Fusarium oxysporum* strains. J Nanobiotechnology 3(1):8

Elcey C, Kuruvilla AT, Thomas D (2014) Synthesis of magnetite nanoparticles from optimized iron reducing bacteria isolated from iron ore mining sites. Int J Curr Microbiol App Sci 3:408–417

El-Shanshoury AE-RR, ElSilk SE, Ebeid ME (2011) Extracellular biosynthesis of silver nanoparticles using *Escherichia coli* ATCC 8739, *Bacillus subtilis* ATCC 6633, and *Streptococcus thermophilus* ESh1 and their antimicrobial activities. ISRN Nanotechnology 2011:7

Fernández-Llamosas H, Castro L, Blázquez ML, Díaz E, Carmona M (2016) Biosynthesis of selenium nanoparticles by *Azoarcus* sp. CIB. Microb Cell Factories 15(1):109

Fesharaki PJ, Nazari P, Shakibaie M, Rezaie S, Banoee M, Abdollahi M, Shahverdi AR (2010) Biosynthesis of selenium nanoparticles using *Klebsiella pneumoniae* and their recovery by a simple sterilization process. Braz J Microbiol 41(2):461–466

Gericke M, Pinches A (2006) Microbial production of gold nanoparticles. Gold Bull 39(1):22–28

Giljohann DA, Seferos DS, Daniel WL, Massich MD, Patel PC, Mirkin CA (2010) Gold nanoparticles for biology and medicine. Angew Chem Int Ed 49(19):3280–3294

Gou Y, Zhou R, Ye X, Gao S, Li X (2015) Highly efficient in vitro biosynthesis of silver nanoparticles using *Lysinibacillus sphaericus* MR-1 and their characterization. Sci Technol Adv Mater 16(1):015004

Govindaraju K, Basha SK, Kumar VG, Singaravelu G (2008) Silver, gold and bimetallic nanoparticles production using single-cell protein (*Spirulina platensis*) Geitler. J Mater Sci Mater Med 43(15):5115–5122

Haefeli C, Franklin C, Hardy KE (1984) Plasmid-determined silver resistance in *Pseudomonas stutzeri* isolated from a silver mine. Res J Recent Sci 158(1):389–392

Hasan S (2015) A review on nanoparticles: their synthesis and types. Res J Recent Sci 2277:9–11

He S, Guo Z, Zhang Y, Zhang S, Wang J, Gu N (2007) Biosynthesis of gold nanoparticles using the bacteria *Rhodopseudomonas capsulata*. Mater Lett 61(18):3984–3987

He S, Zhang Y, Guo Z, Gu N (2008) Biological synthesis of gold nanowires using extract of *Rhodopseudomonas capsulata*. Biotechnol Prog 24(2):476–480

Huang X, Jain PK, El-Sayed IH, El-Sayed MA (2007) Gold nanoparticles: interesting optical properties and recent applications in cancer diagnostics and therapy. Nanomedicine 2(5):681–693

Hulkoti NI, Taranath T (2014) Biosynthesis of nanoparticles using microbes-a review. Colloids Surf B Biointerfaces 121:474–483

Husseiny M, El-Aziz MA, Badr Y, Mahmoud M (2007) Biosynthesis of gold nanoparticles using *Pseudomonas aeruginosa*. Spectrochim Acta A Mol Biomol Spectrosc 67(3–4):1003–1006

Iravani S (2014) Bacteria in nanoparticle synthesis: current status and future prospects. ISRN 2014:18

Iv M, Telischak N, Feng D, Holdsworth SJ, Yeom KW, Daldrup-Link HE (2015) Clinical applications of iron oxide nanoparticles for magnetic resonance imaging of brain tumors. Nanomedicine 10(6):993–1018

Jain N, Bhargava A, Majumdar S, Tarafdar J, Panwar J (2011) Extracellular biosynthesis and characterization of silver nanoparticles using *Aspergillus flavus* NJP08: a mechanism perspective. Nanoscale 3(2):635–641

Jang GG, Jacobs CB, Gresback RG, Ivanov IN, Meyer HM III, Kidder M, Joshi PC, Jellison GE, Phelps TJ, Graham DE (2015) Size tunable elemental copper nanoparticles: extracellular synthesis by thermoanaerobic bacteria and capping molecules. J Mater Chem C 3(3):644–650

Jeevan P, Ramya K, Rena AE (2012) Extracellular biosynthesis of silver nanoparticles by culture supernatant of *Pseudomonas aeruginosa*. Indian J Biotechnol 11:72–76

Jha AK, Prasad K, Kulkarni A (2009) Synthesis of TiO_2 nanoparticles using microorganisms. Colloids Surf B Biointerfaces 71(2):226–229

Jin-Zhou F, Yue-Ying L, Ping-Ying G, Ding-Liang S, Zhong-Yu L, Bing-Xin Y, Sheng-Zhou W (2000) Spectroscopic characterization on the biosorption and bioreduction of Ag (I) by *Lactobacillus* sp. A09. Acta Phys Chim Sin 16(09):779–782

Joerger R, Klaus T, Granqvist CG (2000) Biologically produced silver-carbon composite materials for optically functional thin-film coatings. Adv Mater Res 12(6):407–409

Kalabegishvili TL, Kirkesali EI, Rcheulishvili AN, Ginturi EN, Murusidze IG, Pataraya DT, Gurielidze MA, Tsertsvadze GI, Gabunia VN, Lomidze LG (2012) Synthesis of gold nanoparticles by some strains of *Arthrobacter genera*. Mater Sci Eng A Struct Mater 2(2):164–173

Kalimuthu K, Babu RS, Venkataraman D, Bilal M, Gurunathan S (2008) Biosynthesis of silver nanocrystals by *Bacillus licheniformis*. Colloids Surf B Biointerfaces 65(1):150–153

Kalishwaralal K, Deepak V, Pandian SRK, Gurunathan S (2009) Biological synthesis of gold nanocubes from *Bacillus licheniformis*. Bioresour Technol 100(21):5356–5358

Kessi J, Ramuz M, Wehrli E, Spycher M, Bachofen R (1999) Reduction of selenite and detoxification of elemental selenium by the phototrophic bacterium *Rhodospirillum rubrum*. Appl Environ Microbiol 65(11):4734–4740

Khlebtsov N, Dykman L (2011) Biodistribution and toxicity of engineered gold nanoparticles: a review of in vitro and in vivo studies. Chem Soc Rev 40(3):1647–1671

Kirthi AV, Rahuman AA, Rajakumar G, Marimuthu S, Santhoshkumar T, Jayaseelan C, Elango G, Zahir AA, Kamaraj C, Bagavan A (2011) Biosynthesis of titanium dioxide nanoparticles using bacterium *Bacillus subtilis*. Mater Lett 65(17–18):2745–2747

Klaus T, Joerger R, Olsson E, Granqvist C-G (1999) Silver-based crystalline nanoparticles, microbially fabricated. Proc Natl Acad Sci U S A 96(24):13611–13614

Konishi Y, Ohno K, Saitoh N, Nomura T, Nagamine S (2004) Microbial synthesis of gold nanoparticles by metal reducing bacterium. Trans Mater Res Soc Jpn 29:2341–2343

Konishi Y, Tsukiyama T, Ohno K, Saitoh N, Nomura T, Nagamine S (2006) Intracellular recovery of gold by microbial reduction of $AuCl^{4-}$ ions using the anaerobic bacterium *Shewanella algae*. Hydrometallurgy 81(1):24–29

Konishi Y, Ohno K, Saitoh N, Nomura T, Nagamine S, Hishida H, Takahashi Y, Uruga T (2007) Bioreductive deposition of platinum nanoparticles on the bacterium *Shewanella algae*. J Biotechnol 128(3):648–653

Koren K, Brodersen KE, Jakobsen SL, Kühl M (2015) Optical sensor nanoparticles in artificial sediments – a new tool to visualize O_2 dynamics around the rhizome and roots of seagrasses. Environ Sci Technol Lett 49(4):2286–2292

Kushwaha A, Singh VK, Bhartariya J, Singh P, Yasmeen K (2015) Isolation and identification of *E. coli* bacteria for the synthesis of silver nanoparticles: characterization of the particles and study of antibacterial activity. Eur J Exp Biol 5(1):65–70

Labrenz M, Druschel GK, Thomsen-Ebert T, Gilbert B, Welch SA, Kemner KM, Logan GA, Summons RE, De Stasio G, Bond PL (2000) Formation of sphalerite (ZnS) deposits in natural biofilms of sulfate-reducing bacteria. Science 290(5497):1744–1747

Lee SY (1996) High cell-density culture of *Escherichia coli*. Trends Biotechnol 14(3):98–105

Lengke MF, Fleet ME, Southam G (2006) Morphology of gold nanoparticles synthesized by filamentous cyanobacteria from gold (I) – thiosulfate and gold (III) – chloride complexes. Langmuir 22(6):2780–2787

Lengke MF, Fleet ME, Southam G (2007) Biosynthesis of silver nanoparticles by filamentous cyanobacteria from a silver (I) nitrate complex. Langmuir 23(5):2694–2699

Li X, Xu H, Chen Z-S, Chen G (2011) Biosynthesis of nanoparticles by microorganisms and their applications. J Nanomater 2011:1–16

Lin Z-Y, Fu J-K, Wu J-M, Liu Y-Y, Cheng H (2001) Preliminary study on the mechanism of non-enzymatic bioreduction of precious metal ions. Acta Phys Chim Sin 17(05):477–480

Liu S, Wei L, Hao L, Fang N, Chang MW, Xu R, Yang Y, Chen Y (2009) Sharper and faster "nano darts" kill more bacteria: a study of antibacterial activity of individually dispersed pristine single-walled carbon nanotube. ACS Nano 3(12):3891–3902

Lloyd J, Ridley J, Khizniak T, Lyalikova N, Macaskie L (1999) Reduction of technetium by *Desulfovibrio desulfuricans*: biocatalyst characterization and use in a flowthrough bioreactor. Appl Environ Microbiol 65(6):2691–2696

Lovley DR, Stolz JF, Nord GL, Phillips EJ (1987) Anaerobic production of magnetite by a dissimilatory iron-reducing microorganism. Nature 330(6145):252–254

Malarkodi C, Chitra K, Rajeshkumar S, Gnanajobitha G, Paulkumar K, Vanaja M, Annadurai G (2013) Novel eco-friendly synthesis of titanium oxide nanoparticles by using *Planomicrobium* sp. and its antimicrobial evaluation. Der Pharmacia Sinica 4(3):59–66

Mann S (2001) Biomineralization: principles and concepts in bioinorganic materials chemistry, vol 5. Oxford University Press on Demand, New York

Marshall MJ, Beliaev AS, Dohnalkova AC, Kennedy DW, Shi L, Wang Z, Boyanov MI, Lai B, Kemner KM, McLean JS (2006) C-type cytochrome-dependent formation of U (IV) nanoparticles by *Shewanella oneidensis*. PLoS Biol 4(8):e268

Mohanpuria P, Rana NK, Yadav SK (2008) Biosynthesis of nanoparticles: technological concepts and future applications. J Nanopart Res 10(3):507–517

Mohseniazar M, Barin M, Zarredar H, Alizadeh S, Shanehbandi D (2011) Potential of microalgae and lactobacilli in biosynthesis of silver nanoparticles. Bioimpacts 1(3):149–152

Mokhtari N, Daneshpajouh S, Seyedbagheri S, Atashdehghan R, Abdi K, Sarkar S, Minaian S, Shahverdi HR, Shahverdi AR (2009) Biological synthesis of very small silver nanoparticles by culture supernatant of *Klebsiella pneumonia*: the effects of visible-light irradiation and the liquid mixing process. Mater Res Bull 44(6):1415–1421

Momeni S, Nabipour I (2015) A simple green synthesis of palladium nanoparticles with Sargassum alga and their electrocatalytic activities towards hydrogen peroxide. Appl Biochem Biotechnol 176(7):1937–1949

Morones JR, Elechiguerra JL, Camacho A, Holt K, Kouri JB, Ramírez JT, Yacaman MJ (2005) The bactericidal effect of silver nanoparticles. Nanotechnology 16(10):2346–2353

Mukherjee P, Ahmad A, Mandal D, Senapati S, Sainkar SR, Khan MI, Ramani R, Parischa R, Ajayakumar P, Alam M (2001) Bioreduction of $AuCl4^-$ ions by the fungus, *Verticillium* sp and surface trapping of the gold nanoparticles formed. Angew Chem Int Ed 40(19):3585–3588

Mullen M, Wolf D, Ferris F, Beveridge T, Flemming C, Bailey G (1989) Bacterial sorption of heavy metals. Appl Environ Microbiol 55(12):3143–3149

Nair B, Pradeep T (2002) Coalescence of nanoclusters and formation of submicron crystallites assisted by *Lactobacillus strains*. Cryst Growth Des 2(4):293–298

Nanda A, Saravanan M (2009) Biosynthesis of silver nanoparticles from *Staphylococcus aureus* and its antimicrobial activity against MRSA and MRSE. Nanomedicine 5(4):452–456

Nasrollahzadeh M, Sajadi SM (2015) Green synthesis of copper nanoparticles using *Ginkgo biloba* L. leaf extract and their catalytic activity for the Huisgen [3+2] cycloaddition of azides and alkynes at room temperature. J Colloid Interface Sci 457:141–147

Parikh RY, Singh S, Prasad B, Patole MS, Sastry M, Shouche YS (2008) Extracellular synthesis of crystalline silver nanoparticles and molecular evidence of silver resistance from *Morganella* sp.: towards understanding biochemical synthesis mechanism. Chem Bio Chem 9(9):1415–1422

Peng H-I, Miller BL (2011) Recent advancements in optical DNA biosensors: exploiting the plasmonic effects of metal nanoparticles. Analyst 136(3):436–447

Pereira L, Mehboob F, Stams AJ, Mota MM, Rijnaarts HH, Alves MM (2015) Metallic nanoparticles: microbial synthesis and unique properties for biotechnological applications, bioavailability and biotransformation. Crit Rev Biotechnol 35(1):114–128

Prakash N, Soni N (2011) Factors affecting the geometry of silver nanoparticles synthesis in *Chrysosporium tropicum* and *Fusarium oxysporum*. Am J Biochem 2(1):112–121

Prakash A, Sharma S, Ahmad N, Ghosh A, Sinha P (2011) Synthesis of AgNps by *Bacillus cereus* bacteria and their antimicrobial potential. J Biomater Nanobiotechnol 2(02):155–161

Prasad R (2014) Synthesis of silver nanoparticles in photosynthetic plants. J Nanoparticles:963961. https://doi.org/10.1155/2014/963961

Prasad R (2016) Advances and applications through fungal nanobiotechnology. Springer, Cham. isbn:978-3-319-42989-2

Prasad R (2017) Fungal nanotechnology: applications in agriculture, industry, and medicine. Springer International Publishing. isbn:978-3-319-68423-9

Prasad K, Jha AK, Kulkarni A (2007) Lactobacillus assisted synthesis of titanium nanoparticles. Nanoscale Res Lett 2(5):248–250

Prasad NVKV, Subba Rao Kambala V, Naidu R (2011) A critical review on biogenic silver nanoparticles and their antimicrobial activity. Curr Nanosci 7(4):531–544

Prasad R, Kumar V, Prasad KS (2014) Nanotechnology in sustainable agriculture: present concerns and future aspects. Afr J Biotechnol 13(6):705–713

Prasad R, Pandey R, Barman I (2016) Engineering tailored nanoparticles with microbes: quo vadis. WIREs Nanomed Nanobiotechnol 8:316–330. https://doi.org/10.1002/wnan.1363

Prasad R, Bhattacharyya A, Nguyen QD (2017) Nanotechnology in sustainable agriculture: recent developments, challenges, and perspectives. Front Microbiol 8:1014. https://doi.org/10.3389/fmicb.2017.01014

Prasad R, Kumar V, Kumar M, Shanquan W (2018) Fungal nanobionics: principles and applications. Springer, Singapore. isbn:978-981-10-8666-3. https://www.springer.com/gb/book/9789811086656

Priyadarshini S, Gopinath V, Priyadharsshini NM, Mubarak Ali D, Velusamy P (2013) Synthesis of anisotropic silver nanoparticles using novel strain, *Bacillus flexus* and its biomedical application. Colloids Surf B Biointerfaces 102:232–237

Rai M, Gade A, Yadav A (2011) Biogenic nanoparticles: an introduction to what they are, how they are synthesized and their applications. In: Metal nanoparticles in microbiology. Springer, Berlin, Heidelberg, pp 1–14

Rai M, Ingle AP, Birla S, Yadav A, Santos CAD (2016) Strategic role of selected noble metal nanoparticles in medicine. Crit Rev Microbiol 42(5):696–719

Ramanathan R, O'Mullane AP, Parikh RY, Smooker PM, Bhargava SK, Bansal V (2010) Bacterial kinetics-controlled shape-directed biosynthesis of silver nanoplates using *Morganella psychrotolerans*. Langmuir 27(2):714–719

Ranganath E, Rathod V, Banu A (2012) Screening of *Lactobacillus* spp. for mediating the biosynthesis of silver nanoparticles from silver nitrate. IOSR PHR 2(2):237–241

Roh Y, Lauf R, McMillan A, Zhang C, Rawn C, Bai J, Phelps T (2001) Microbial synthesis and the characterization of metal-substituted magnetites. Solid State Commun 118(10):529–534

Saif Hasan S, Singh S, Parikh RY, Dharne MS, Patole MS, Prasad B, Shouche YS (2008) Bacterial synthesis of copper/copper oxide nanoparticles. J Nanosci Nanotechnol 8(6):3191–3196

Saifuddin N, Wong C, Yasumira A (2009) Rapid biosynthesis of silver nanoparticles using culture supernatant of bacteria with microwave irradiation. J Chem 6(1):61–70

Saklani V, Suman JV, Jain K (2012) Microbial synthesis of silver nanoparticles: a review. J Biotechnol Biomaterial 13:007

Samadi N, Golkaran D, Eslamifar A, Jamalifar H, Fazeli MR, Mohseni FA (2009) Intra/extracellular biosynthesis of silver nanoparticles by an *Autochthonous Strain* of *Proteus mirabilis* isolated from photographic waste. J Biomed Nanotechnol 5(3):247–253

Saravanan M, Vemu AK, Barik SK (2011) Rapid biosynthesis of silver nanoparticles from *Bacillus megaterium* (NCIM 2326) and their antibacterial activity on multi drug resistant clinical pathogens. Colloids Surf B Biointerfaces 88(1):325–331

Schröfel A, Kratošová G, Bohunická M, Dobročka E, Vávra I (2011) Biosynthesis of gold nanoparticles using diatoms-silica-gold and EPS-gold bionanocomposite formation. J Nanopart Res 13(8):3207–3216

Schübbe S, Kube M, Scheffel A, Wawer C, Heyen U, Meyerdierks A, Madkour MH, Mayer F, Reinhardt R, Schüler D (2003) Characterization of a spontaneous nonmagnetic mutant of *Magnetospirillum gryphiswaldense* reveals a large deletion comprising a putative magnetosome island. J Bacteriol 185(19):5779–5790

Selid PD, Xu H, Collins EM, Striped Face-Collins M, Zhao JX (2009) Sensing mercury for biomedical and environmental monitoring. Sensors 9(7):5446–5459

Shah M, Fawcett D, Sharma S, Tripathy SK, Poinern GEJ (2015) Green synthesis of metallic nanoparticles via biological entities. Materials 8(11):7278–7308

Shahverdi AR, Fakhimi A, Shahverdi HR, Minaian S (2007a) Synthesis and effect of silver nanoparticles on the antibacterial activity of different antibiotics against *Staphylococcus aureus* and *Escherichia coli*. Nanomedicine 3(2):168–171

Shahverdi AR, Minaeian S, Shahverdi HR, Jamalifar H, Nohi A-A (2007b) Rapid synthesis of silver nanoparticles using culture supernatants of Enterobacteria: a novel biological approach. Process Biochem 42(5):919–923

Shantkriti S, Rani P (2014) Biological synthesis of copper nanoparticles using *Pseudomonas fluorescens*. Int J Curr Microbiol App Sci 3(9):374–383

Sharma VK, Yngard RA, Lin Y (2009) Silver nanoparticles: green synthesis and their antimicrobial activities. Adv Colloid Interf Sci 145(1–2):83–96

Sharma N, Pinnaka AK, Raje M, Ashish F, Bhattacharyya MS, Choudhury AR (2012) Exploitation of marine bacteria for production of gold nanoparticles. Microb Cell Factories 11(1):86

Shedbalkar U, Singh R, Wadhwani S, Gaidhani S, Chopade B (2014) Microbial synthesis of gold nanoparticles: current status and future prospects. Adv Colloid Interf Sci 209:40–48

Shim H-W, Jin Y-H, Seo S-D, Lee S-H, Kim D-W (2010) Highly reversible lithium storage in bacillus subtilis-directed porous Co_3O_4 nanostructures. ACS Nano 5(1):443–449

Shivaji S, Madhu S, Singh S (2011) Extracellular synthesis of antibacterial silver nanoparticles using psychrophilic bacteria. Process Biochem 46(9):1800–1807

Shivakrishna P, Krishna MRPG, Charya MS (2013) Synthesis of silver nano particles from marine bacteria *Pseudomonas aeruginosa*. Octa J Biosci 1(2):108–114

Shobha G, Moses V, Ananda S (2014) Biological synthesis of copper nanoparticles and its impact-a review. Int J Pharm Sci Invent 3(8):28–38

Singh S, Bhatta UM, Satyam P, Dhawan A, Sastry M, Prasad B (2008) Bacterial synthesis of silicon/silica nanocomposites. J Mater Chem 18(22):2601–2606

Singh P, Kim Y-J, Zhang D, Yang D-C (2016) Biological synthesis of nanoparticles from plants and microorganisms. Trends Biotechnol 34(7):588–599

Slawson RM, Van Dyke MI, Lee H, Trevors JT (1992) Germanium and silver resistance, accumulation, and toxicity in microorganisms. Plasmid 27(1):72–79

Slawson R, Lohmeier-Vogel E, Lee H, Trevors J (1994) Silver resistance in *Pseudomonas stutzeri*. Biometals 7(1):30–40

Sneha K, Sathishkumar M, Mao J, Kwak I, Yun Y-S (2010) *Corynebacterium glutamicum*-mediated crystallization of silver ions through sorption and reduction processes. Chem Eng J 162(3):989–996

Southam G, Beveridge TJ (1996) The occurrence of sulfur and phosphorus within bacterially derived crystalline and pseudocrystalline octahedral gold formed in vitro. Geochim Cosmochim Acta 60(22):4369–4376

Sunkar S, Nachiyar CV (2012) Microbial synthesis and characterization of silver nanoparticles using the endophytic bacterium *Bacillus cereus*: a novel source in the benign synthesis. Global J Med Res 12:43–50

Sweeney RY, Mao C, Gao X, Burt JL, Belcher AM, Georgiou G, Iverson BL (2004) Bacterial biosynthesis of cadmium sulfide nanocrystals. J Chem 11(11):1553–1559

Syed B, Prasad NM, Satish S (2016) Endogenic mediated synthesis of gold nanoparticles bearing bactericidal activity. J Microsc Ultrastruct 4(3):162–166

Taylor DE (1999) Bacterial tellurite resistance. Trends Microbiol 7(3):111–115

Thakkar KN, Mhatre SS, Parikh RY (2010) Biological synthesis of metallic nanoparticles. Nanomedicine 6(2):257–262

Thomas R, Janardhanan A, Varghese RT, Soniya E, Mathew J, Radhakrishnan E (2014) Antibacterial properties of silver nanoparticles synthesized by marine *Ochrobactrum* sp. Braz J Microbiol 45(4):1221–1227

Torres S, Campos V, León C, Rodríguez-Llamazares S, Rojas S, Gonzalez M, Smith C, Mondaca M (2012) Biosynthesis of selenium nanoparticles by *Pantoea agglomerans* and their antioxidant activity. J Nanopart Res 14(11):1236

Tsibakhashvil N, Kalabegishvili T, Gabunia V, Gintury E, Kuchava N, Bagdavadze N, Pataraya D, Gurielidzse M, Gvarjaladze D, Lomidze L (2010) Synthesis of silver nanoparticles using bacteria. Nano Studies 2:179–182

Vanaja M, Rajeshkumar S, Paulkumar K, Gnanajobitha G, Malarkodi C, Annadurai G (2013) Kinetic study on green synthesis of silver nanoparticles using *Coleus aromaticus* leaf extract. Adv Appl Sci Res 4(3):50–55

Varshney R, Bhadauria S, Gaur M, Pasricha R (2011) Copper nanoparticles synthesis from electroplating industry effluent. Nano Biomed Eng 3(2):115–119

Visha P, Nanjappan K, Selvaraj P, Jayachandran S, Elango A, Kumaresan G (2015) Biosynthesis and structural characteristics of selenium nanoparticles using *Lactobacillus Acidophilus* bacteria by wet sterilization process. J Adv Vet Anim Res 4:178–183

Visweswara Rao P, Hua Gan S (2015) Recent advances in nanotechnology-based diagnosis and treatments of diabetes. Curr Drug Metab 16(5):371–375

Waghmare S, Deshmukh A, Kulkarni S, Oswaldo L (2011) Biosynthesis and characterization of manganese and zinc nanoparticles. Univers J Environ Res Technol 1(1):64–69

Waki M, Sugiyama E, Kondo T, Sano K, Setou M (2015) Nanoparticle-assisted laser desorption/ionization for metabolite imaging. In: Mass spectrometry imaging of small molecules. Humana Press; Copyright Holder: Springer, New York, pp 159–173

Watson J, Ellwood D, Soper A, Charnock J (1999) Nanosized strongly-magnetic bacterially-produced iron sulfide materials. J Magn Magn Mater 203(1–3):69–72

Whiteley C, Govender Y, Riddin T, Rai M (2011) Enzymatic synthesis of platinum nanoparticles: prokaryote and eukaryote systems. In: Metal nanoparticles in microbiology. Springer, Berlin, Heidelberg, pp 103–134

Wrótniak-Drzewiecka W, Gaikwad S, Laskowski D, Dahm H, Niedojadło J, Gade A, Rai M (2014) Novel approach towards synthesis of silver nanoparticles from *Myxococcus virescens* and their lethality on pathogenic bacterial cells. Austin J Biotechnol Bioeng 1(1):7

Yadav A, Theivasanthi T, Paul P, Upadhyay K (2015) Extracellular biosynthesis of silver nanoparticles from plant growth promoting rhizobacteria *Pseudomonas* sp. Int J Curr Microbiol App Sci 4(8):1057–1068

Yates MD, Cusick RD, Logan BE (2013) Extracellular palladium nanoparticle production using *Geobacter sulfurreducens*. ACS Sustain Chem Eng 1(9):1165–1171

Yong P, Rowson NA, Farr JPG, Harris IR, Macaskie LE (2002) Bioreduction and biocrystallization of palladium by *Desulfovibrio desulfuricans* NCIMB 8307. Biotechnol Bioeng 80(4):369–379

Zahir AA, Chauhan IS, Bagavan A, Kamaraj C, Elango G, Shankar J, Arjaria N, Roopan SM, Rahuman AA, Singh N (2015) Green synthesis of silver and titanium dioxide nanoparticles using *Euphorbia prostrata* extract shows shift from apoptosis to G0/G1 arrest followed by necrotic cell death in *Leishmania donovani*. Antimicrob Agents Chemother 59(8):4782–4799

Zarina A, Nanda A (2014) Green approach for synthesis of silver nanoparticles from marine Streptomyces-MS 26 and their antibiotic efficacy. J Pharm Sci Res 6(10):321–327

Zhang W, Chen Z, Liu H, Zhang L, Gao P, Li D (2011) Biosynthesis and structural characteristics of selenium nanoparticles by *Pseudomonas alcaliphila*. Colloids Surf B Biointerfaces 88(1):196–201

Chapter 4
Mushrooms: New Biofactories for Nanomaterial Production of Different Industrial and Medical Applications

Hesham Ali El Enshasy, Daniel Joel, Dhananjaya P. Singh, Roslinda Abd Malek, Elsayed Ahmed Elsayed, Siti Zulaiha Hanapi, and Kugen Kumar

Contents

H. A. El Enshasy (✉)
Institute of Bioproduct Development (IBD), Universiti Teknologi Malaysia (UTM), Johor Bahru, Johor, Malaysia

Faculty of Chemical and Energy Engineering, Universiti Teknologi Malaysia (UTM), Johor Bahru, Johor, Malaysia

City of Scientific Research and Technology Application, New Burg Al Arab, Alexandria, Egypt
e-mail: henshasy@ibd.utm.my

R. Prasad (ed.), *Microbial Nanobionics*, Nanotechnology in the Life Sciences,
https://doi.org/10.1007/978-3-030-16383-9_4

4.1 Introduction

Nanotechnology is one of the most rapidly growing sectors of multidisciplinary sciences during the last four decades. It is based on exploiting new chemical and physical properties of matter depending on having their particle sizes within the nanoscale. These properties are extremely different owing to their small sizes and large ratios of surface area to volume. Nanobiotechnology, is a sector of nanotechnology with growing interest because of its enormous benefit (Khan et al. 2017a, b). This technology deals with synthesizing and preparing nanoparticles (NPs) using biological sources, i.e., bacteria, fungi, plant, etc., and testing their properties on biological systems. Biologically prepared NPs have different advantages over other chemically synthesized NPs. They are characterized by their eco-friendly, cost-effective, and easily scalable production processes (Prasad 2014; Prasad et al. 2016; Borthakur et al. 2017). Biologically synthesized NPs have found many important applications in pharmaceutical industries (Ahmad et al. 2003). Traditionally, NPs have been used as antimicrobial and anticancer agents (Nithya and Rangunathan 2012). Recently, NPs have gained more interest in new fields and applications, such as drug delivery and medical imaging systems (Schrofel et al. 2014), gene and cancer therapy (Can 2011), antiviral agents (Elechiguerra et al. 2005), and coatings for medical instruments and tools (Li et al. 2006) tissue engineering and bone replacements (Xing et al. 2010). Besides medically oriented applications, the use of NPs has spread to applications in water purification systems, environmental pollution control (Dankovich and Gray 2011), the electronics industry, agriculture, and biosensors (Shulka et al. 2012; Prasad et al. 2014, 2017).

Mushrooms are macrofungal biological systems that have been traditionally consumed for food and medicinal purposes. They contain carbohydrates, proteins, fats, fibers, minerals, and vitamins (Owaid et al. 2017a, b, c, d, e). Moreover, they

D. Joel
Institute of Bioproduct Development (IBD), Universiti Teknologi Malaysia (UTM), Johor Bahru, Johor, Malaysia

Faculty of Chemical and Energy Engineering, Universiti Teknologi Malaysia (UTM), Johor Bahru, Johor, Malaysia

D. P. Singh
ICAR-National Bureau of Agriculturally Important Microorganisms, Kushmaur, Uttar Pradesh, India

R. A. Malek · S. Z. Hanapi · K. Kumar
Institute of Bioproduct Development (IBD), Universiti Teknologi Malaysia (UTM), Johor Bahru, Johor, Malaysia

E. A. Elsayed
Bioproducts Research Chair, Zoology Department, Faculty of Science, King Saud University, Riyadh, Kingdom of Saudi Arabia

Chemistry of Natural and Microbial Products Department, National Research Centre, Cairo, Egypt contain polyphenols, terpenoids, and lectins, and are also famous for their wide range of biological activity (Bernardshaw et al. 2005). As macrofungi, they can tolerate higher concentrations of metals used for NP synthesis. Additionally, their metabolic machinery depends on secreting various extracellular enzymes that can be used in the biosynthesis of NPs (Parikh et al. 2011; Prasad 2016, 2017; Prasad et al. 2018). Accordingly, NPs that are synthesized by mushrooms have been found in many applications in various fields. AgNPs prepared by *Agaricus bisporus* have been extensively used for their antibacterial, antifungal, antioxidant, and drug delivery applications (Sudhakar et al. 2014: Dhanasekaran et al. 2013: Majumder 2017). Gold NPs (AuNPs) have been prepared from *P. bisporus* and the yellow oyster mushroom *Pleurotus cornucopiae* var. *citrinopileatus* and have been used for cancer cells as well as HIV treatment (Owaid et al. 2017a, b, c, d, e). Furthermore, the milky mushroom, *Calocybe indica*, has been used for the preparation of AgNPs, which have been found to exert a potential effect on the development of apoptotic cell death in MDA-MB-231 human breast cancer cells (Gurunathan et al. 2015). Recently, nano-protein extracts synthesized by *Ganoderma lucidum* and *Pleurotus ostreatus* mushrooms showed an inhibitory potential against acute and chronic inflammation and exhibited a great potential for application in a variety of orthopedic illnesses (Zhang et al. 2017). Moreover, polysaccharides extracted from *Dictyophora indusiata* have been used in the coating and bio-functionalization of selenium nanoparticles (SeNPs) to improve their semiconducting/metallic properties, and hence, their application in cancer diagnostic and healing therapy (Liao et al. 2015). Additionally, application of curcumin, a potential pharmaceutical therapeutic agent, has been limited because of its insolubility in aqueous solutions. Glucans extracted from the medicinal mushrooms *Hericium erinaceus* and *G. lucidum* have been used for curcumin encapsulation, increasing their potential as a drug delivery system with controlled drug release (Le et al. 2016).

4.2 Biosynthesis and Characterization of Metal Nanoparticles Using Mushrooms

Review of the discursive literature on NP research presents three emerging themes: (1) synthesis and assembly of metal particles of well-defined size and geometry; (2) structure and surface chemistry effects on single electron charging; and (3) size, shape, and surface chemistry effects on particle optical properties (Fedlheim and Foss 2001). In the biosynthesis of metal NPs by eukaryotes such as fungi, one or more enzymes or metabolites are produced to reduce its metallic solid NPs through the catalytic effect (Oksanen et al. 2000). However, the synthesis to generate the NPs occurred at a much slower rate; therefore, it must be corrected to compete with other methods.

In the microbial synthesis of NPs, the target ions from the environment were taken by the microorganism and synthesized into the element metals by the enzymes generated from the cellular activities (Mishra et al. 2015). Furthermore, the presence

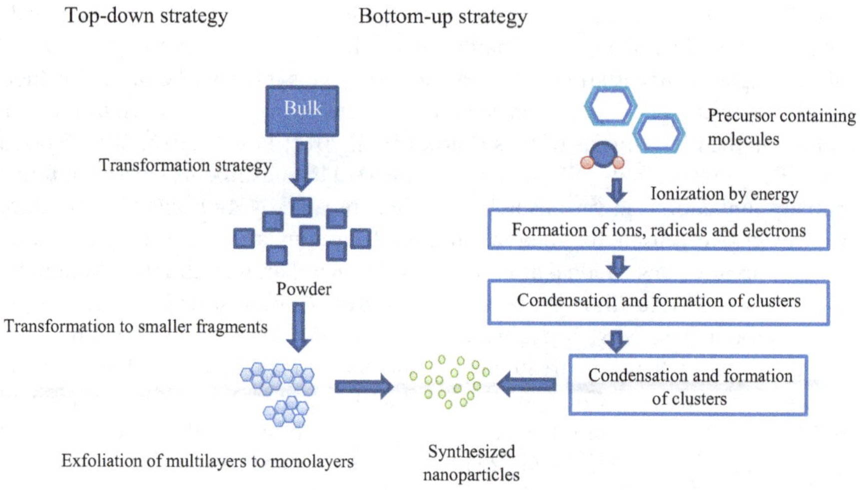

Fig. 4.1 Two strategies for the synthesis of nanoparticles. (Adapted from Aleksandra and Magdalena (2017))

of polysaccharide/oligosaccharide in the broth could be responsible for the reduction of metal ions to the corresponding metal NPs. It is also possible that proteins/enzymes play a role in the reduction of metal ions by the oxidation of benzaldehyde (aldehyde groups) to carboxylic acids. Anthony et al. (2014) had suggested that the carbonyl group of amino acid residue and peptides of proteins might have a strong ability to bind with the metal, and led to the formation of a layer on the AgNPs. This layer prevents the agglomeration of the particles, and thus stabilizes the NPs in the medium.

Depending on the location in which the NPs are formed, NP synthesis is categorized into intracellular and extracellular processes (Tiquia-Arashiro and Rodrigues 2016). The intracellular process involves the transportation of ions into the microbial cells, whereas the extracellular process leads the synthesized NPs to accumulate on the surface of the cell wall of the microorganisms. Various factors have greatly influenced biological production in both intracellular and extracellular synthesis, such as the types of the metal NPs used, growth temperature, synthesis time, extraction methods, and percentage of synthesized versus percentage removed from the sample ratio (Pantidos and Horsfall 2014; Prasad et al. 2016). According to Kumar and Ghosh (2016), synthesis via the extracellular mode offers great advantages over intracellular synthesis from an application point of view. The synthesis of NPs using fungi is typically a bottom-up strategy (Fig. 4.1), where the major reactions involved are the reduction and oxidation of the substrate to increase the colloidal structures (Kashyap et al. 2013). During biosynthesis, the fungal mycelium, which is treated with a metal salt solution, produces enzymes in addition to metabolites as a response to their survival. Therefore, the catalytic effect due to the enzymatic activity of metabolites resulted in a reduction of toxic metal ions from non-toxic metallic solid NPs.

The other strategy is top-down, of which the reduction of materials is completed from large scale to no scale dimension using cutting, polishing, and etching techniques, excluding scaling down to the atomic level. Kashyap et al. (2013) agreed that the bottom-up strategy is preferable for obtaining the level of nanostructures, such as nanorods or nanosheets, which can maximize the homogeneity of the chemical content. In addition, this method also provides fewer defective NPs compared with the top-down approach. Meanwhile, a bottom-up strategy, such as preservation technique, maceration or heating the reaction mixture, are time-consuming and laborious.

Once the NPs are synthesized, precise particle characterization is necessary before application, especially with the purpose of human welfare, in nanomedicines, or in the health care industry. The ultimate goal of developing a safer, more effective, and stabilized method is achieved via a successful measurement method during the characterization process. No standard method exists for the characterization of NPs; however, the formation and characterization of NPs were always confirmed by UV-vis spectroscopy, X-ray diffraction (XRD), Fourier transform infrared spectroscopy (FTIR), scanning electron microscopy (SEM), atomic force microscopy (AFM), etc. (Kumar and Ghosh 2016). Excitation of blueshift in UV-vis spectroscopy reveals the formation of NPs. Meanwhile, transmission electron microscopy (TEM), XRD, and SEM were used to study the morphology, distribution, crystallinity, and size of the NPs. FTIR spectra were recorded to identify the functional groups involved in the synthesis of NPs. More information gathered in the characterization process enhances the bioavailability of therapeutic agents (Zhang et al. 2016). Table 4.1 represents the cumulative studies of the recent characterization of different mushrooms using various fundamental approaches in metal NP research.

4.2.1 UV-Visible Absorption Spectrum

The UV-visible absorption spectrum (UV-vis) was considered a novel technique for the characterization of NPs, and provides information on morphology, size, and stabilization of metal NPs (Kanmani and Lim 2013). During analysis, absorbance between 200 and 700 nm is authenticated to the sample to depict the formation of metal NPs during the characterization process, based on the analysis of localized surface plasmon resonance (SPR) peak. The position and width of the SPR absorption peak depend on the size and agglomeration of the NPs. As the absorbance shifts to longer wavelengths and the SPR peak broadens, the size of the NPs increases whereas the distance between particles decreases owing to aggregation (Zuber et al. 2016). The increase in bandgap can be experimentally shown by the blue shift in the absorption spectrum or else visualized through the color change of the samples. For absorption peaks that exhibited rays, the blue shift showed strong intensity, which is known to arise from the quantum confinement effect as the particle size becomes comparable with or smaller than the Bohr radius of excitation (Suganya and Mahalingam 2017). Although the color change is due to the excitation of free electrons present in NPs (Manzoor-ul-haq et al. 2014), the stabilization of the color intensity of the cell filtrate with metal NPs demonstrated the good distribution of the particles in the solution, with no obvious aggregation.

Table 4.1 Biosynthesis of nanoparticles (NPs) by different mushrooms

Fungus	NPs	Characterization method	NP characteristics	References
Pleurotus ostreatus	A	UV-vis	AgNP SPR band occurs at 420 nm	Al-Bahrani et al. (2017)
		SEM	Spherical (10–40 nm)	
		EDX	Silver, carbon, and oxygen atoms detected at 2.8–3.2 keV	
		FTIR	Functional groups: N–H (3318 cm^{-1}), C–H (2944 cm^{-1}), C=O (1616 cm^{-1}) and carbohydrate group (1411 cm^{-1}).	
		MIC	*B. subtilis* (8–14 mm), *B. cereus* (9–12 mm), *S. aureus* (8–11 mm), *E. coli* (9–12 mm), and *P. aeruginosa* (9–13 mm)	
White mushroom	Mg	XRD	Cubic, diffraction peaks at 36.93, 42.93, 62.27, 74.77, and 78.69 along with the Miller indices (hkl) values (111), (200), (220), (311), and (222) crystal planes	Jhansi et al. (2017)
		Debye–Scherrer equation	Average crystallite size: 10 ml (20 nm), 20 ml (18.5 nm), 40 ml (18 nm), 60 ml (16.5 nm), and 80 ml (16 nm)	
		Particle size distribution analysis	Volume of extract: 10 ml (38.6 nm), 20 ml (35.2 nm), 40 ml (34.1 nm), 60 ml (32.4 nm), and 80 ml (29.6 nm)	
		SEM	Average size 1 μm	
		EDX	Mg: O = 1.2:1	
		TG/DTA	Weight losses; 10 ml (3.8%), 20 ml (4.6%), 40 ml (5%), 60 m (3.6%), and 80 ml (4.1%)	
Pleurotus florida	ZnS	UV-vis	ZnSNP SPR band occurs at 445 nm	Suganya and Mahalingam (2017)
		FTIR	Functional groups: O–H (3448 cm^{-1}), C–H (2926 cm^{-1}), C–O (2856 cm^{-1}), C=O (1720 cm^{-1}), N–H (1562 cm^{-1}), and N–H (1543 cm^{-1})	
		XRD	Cubic lattice structure, diffraction peaks at 25.879, 27.958, 31.449, 32.585, 33.324, and 33.324 units along with the Miller indices (hkl) values (111), (220), and (311)	
		SEM	Spherical (10–20 nm) Zn: sulfur = 48.67:50.73	
		TEM	Spherical with ±1.7207 ± 0.064 nm	

Penicillium oxalicum	Ag	UV-vis	AgNP SPR band occurs at 452 nm	Bhattacharjee et al. (2017)
		TEM	Spherical (10–50 mm)	
		SAED	Crystalline nature	
		FTIR	Absorption peaks located at 3427 cm^{-1}, 2811 cm^{-1}, 1596 cm^{-1}, 1383 cm^{-1}, 1122 cm^{-1}, 921 cm^{-1}, and 612 cm^{-1}	
		MIC	*P. aeruginosa* (18.25 ± 0.75 mm), *E. coli* (18.50 ± 1.00 mm), *S. aureus* (17.25 ± 2.25 mm), and *B. subtilis* (17.75 ± 2.75 mm)	
GL and *AB*	Ag	UV-vis	AgNP SPR band occurs at 434 nm (GL) and 422 nm (AB)	Sriramulu and Sumathi (2017)
		FTIR	Presence of –OH group (3304–3340), C–H stretching (2923), amide C=O stretch (1620), alkane –C–H-bond (1350–1400), and C–O–C (1000–1050)	
		XRD	Cubic centered, diffraction peaks at 38.1°, 44.6°, 64.3°, and 78.5°, which corresponds to (111), (200), (220), and (311)	
		SEM	GL: small rod-shaped and agglomerated needle-like structure AB: agglomerated with a sponge-like structure	
		MIC	GL: *S. aureus* (24 mm) and *E. coli* (28 mm) AB: *S. aureus* (22 mm) and *E. coli* (26 mm)	
Pleurotus ostreatus, Pleurotus cornucopiae var. citrinopileatus and *Pleurotus salmoneostramineus*	Ag	UV-vis	Spherical (10–20 mm)	Owaid et al. (2017b)
Pleurotus cornucopiae var. *citrinopileatus* (fresh and dry)	Au	UV-vis	AgNP SPR band occurs at 550 nm (fresh) and 540 nm (dried)	Owaid et al. (2017a)
		FESEM/HRTEM	Fresh: spherical (23–100 nm) Dried: spherical (16–91 nm)	
Avena sativa	Ag	SEM	Negative without peak Agglomeration in varying sizes (60–100 nm)	Amini et al. (2017)
Reishi mushroom	CdS	UV-vis	CdSNP SPR band occurs at 255.5 nm	Raziya et al. (2016)
		XRD	Diffraction peaks at 12.15, 26.81, 32.91, 33.46, 36.71, 43.93, and 48.08	

(continued)

Table 4.1 (continued)

Fungus	NPs	Characterization method	NP characteristics	References
		SEM	Spherical with crystalline structure	
		TEM	Spherical (±10 nm)	
		Debye–Scherrer equation	21.00784 nm	
		EDS	Confirmed the presence of characteristic elements Cd and S	
Flammulina velutipes	Au	TEM	≤20 nm	Narayanan et al. (2015)
		ICP-OES	64.4–330.5 mg kg^{-1} dry weight	
		XRD	Crystalline AuNPs	
Lentinus edodes	Ag	UV-vis	AgNP SPR occurs at 430 nm	Lateef and Adeeyo (2015)
		FTIR	Functional groups: OH (1039 cm^{-1}; 1242 cm^{-1}), –OH (1394 cm^{-1}), C=O (1459 cm^{-1}), secondary amine (1546 cm^{-1}; 1633 cm^{-1}; 2923 cm^{-1}), and primary amines (3354 cm^{-1})	
		SEM	Walnut-shaped (50–100 nm)	
		XRD	Crystal Au structure	
Tricholoma matsutake	Ag	UV-vis	Ag SPR band occurs at 430 nm	Anthony et al. (2014)
		TEM	Spherical (10 ± 5 nm)	
		XRD	Diffraction peaks to (111), (200), (220), (311), and (222) confirmed the crystalline structure of AgNPs	
		Disc diffusion	*E. coli* – <21 mm zone of inhibition *B. subtilis* – 18.0 mm zone of inhibition	
		DLS	Average size AgNPs ~15 nm	
Agaricus bisporus *Pleurotus florida*	Ag	UV-vis	Ag SPR band occurs at 420–430 nm	Manzoor-ul-haq et al. (2014)
		TEM	Spherical (20–44 nm)	
		FTIR	C=C alkyne (2250–2100 cm^{-1}), C=C amine (1500–1700 cm^{-1}), and C–O alcohols (1000 cm^{-1})	

Schizophyllum commune	Ag	SEM	Spherical (54–99 nm)	Arun et al. (2014)
		UV-vis	Ag SPR band occurs at 440 nm	
		FTIR	Functional groups: –C–OC (1020 cm^{-1}), primary amide (1538 and 1638 cm^{-1})	
		PAGE analysis	Synthesized NPs with 30 and 70 kDa	
		MIC	*K. pneumoniae* (3.1 cm); *B. subtilis* (3.0 cm); *E. coli* (2.5 cm); *P. fluorescens* (2.3 cm)	
		Antifungal (25–100 μg)	*Trichophyton simii* (12–27 mm); *Trichophyton mentagrophytes* (13–26 mm); *Trichophyton rubrum* (17–21 mm)	
		Anticell proliferation activity (10–100 μg)	Cell viability in descending order; 10 μg (70.8%); 20 μg (73.6%); 50 μg (46.8%); 100 μg (35.6%)	
Pleurotus florida	Ag	TEM	Monodispersed, almost spherical and very small (±2.445 ± 1.08 nm)	Sen et al. (2013a)
		XRD	Diffraction peaks to (1 1 1), (2 0 0), (2 2 0), and (3 1 1) confirmed the crystalline face-centered cubic structure of AgNPs	
		MIC (turbidimetric growth assay)	*K. pneumoniae* – CFU count reduced to 50% (from 4.9×10^8 CFU to 2.4×10^8 CFU) at 15 μg/ml and reduced to 0.02% (from 4.9×10^8 CFU to 1.0×10^5 CFU) at 40 μg/ml of AgNP–glucan conjugates	
		FTIR	Functional groups: OH (3392 cm^{-1})	
Agaricus bisporus	Ag	UV-vis	Ag SPR band occurs at 420 nm	Narasimha et al. (2011a, b)
		XRD	Diffraction peak to (111) confirmed presence of NPs	
		Debye–Scherrer equation	Average crystallite size is 8.5 nm	
		TEM	Particle sizes and average size vary and range from 6 nm to 50 nm	

UV-vis UV-visible absorption spectroscopy, *AgNP* silver nanoparticle, *SPR* surface plasmon resonance, *SEM* scanning electron microscopy, *EDX* energy dispersive X-ray analysis, *FTIR* Fourier transform infrared spectroscopy, *MIC* minimum inhibitory concentration, *XRD* X-ray diffraction, *TG/DTA* thermo gravimetric and differential thermal analysis, *ZnSNP* zinc sulfide nanoparticle, *TEM* transmission electron microscopy, *SAED* selected area diffraction, *GL Ganoderma lucidum*, *AB Agaricus bisporus*, *FESEM* field emission scanning electron microscope, *CdSNP* cadmium sulfide nanoparticle, *HRTEM* high-resolution transmission electron microscopy, *EDS* energy dispersive X-ray spectroscopy, *ICP-OES* inductively coupled plasma-optical emission spectrometry, *DLS* dynamic light scattering, *PAGE* polyacrylamide gel electrophoresis, *CFU* colony-forming unit

4.2.2 Scanning Electron Microscopy and Transmission Electron Microscopy

The size and shapes of the metal NPs are affected by various factors, including pH, temperature, incubation time, reductant concentration, and the method of preparation (Sriramulu and Sumathi 2017). A special technique is needed to provide an insight into the structure, size, and morphology of the biogenic NPs (Jhansi et al. 2017). SEM and TEM are widely known to accomplish this task. As shown by Al-Bahrani et al. (2017), SEM and TEM analysis are the most reliable methods of defining the size and numbers of the NPs affected by various temperatures. Other image analysis techniques are selected area diffraction (SAED), which describes the polycrystalline nature of the metal NPs (Suganya and Mahalingam 2017).

4.2.3 Fourier Transform Infrared Spectroscopy

The measurement of NPs using Fourier transform infrared (FTIR) spectroscopy is a function to identify the possible biomolecules that are responsible for the stabilization of the synthesized metal NPs such as AgNPs in the solution (Lateef and Adeeyo 2015). This method also highlights the interactions occurring between organic molecules and metal NPs during bioreduction (Al-Bahrani 2017). The strong peaks shown in FTIR spectroscopy confirm the presence of some functional groups such as carbonyl groups, esters, amines, aldehydes, and alkanes, which are responsible for the synthesis of NPs (Narasimha et al. 2011a, b). Different classes of functional groups such as amines, carboxylic acid,, aldehydes, and alkanes are mainly found in plants and microorganisms during the synthesis of metal NPs (Sriramulu and Sumathi 2017).

4.2.4 X-ray Diffraction

The "structure-driven properties" using X-ray diffraction (XRD) crystallography is a widely accepted technique for analyzing NPs of native metals and *a fortiori* the properties of the latter. Giannini et al. (2016) reported that XRD and Bragg diffraction can quantitatively investigate the morphological and structural information of nanomaterials: (1) analyzing the crystal's atomic structure (positions/symmetry of the atoms in the unit cell, unit size, size/shape of the nanocrystalline domain); (2) crystalline mixture (identification of crystalline phases and determination of weight fractions); and (3) nanoscale assembly (position/symmetry of the NPs/nanocrystals in the assembly and extension of the assembly). Recently, this technique has created noteworthy commercial interest in nanostructured research owing to its electrical conductivity, high strength, structure, electron affinity, and versatility.

4.2.5 Debye–Scherrer

The term Debye–Scherrer is erroneously named after the originators, Debye, Scherrer, and Hull; it is used in X-ray diffraction and crystallography to determine the size of crystal particles in the form of powder. The Debye–Scherrer equation relates the size of sub-micrometer particles, or crystallites in a solid to the broadening of a peak in a diffraction pattern. The average crystallite size is calculated using the Debye–Scherrer formula, i.e., $D = K \lambda / \beta \cos\theta$, where K is the constant ($K = 0.9$), λ is the wavelength of the X-ray ($\lambda = 1.54060$ A0), β is the width of the XRD peak at half height, and θ is the Bragg diffraction angle (Suganya and Mahalingam 2017). The results of this equation may describe the formation of metal NPs, which are strongly dependent on the species of mushrooms, method of extraction, the concentration of extract, and the conditions of incubation (such as light and temperature) (Owaid et al. 2017b).

4.2.6 Dynamic Light Scattering

Dynamic light scattering (DLS) was previously known as photon correlation spectroscopy and quasi-elastic light scattering in the older literature. This technique is useful for the determination of size distribution profile and particle size by measuring the random changes in the intensity of light scattered from the suspension or solution (Amini et al. 2017).

4.3 Advantages of Mushroom as a Biofactory for Biosynthesis in Nanoparticles Compared with the Biosynthesis of Other Microorganisms

Green nanotechnology using a mushroom as a bio-manufacturing unit results in the discovery of lighter, stronger, cleaner, less expensive, sustainable, and more precise products compared with other microbes. It is a gateway to the synthesis of NPs of variable size, shape, chemical composition, and controlled dispersal without the presence of toxic chemicals. Biosynthesis of metals such as using AgNPs with microbes is considered to be environmentally friendly and is becoming more popular because of the choice of the solvent medium, reducing agent, and a non-toxic material for the stabilization of the NPs (Bhattacharjee et al. 2017; Aziz et al. 2015, 2016). Compared with plants and other organisms, metallic NPs via fungi (mushrooms) are more stable and nontoxic (Sriramulu and Sumathi 2017; Prasad 2016, 2017; Prasad et al. 2018).

Mushrooms can be used as a source for the production of larger amounts of NPs because of its capability of producing high volumes of protein that directly increase NP formation (Aleksandra and Magdalena 2017). For large-scale production, fungi or mushrooms become ideal biocatalysts owing to their resistance to flow pressure, agitation, and harsh conditions in chambers such as bioreactors (Aleksandra and Magdalena 2017). The exploitation of this organism in a bioreactor exudes extracellular reductive proteins, which are useful in subsequent process steps, and the cell itself can be manipulated in the manifold ways of nanoscience and nanotechnology. Furthermore, mushroom extract is very rich in proteins, amino acids, polysaccharides, and vitamins, which act as the reducing, stabilizing, and capping agents to the process of NPs compared with other biosynthesis routes already reported (Philip 2009). Recent studies have strengthened the link between the mushroom biosynthesis of metal NPs and antimicrobial activities. The synthesis of AgNPs with the biomolecules present in the microbial cells exudes the inhibition of growth and multiplication of many bacteria such as *Bacillus cereus*, *Staphylococcus aureus*, *Citrobacter koseri*, *Salmonella typhi*, *Pseudomonas aeruginosa*, *Escherichia coli*, *Klebsiella pneumonia, Vibrio parahaemolyticus*, and the fungus *Candida albicans*. It is believed that that AgNPs produce reactive oxygen species (ROS) and free radicals, which cause apoptosis, leading to cell death, which eventually prevents their replication (Siddiqi et al. 2018). These unique, unusual physiochemical properties and biological activities extend the application of AgNPs to a wide area, including antiviral, anti-inflammatory, and cancer therapy (Owaid et al. 2017a, b, c, d, e).

Although research has shown that particularly with regard to monodispersity, NPs with well-defined dimensions can be obtained using fungi (Yadav et al. 2017), there is still some debate about the effect of NPs varying from mushroom to mushroom, and its mechanism is not well known; thus, there is more about the condition to be discovered (Manzoor-ul-haq et al. 2014). However, with an established research field, a roadmap for research in this area will continue to grow.

4.3.1 Metal Nanoparticle Biosynthesis by Mushrooms

4.3.1.1 Characteristics of Mushrooms as a Biomanufacturing Unit in Nanotechnology

Biosynthesis of metallic NPs has been established by utilizing edible and medicinal mushrooms as a biomanufacturing unit. They have great potential and advantages over a conventional approach using chemicals. This method is also known to be eco-friendly, cost-effective, easily scaled up, single-step, and no requirement for high energy, pressure, temperature, and toxic chemicals (Khan et al. 2017a, b). Mushrooms are categorized as eukaryotic microorganisms with rigid cell walls and are nonphototrophic. They have advantages with regard to the green synthesis of NPs, especially their ability to secrete enzymes extracellularly and with easy isolation (Prasad et al. 2015). The use of organisms such as mushrooms as a bio-manufacturing unit in nanotechnology is a smart approach, as they secrete many enzymes and are also easy to manage (Poudel et al. 2017). It is known that mushrooms are able to produce NPs extracellularly or intracellularly. The rate of synthesis, especially of

Table 4.2 The list of nanoparticles synthesized by different mushrooms

NP type	Mushroom name	References
Silver	*Agaricus bisporus*	Al-Hamadani and Kareem (2017)
	Ganoderma lucidum	Sriramulu and Sumathi (2017)
	Ganoderma applanatum	Jogaiah et al. (2017)
	Helvella lacunosa	Manzoor-ul-haq et al. (2014)
	Pleurotus ostreatus	Al-Bahrani et al. (2017)
	Pleurotus citrinopileatus	Maurya et al. (2016)
	Schizophyllum commune	Arun et al. (2014)
	Tricholoma matsutake	Anthony et al. (2014)
	Volvariella volvacea	Singha et al. (2017)
Gold	*Flammulina velutipes*	Narayanan et al. (2015)
	Inonotus obliquus	Lee et al. (2015)
	Pleurotus cornucopiae var. *citrinopileatus*	Owaid et al. (2017a, b, c, d, e)
	Pleurotus sapidus	Sarkar et al. (2013)
	Pleurotus florida	Bhat et al. (2013)
	Tremella fuciformis	Tang et al. (2018)
Zinc sulfide	*Agaricus bisporus*	Senapati et al. (2015)
	Pleurotus florida	Suganya and Mahalingam (2017)
	Pleurotus ostreatus	Senapati and Sarkar (2014)
Iron	*Pleurotus* sp.	Mazumdar and Haloi (2011)
Selenium	*Catathelasma ventricosum*	Liu et al. (2018)

AgNPs, is much more rapid compared with chemical synthesis. Some species grow fast and therefore culturing becomes easier (Castro-Longoria et al. 2011). The use of fungi in the form of mushroom extract has attracted much attention regarding research into the biological production of metallic NPs owing to their tolerance and their capability to accumulate metal (Blackwell 2011). Table 4.2 provides the list of NPs synthesized by different types of mushrooms.

Another interesting aspect is their high metal binding to the surface of the cell wall and intracellular metal uptake capacities (Gupta et al. 2015). In addition, they acquire one or more metal tolerance strategies, including extracellular metal sequestration and precipitation, suppressed influx, enhanced metal efflux, intracellular sequestration, and complexation (Oladipo et al. 2018). Furthermore, they have the ability to produce metal NPs and nanostructures by reducing enzymes and the procedure of biomimetic mineralization (Duran et al. 2005). The mycelial mesh used for NP synthesis can bear flow pressure, agitation, and other conditions in bioreactors compared with bacteria (Saglam et al. 2016). It can be easily handled in downstream processing as the reductive proteins secreted are well known (Prasad 2016). Another characteristic is that mushrooms exhibit tolerance to a range of toxicants and heavy metals in *in vitro* conditions (Oladipo et al. 2018). Their cell wall is a dynamic structure that protects the cell from changes in osmotic pressure due to other environmental stresses (Bowman and Free 2006). With this, mushrooms will be able to produce NPs under conditions of high metal concentrations.

4.3.1.2 Extracellular and Intracellular Biosynthesis of Metal Nanoparticles

The clear mechanism of NP synthesis using mushrooms is somehow not well known. However, the formation of NPs is believed to be responsible for the enzyme reductase, which exists in the culture filtrate intermingled with metal ions (Phanjom and Ahmed 2017). In general, there are three steps involved in the mechanisms for extracellular synthesis of NPs, which are the nitrate reductase process, electron shuttle quinones, or both (Alghuthaymi et al. 2015). Molnár et al. (2018) suggested that the synthesis of NPs might consist of two simultaneous processes, which are the reduction of Au(III)- into Au(0)-producing nuclei followed by the growth of particles and the stabilization by capping ligands of the NPs produced. The protecting ligand prevents further growth and the coagulation of particles owing to electrostatic or steric stabilization (Molnár et al. 2018).

Nitrate reductase was suggested to initiate NP formation by many fungi, including mushrooms. Chan and Mashitah (2012) reported that the reduction of Ag ions was probably caused by the presence of diketone compound. Several enzymes, such as α-NADPH-dependent reductases, nitrate-dependent reductases, and an extracellular shuttle quinone, were associated with AgNP synthesis (Alghuthaymi et al. 2015). It was also reported that the synthesis of NPs may be due to the transfer of electrons from free amino acids to metal ions (Birla et al. 2013). The ninhydrin test was done to confirm the presence of free amino acids in heat-denatured fungal filtrate, which was perhaps responsible for the synthesis of NPs.

The enzymes found in the extract could play a role in the reduction of metal ions by the oxidation of benzaldehyde (aldehyde groups) to carboxylic acids (Mandal et al. 2006). A recent report shows that a polysaccharide called glucan isolated from the edible mushroom *Pleurotus florida* acts as both a reducing and a stabilizing agent (Sen et al. 2013a). A similar study using the pine mushroom extract *Tricholoma matsutake* suggested the possibility of the reduction of metal ions to the corresponding NPs because of the presence of polysaccharide (Anthony et al. 2014). Bhat et al. (2011) proposed that flavins (flavoproteins) existing in the mushroom extract are responsible for the reduction process of Ag ions into AgNPs.

Limited studies have been carried out on the intracellular synthesis of NPs using mushroom species. The proposed mechanisms during intracellular synthesis involve the binding of heavy metals to the fungal cell wall by proteins or enzymes, which produce electrostatic interactions that lead to the transfer of electrons and the reduction of metal ions (Khan et al. 2017a, b). Fungal cell wall and cell wall sugars are likely to play an important role in the absorption and reduction of metal ions (Alghuthaymi et al. 2015). The intracellular synthesis of NPs can be explained using a stepwise mechanism. In the preliminary step of bioreduction, the trapping of metal ions takes place at the fungal cell surface. This is probably due to the electrostatic interaction of the positively charged groups in enzymes present on the cell wall mycelia. In the next step, the metal ions are probably reduced by the enzymes within the cell wall, which leads to the aggregation of metal ions and formation of NPs. Glucose-containing medium can also act as a reducing agent and can reduce Au(III) to Au(0).

4.3.1.3 Factors Affecting Fungal Synthesis of Metallics Application

The challenges frequently encountered in the biosynthesis of NPs are the control of the shape and size of the NPs in addition to achieving the monodispersity in the solution phase. There are several factors affecting the fungal synthesis of metals by mushrooms, e.g., media type, pH, temperature, incubation time, and biomass concentration.

Type of Medium

Biosynthesis of NPs can be varied depending on the type of media used. Saxena et al. (2016) showed that potato dextrose broth (PDB) was the best for maximal protein concentration among other media studied, such as Sabouraud dextrose agar (SDA), protease production (PP) media, Czapek Dox (CZAPEK), Richard's medium (RM), and glucose yeast extract peptone (GYP). Birla et al. (2013) has studied ten different types of media for synthesis of NPs by *Fusarium oxysporum*, including CZAPEK, GYP, gluten glucose broth, lipase assay medium, malt glucose yeast peptone broth, PDB, Richard's broth, SDA, and sucrose peptone yeast broth. They found that the highest protein concentration was produced in PP medium. Another study by Saxena et al. (2016) described that the fungal biomass *Sclerotinia sclerotiorum* grown in PDB has shown enhanced AgNP synthesis followed by SDA, RM, CZAPEK, PP, and GYP. Another study reported the use of PDB broth and modified CZAPEK medium for the production of AuNPs (Molnár et al. 2018). They found that the PDB medium could act as a reducing agent and form stable AuNPs without the inoculation of fungi, whereas no NPs were observed on the modified CZAPEK medium. PDB is a complex medium, containing biomolecules ranging from small molecules such as sugars and amino acids to polymers such as potato-starch, which can act as both a reducing and a capping agent, whereas no capping agent was present in CZAPEK medium (Molnár et al. 2018).

pH

The impact of the pH value of the reaction solution was also found to play an important role in the biosynthesis of NPs by mushrooms. The pH of the solution determines the rate, shape, and size of the synthesized NPs, which is due to the formation of nucleation centers, which increases with the increase in pH value of the solution (Vijayaraghavan and Ashokkumar 2017). The reduction of metallic ion to metal NPs increases as the nucleation center increases. The pH of the solution was reported to be an important factor affecting the morphology of the NPs obtained. Another study using the fungus *Neurospora crassa* shows that the optimal pH for the production of small AgNPs is pH 3. However, by using pH 10, the particles are able to maintain the same size range after storage of 10 months (Quester et al. 2016). Maximum synthesis of AgNPs by *Fusarium oxysporum* were obtained under highly alkaline conditions of pH 9 and 11. Another study also reported the high synthesis of AgNPs by the same species under alkaline conditions of pH 8 (Khan and Jameel 2016a, b). At neutral pH, there were fewer NPs being synthesized compared with the alkaline condition. No flocculation was observed at alkaline pH whereas aggregates were formed at acidic pH, which indicates that NPs were monodispersed and stable at alkaline pH (Birla et al. 2013). A fungus named *Penicillium fellutanum* produced AgNPs rapidly at pH 8.0 (Khan and Jameel 2016a, b). However, in the synthesis of AuNPs by *Trichoderma viride*, the size of the NPs was reduced as pH increased. Repulsion between negatively charged ions and the carboxylic group of extract was reduced at lower pH, resulting in uncontrolled nucleation of seeds and formation of larger particles (Starnes et al. 2010). The formation of AgNPs was inhibited when below pH 6 and stopped completely below pH 5 (Ghareib et al. 2016). The study also reported an enhancement effect of particle formation as pH was increased up to pH 9. A similar study also reported using *Aspergillus oryzae* where alkaline conditions are necessary for the reduction of metal ions, and maximum synthesis happened at pH 10 with a reduction of time and no synthesis observed at pH 4–pH 5 (Phanjom and Ahmed 2017). NPs that synthesized under alkaline conditions promoted competition between protons and metal ions for a negatively charged binding site (Sintubin et al. 2009).

Temperature

Temperature is another crucial factor that influences the rate, size, and shape of NPs produced by mushrooms. Small particles within the range 2–9 nm were obtained at a low temperature of 4 °C and the size range increased from 2–22 nm when synthesized at 25 °C (Quester et al. 2016). A similar study was also conducted by Sani et al. (2011) showing that at a temperature of 25 °C, the majority of AgNPs were smaller and larger ones were observed at a higher temperature of 30 °C. The rate of NP formation is related to the incubation temperature and an increase in temperature allowed particles to grow at faster rate. The size of individual particles and rate of synthesis increased with an increase in temperature up to 60 °C (Birla et al. 2013). They found that the maximum protein was excreted at this temperature owing to heat shock. The study conducted for the synthesis of AuNPs by *Trichoderma viride* shows that lower temperature produces smaller particles (Kumari et al. 2016). Another study reported that complete reduction of AgNPs by *Cunninghamella phaeospora* was 4 min faster at 100 °C compared with a temperature of 30 °C, where particles were produced after about a day of reaction (Ghareib et al. 2016). The rapid production of NPs is an important parameter for scaling up on a commercial scale, which may lead to cost-effective production. In a further study using *Aspergillus oryzae*, the rapid synthesis of AgNPs observed within less time when the temperature increased from 30 to 90 °C. However, the size of the synthesized particles decreased with an increase in reaction time and were more uniform in spherical sizes. Yet another study revealed that the maximal synthesis of AgNPs by

Fusarium oxysporum was at 40 °C (Khan and Jameel 2016a, b). Enzymatic activity of nitrate reductases was maximal at this temperature, resulting in increased production of NPs. Synthesis of NPs using *Arthroderma fulvum* showed an optimal temperature at 55 °C and synthesis of NPs was inhibited at temperatures beyond 65 °C (Xue et al. 2016). A report using *Cladosporium sphaerospermum* revealed a sharp narrow UV spectra peak at a lower wavelength region (412 nm at 70 °C) indicating formation of smaller NPs and higher wavelength regions (440 nm at 30 °C) indicating formation of larger particles (Abdel et al. 2016). These findings are in agreement that the reactants are consumed rapidly as the temperature increased, leading to the development of smaller NPs.

Biomass Concentration

It is well understood that the enzymes produced by organisms are the major factor in the biosynthesis of NPs. An increasing amount of AgNP synthesis by *Fusarium oxysporum* was observed when the quantity of biomass increased from 0.2 g to 6.0 g (Birla et al. 2013). Another study using the same fungi species reported synthesis of maximal NPs at a high biomass of 20 g (Khan and Jameel 2016a, b). The amount of biomass plays a key role in the synthesis or complete reduction of Ag^+ to Ag^0. A similar result was obtained for the biosynthesis of AgNPs by using different types of fungi where the amount NP synthesis increased with an increase in biomass quantity from 0.5 to 10 g (Saxena et al. 2016). A further study using *Trichoderma viride* for the synthesis of AuNPs showed that the size of particles increased when the biomass concentration increased (Kumari et al. 2016). Yet another study reported that fungi *Sclerotinia sclerotiorum* showed an increased synthesis of NPs when the biomass increased from 0.5 to 10 g (Saxena et al. 2016). A higher biomass concentration probably increased the catalytic activity, which caused a faster reaction. It was also reported that *Trichoderma longibrachiatum* was able to synthesize NPs at an optimal biomass concentration of 10 g when the biomass was varied between 1 and 20 g (Elamawi et al. 2018). This was most probably due to the presence of a higher number of enzymes than the NPs ion for the reaction to take place.

Incubation Time

The length of incubation time shows an effect toward the amount of NPs synthesized. It was reported that the fungal suspension of *Chrysosporium tropicum* and *Fusarium oxysporum* produced an increasing amount of NPs with increased of incubation time from day 1 to day 5. Another study reported the continuous synthesis of AgNPs up to 1 week and thereafter no production was observed after 1 month of incubation (Abdel et al. 2016). It was reported that the formation time for AgNPs by *Arthroderma fulvum* happened after about 5-min incubation and more NPs were produced as incubation time increased to 10 h (Xue et al. 2016). Another study found that the production of AgNPs by *Aspergillus terreus* was optimal at an incubation period of 55 h for the interaction of proteins with aqueous Ag metal ions (Khan and Jameel 2016a, b). The stability of the synthesized AgNPs was affected if the incubation period was beyond 55 h. *Nigrospora sphaerica* started to synthesize AgNPs after an incubation time of 72 h (Muhsin and Hachim 2014). A further study reporting the synthesis of AgNPs by *Aspergillus terreus* showed formation of AgNPs after 12 h of incubation (Balakumaran et al. 2016). In the case of AuNPs, a previous study reported that the fungal species of *Aspergillus terreus* was able to synthesize AuNPs at a faster rate (Balakumaran et al. 2016) compared with the study using *Aspergillus fumigatus* and *Aspergillus flavus* (Gupta and Bector 2013). Different types of fungi have different rates of synthesis; therefore, the selection of fast-rate NP producers is important for use in industry.

4.4 Medical Application of Metal Nanoparticles Produced by the Mushroom

Edible mushrooms have long been known to have medicinal properties. Many bioactive substances with immunomodulating effects have been isolated and further research was conducted. The bioactive substances present in the mushroom include polysaccharides (high and low molecular polysaccharides), glycoproteins, triterpenoids, and fungal immunomodulator (Valverde et al. 2015; Sonawane et al. 2014). The whole mushroom, its fruiting bodies, mycelium, and culture filtrate, has been demonstrated to be beneficial owing to its proactive polysaccharides and its ability to contribute toward antibacterial, antiviral, anticancer, antioxidant, hematological, and antitumor effects. By definition, nanotechnology is the engineering of functional systems at the molecular scale and NPs are the domain for essential elements of nanotechnology. The unique features of NPs such as size, morphology, and size-dependent properties of NPs (Khan et al. 2017a, b; Hoshyar et al. 2016) may lead to its applications in biomedicine, healthcare, chemicals, textile industries, medical diagnostics, drug and gene delivery, artificial implants, tissue engineering, biochemical sensors, medical imaging, and computing (Valverde et al. 2015; Sonawane et al. 2014; Prasad et al. 2016). Biosynthesis is provided with a platform for important properties of metal NPs with the use of the mushroom as a biofactory. In fact, the use of edible mushrooms in NP synthesis plays a promising role in large-scale metal NP production, due to advantages such as shorter production time, biocompatibility, cost-effectiveness, and simple and industrial scale with a high yield. Currently, metal NPs are synthesized from various species of edible mushroom and their biological activity studied. Results from this study (antioxidant activity, antifungal, anticancer, antiproliferative) will be further used to produce novel therapeutic, imaging, and biomedical applications.

4.4.1 Antioxidant Activity

The ability of edible mushrooms to prevent oxidative damage to cellular deoxyribonucleic acid has been evaluated and reported (Sánchez 2017). Antioxidants play a role in the treatment of hypertension and ischemic heart disease (Pellegrino 2016; Leopold 2015). There is also information on the function of flavonoid, a potent antioxidant, in an analgesic action mainly targeting prostaglandins (Kumar 2011). Therefore, interest in metal NPs has increased in therapeutics used to prevent oxidative damage, thus considered to reduce the apoptotic pathway (Arvizo et al. 2012). Therefore, the antioxidant potential of NPs such as Au and Ag has been investigated and reported by several researchers (Table 4.3). The well-known methods of investigating antioxidant properties of edible mushroom are di(phenyl)-(2,4,6-trinitrophenyl)iminoazanium (DPPH), ROS, and 2,2′-azino-bis(3-ethylbenzothiazoline-6-sulphonic acid (ABTS). With regard to antioxidant study, a researcher reported that the biosynthesis AuNPs and AgNPs was shown to be significantly higher in antioxidant activities. The present review demonstrates an eco-friendly and low-cost method for the biosynthesis of AuNPs and AgNPs using basidiomycetes mushroom fungal strains *Agaricus bisporus, Ganoderma lucidum, Ganoderma*

Table 4.3 Antioxidant activities of metal nanoparticle synthesis by several species of mushroom

Mushroom	NPs	Method used	Phytochemical study	Antioxidant activity	References
Agaricus bisporus	Ag	DPPH assay	Phenolic, flavonoid	75% ± 0.24	Sriramulu and Sumathi (2017)
Ganoderma lucidum	Ag	DPPH assay	Phenolic, flavonoid	60% ± 0.17	Sriramulu and Sumathi (2017)
Ganoderma sessiliforme	Ag	DPPH assay	Flavonoids, phenolics, protein, tannins, sugar	40–90%	Mohanta et al. (2018)
Ganoderma lucidum	Ag	DPPH assay	Phenolic, flavonoid, polysaccharides	82.59%	Poudel et al. (2017)
Pleurotus eous,	Ag	DPPH assay	Flavonoids, phenolics	76.79%	Madhanraj et al. (2017)
Pleurotus pulmonarius	Au	DPPH assay	Flavonoids, phenolics	76.79%	Madhanraj et al. (2017)
Ganoderma applanatum	Ag	DPPH assay	Phenolic, protocatechuic acid, catechin, apigenin, ferulic acid	98.73% ± 3.45	Jogaiah et al. (2018)
Inonotus obliquus	Au	ABTS assay	Phenols, amines	60–70%	Lee et al. (2015)
Agaricus bisporus	Chi	DPPH assay	Terpenoid, alkaloid Steroid, carbohydrates, tannins, proteins, flavonoids	27.28%	Dhamodharan and Mirunalini (2013)

DPPH di(phenyl)-(2,4,6-trinitrophenyl)iminoazanium, *ABTS* 2,2’-azino-bis(3-ethylbenzothiazoline-6-sulphonic acid, *Chi* chitosan *sessiliforme*, *Pleurotus eous, Pleurotus pulmonarius*, *Ganoderma applanatum*, and *Inonotus obliquus* and antioxidant activities (Table 4.3).

It has been reported that antioxidant activity (DPPH radical scavenging activity) of *G. sessiliforme* extract recommends the plausible commitment of antioxidant molecules from the mushroom concentrate to the biosynthesis of AgNPs. The production of phenolics and flavonoids in *G. sessiliforme* promotes high antioxidative proficiency and these mixes are viewed as effective free-radical scavengers. Furthermore, because such phytochemicals (flavonoids, phenolics, protein, tannins, and sugar) are the main molecules in charge of stabilizing the AgNPs, the NPs are likewise anticipated to have antioxidant potential, which can be extremely helpful for biomedicine. Moreover, antioxidant activity is emphatically ascribed to the presence of flavonoids and phenolic mixes in the concentrates of plants or mushrooms. Phenolic compounds are considered to generate significant antioxidant and antidiabetic activity. There is increased interest in antioxidants, especially in those planned to keep the assumed injurious impacts of free radicals in the human body and to maintain the deterioration of fats and different constituents of foodstuffs (Madhanraj et al. 2017).

4.4.2 Antibacterial Activity

In recent years, microbial infection and contamination are some of the major problem in healthcare and agriculture sectors worldwide. Microbial contaminations are responsible for most common clinical diseases around the world, conveying huge dangers to the general well-being of humans. Presently, the antimicrobial agents in the market are ammonium salt, metal salt arrangements, and antibiotics (Jennings et al. 2015). Unfortunately, the poor efficiency and misapplication of these agents have led to the development of multidrug resistance (MDR) of pathogenic bacteria, yeast, and fungal strains. The major contaminations that have happened because of MDR strains of *Acinetobacter baumannii, Klebsiella* sp., and *Pseudomonas sp.* are increasing at an alarming signal (Manzoor-ul-haq et al. 2015a, b). Contaminations caused by these strains are harder to prevent and cure. In this way, it is critical to discover novel antimicrobial specialists with low toxic qualities and high productivity or elective treatments to solve these issues. Since the 1920s, when AgNPs were officially approved by the FDA to be used in wound therapy as an antibacterial agent, the exploration of metal NPs in the antibacterial field has increased rapidly. Nowadays, numerous kinds of MNPs have been synthesized and demonstrated to have significant antimicrobial activity (Table 4.4). Various methods have been applied to evaluate the antimicrobial activity of MNPs, of which the agar well diffusion method is the most widely used to measure the antimicrobial activity of MNPs against different bacteria and fungi because of its easy, quick and intuitive properties.

Therefore, through NPs, we need to develop new antimicrobial agents with characteristics such as antimicrobial potency, lower toxicity, and compatibility. NPs have been considered a potential alternative because of their electrostatic attraction

Table 4.4 Antibacterial activity of metal NP synthesis by several species of mushroom

Mushroom sp.	NPs	Shapes/size (nm)	Pathogenic microorganism	Inhibition zone (mm)	Activity	References
Agaricus bisporus	Ag	Spherical/10–80	*E. coli*, *S. aureus*	26 22	100 µl of AgNPs	Sriramulu and Sumathi (2017)
Pleurotus ostreatus	Ag	Spherical/>40	*B. subtilis*, *B. cereus*, *S. aureus*, *E. coli*, *P. aeruginosa*	8.8–14 9.0–12 8.0–11 9.0–12 9.0–13	MIC = 13–27 µg/ml	Al-Bahrani et al. (2017)
Volvariella volvacea	Ag	–/7	*E. coli*, *S. flexneri*	15 18	Not reported	Singha et al. (2017)
Ganoderma sessiliforme	Ag	Spherical/45.26	*E. coli*, *B. subtilis*, *S. faecalis*, *L. innocua*, *M. luteus*	11 ± 0.50 20 ± 1.00 16 ± 1.00 22 ± 1.15 21 ± 1.15	500 µg/ml of AgNPs	Mohanta et al. (2018)
Agaricus bisporus	Ag	Spherical/5–35	MRSA *E.coli*, *P. aeruginosa*, *P. mirabilis*	14 20 18 15	50 µl of AgNPs	Al-Hamadani and Kareem (2017)
Agaricus bisporus	Ag	Spherical/15–20	*E. coli*	19	5 µl of AgNPs	Sudhakar et al. (2014)
Trametes ljubarskyi	Ag	–/15–25	*B. subtilis*, *S. aureus*, *M. luteus*, *Staphylococcus*, *E. coli*, *P. putida*, *K. pneumoniae*, *K. aerogenes*	26 26 24 28 25 28 24 27	60 µl of AgNPs	Gudikandula et al. (2017)

(continued)

Table 4.4 (continued)

Mushroom sp.	NPs	Shapes/size (nm)	Pathogenic microorganism	Inhibition zone (mm)	Activity	References
Ganoderma enigmaticum	Ag	–/15–25	*B. subtilis*, *S. aureus*, *M. luteus*, *Staphylococcus*, *E. coli*, *P. putida*, *K. pneumoniae*, *K. aerogenes*	28 24 26 26 26 28 26 26	60 µl of AgNPs	Gudikandula et al. (2017)
Pleurotus citrinopileatus	Ag	Round/6–10	*E.coli*, *S. aureus*	*Data not mention*	0.1 M of AgNPs	Maurya et al. (2016)
Agaricus bisporus	Ag	Spherical/5–50	MRSA 1 MRSA 2 MRSA 3	23 14 16	80 µl of AgNPs	Manzoor-ul-haq et al. (2015a, b)
Agaricus bisporus	Ag	Spherical/5–50	*Klebsiella* sps-1, *Pseudomonas* sp., *Acinetobacter* sp.	17 17 17	80 µl of AgNPs	Manzoor-ul-haq et al. (2015a, b)
Cordyceps sinensis	Ag	ND/50	*E. coli*, *S. aureus*	MIC = 1.6 mg/ml MIC = 0.8 mg/ml	3.0 g/L EPS	Chen et al. (2016)
Agaricus bisporus	Ag	Irregular shape/500 nm–10 µm	*P. vulgaris*, *E. coli*, *P. aeruginosa*, *K. pneumoniae*	12 11 13 10	Not reported	Nath et al. (2015)
Reishi mushroom	CdS	Spherical/10	*S. aureus*	Data not shown	Not reported	Raziya et al. (2016)

Microporus xanthopus	Ag	Spherical/30–50	*P. aeruginosa*, *S. aureus* *E. coli*, *Shigella* sp., *B. subtilis*, *K. pneumoniae*	18 12 11 22 10 17	MIC = 20 µl MIC = 20 µl MIC = 10 µl MIC = 20 µl MIC = 40 µl MIC = 20 µl	Balashanmugam et al. (2007)
Schizophyllum commune	Ag	Spherical/54–99	*E. coli*, *K. pneumonia*, *P. fluorescence*, *B. subtilis*	20–25 25–31 18–23 25–30	1 mM AgNPs	Arun et al. (2014)
Ganoderma lucidum	Ag	Spherical/5–30	*S. aureus*, *E. coli*, *S. mutans*, *K. pneumoniae*, *P. aeruginosa*	29.4 28.12 30.65 22.27 30.28	Not reported	Kannan et al. (2014)
Ganoderma lucidum	Ag	Spherical/10–30	*B. subtilis*, *E.coli*, *S. typhi*, *K. pneumoniae*, *S. aureus*, *B. cereus*	17.0 10.1 12.2 12.1 13.9 13	200 µg/ml (W/v)	Poudel et al. (2017)
Ganoderma applanatum	Ag	Spherical/20–25	*S. aureus*, *E. coli*	16.33 13.33	50 µg ml^{-1}	Jogaiah et al. (2018)
Inonotus obliquus	Au	Spherical, Triangle, hexagonal rod/ 25	*B. subtilis*, *S. aureus*, *E. coli*	12 16 14	30 µl	Lee et al. (2015)
Lentinus edodes	Ag	Walnut-shapes/50–100	*E. coli*, *P. aeruginosa*, *K. pneumoniae*	15–20 12–14 11–12	141 µ g/ml for UV_{10} AgNPs	Lateef and Adeeyo (2015)

(continued)

Table 4.4 (continued)

Mushroom sp.	NPs	Shapes/size (nm)	Pathogenic microorganism	Inhibition zone (mm)	Activity	References
Agaricus bisporus	Ag	Crystalline structure of metallic silver/ 420–444	*E. coli*,	14	Not reported	Dhanasekaran et al. (2013)
			Klebsiella sp.,	15		
			Pseudomonas sp.,	–		
			Enterobacter sp.,	18		
			Proteus sp.,	20		
			S. aureus,	17		
			S. typhi,	22		
			S. paratyphi	17		
Pleurotus platypus	Ag	Spherical/5–50	*E. aerogenes*,	18	14	Sujatha et al. (2013)
			K. pneumoniae,	14	12	
			E. coli,	16	12	
			S. aureus,	12	11	
			P. putida	14	12	

MIC minimum inhibitory concentration, *MRSA* methicillin-resistant *Staphylococcus aureus*, *EPS* exopolysaccharide, *UV* ultraviolet between the positive charge (NPs) and negative charge (microbial cells), and surface to volume ratio. The reported antimicrobial properties of NPs were divided into two categories: antibacterial and antifungal. Ag- and AuNPs are the favorite NPs and have unique properties. They have great potential in antimicrobial activity (inhibition zone against pathogenic bacteria or fungi), in the stability of the chemicals, and in high thermal and electrical conductivity. Applications of metal NPs have been considered to have the potential to inhibit microorganisms in different products, including soaps, plastics, food, biomedical apparatus, and textiles using mushroom species such as *A. bisporus*, *G. lucidum*, *Trametes ljubarskyi*, *Ganoderma enigmaticum*, *Pleurotus citrinopileatus*, *Cordyceps sinensis*, and *Microporus xanthopus*. Use of the consolidation of AgNPs into or on the surface of items such as cleaning splashes, skin creams, ATM buttons, and clothing influences it to accomplish antibacterial properties (Sudhakar et al. 2014) The most important use of AgNPs is in topical balms to prevent contagion against burns and open injuries. The synthesis of AgNPs showed a good inhibitory activity against pathogenic bacteria species such as *E. coli*, *S. aureus*, *P. aeruginosa*, *B. subtilis*, *Klebsiella* sp., *Pseudomonas* sp., *Acinetobacter* sp., *Shigella*, and *Proteus* sp. (Table 4.4).

The spherical shapes of NPs were extended in measurement from 10 nm to 50 nm and indicated a brilliant antimicrobial action against *S. aureus* and *E. coli* (Sriramulu and Sumathi 2017). The free Ag particle has two particular antimicrobial instruments. The first includes denaturation of the disulfide obligations of bacterial proteins, which shapes the basic components of the bacterial structure and ends up incoherent to the synergist impact of the Ag particle. The second includes oxidation, where ionized Ag produces responsive oxygen that connects around the cell, preventing proliferation, and causing cell passage (Gurunathan et al. 2009; El-Sonbaty 2013). Therefore, AgNPs have the capability to serve as another option to antibiotics to control the MDR contamination.

The antibacterial activity of cotton fabrics with and without AgNPs was assessed and investigated by Paul et al. (2015). The cotton fabrics incorporating AgNPs exhibited strong antibacterial activity against three pathogens (*Proteus* sp., *Streptococcus aureus,* and *Pseudomonas* sp.). The results showed the percentage of bacterial colonies present in antibacterial treatment. Thus, it is shown that a dressing material incorporating AgNPs can be utilized commercially as a sterile fabric for wounds and infections.

The antibacterial activity was performed on gram-positive and gram-negative microbes by using the disk diffusion method. The diameter of the inhibition zone was measured and the results were recorded in millimeters (mm). The positive charge on Au particles is basic for its antimicrobial function through electrostatic attractions between the negatively charged cell layer of microorganism and the positively charged NPs (Dakal et al. 2016; Wang et al. 2017). Another system is that the AuNPs produce gaps in the bacterial cell wall by expanding the penetrability of the cell wall, bringing about the spillage of the cell substance, which leads to cell demise.

It is additionally conceivable that Ag-NPs not only interact with the surface of the membrane, it can also penetrate the thick wall of the Gram-positive bacteria

Table 4.5 Application of metal synthesis by the fungus

Applications	Material
Antibacterial coating of the implantable device	Hard valves, dental implants, the catheter, vascular grafts, orthopedic, hernia meshes, dental material
Wound dressing	Diabetic foot ulcer, trauma, burns, chronic skin ulcers
Bone cement	Use to fix joint prostheses

(Poudel et al. 2017). This guarantees that a huge surface region of the molecule is in contact with the bacterial cell surface, which is relied upon to improve the degree of bacterial exclusion (Poudel et al. 2017). A more probable cause of the synergistic effect may be due to the drug delivery of AgNPs to the cell. To the best of author's knowledge, cell membrane consists of phospholipids and glycoprotein, which are hydrophobic. Gentamicin and streptomycin are hydrophilic, but AgNPs are hydrophobic. Thereby, AgNPs, unlike antibiotics, can easily pass through cellular membrane. It can be concluded that AgNPs that bond with antibiotics can be easily delivered to the cell (Hwang et al. 2012). Table 4.5 shows the application of metal synthesis by the fungus.

4.4.3 *Antifungal Activity*

Recently, the resistance of fungal pathogens to antifungal drugs has received worldwide recognition. Thus, there is a growing demand for new antifungal compounds. Natural products from fungi are considered an important source of novel antifungals. This is because fungal species are abundant and diverse, rich in secondary metabolites, and there are improvements in their genetic breeding and fermentation processes (Deyá and Bellotti 2017). The antimicrobial activities of fungi living in distinctive environments are being investigated to discover new antifungal compounds. Few examinations have detailed the intense antifungal action of AgNPs against pathogenic growth, and the vast majority have been centered around the impact of AgNPs on *Candida* species. *Candida albicans* is a commensal yeast species that coexists with humans in distinct niches, such as the oral cavity, mucous membranes, the vagina, and the gastrointestinal tract. Being an opportunistic yeast, it can become pathogenic as the host's immune system weakens. PSC-AgNPs with a normal molecule size of 11.68 nm repressed the development of the pathogenic yeast *C. albicans*. Qualities for least inhibitory focus and lowest fungicidal concentrations were at 250 and 500 mg L^{-1} respectively. TEM pictures discovered that the molecular size of PSC-AgNPs was 16.8 nm, with the qualities of zeta potential and polydispersity record being −8.54 mV and 0.137 respectively (Musa et al. 2017).

Silver NPs within an IC_{80} range of 5–28 mg ml^{-1} indicated very good antifungal activity against *Candida* species. AgNPs showed a comparative action with amphotericin B toward all strains tried with IC_{80} estimations of 5–8 mg ml^{-1}. However, it had more potent activity than fluconazole, with IC_{80} values of 13–33. Yet, the latter

Table 4.6 Antifungal activity of metal NP synthesis by several species of mushroom

Mushroom sp.	NPs	Shapes	Size (nm)	Pathogenic microorganism	Inhibition zone (mm)	References
Ganoderma applanatum	Ag	Spherical	20–25	*Botrytis cinerea*	Data not shown	Jogaiah et al. (2018)
Pleurotus sajor-caju	Ag	Cubic crystalline structure	16.8	*Candida albicans*	11.68	Musa et al. (2017)
Agaricus bisporus	Au	Spherical	10–50	*Aspergillus flavus*	1–3	Eskandari-Nojedehi et al. (2017)
Reishi mushroom	Cd	–	–	*Aspergillus niger*	Not reported	Raziya et al. (2016)
Pleurotus ostreatus	Ag	–	4–15	*Candida albicans*	IC^{80} scope of 5–28 mg ml^{-1}	Yehia and Al-Sheikh (2014)
Schizophyllum commune	Ag	Spherical	54–99	*Trichophyton simii*, *Trichophyton mentagrophytes*, *Trichophyton rubrum*	27 26 21	Arun et al. (2014)
Agaricus bisporus	Ag	–	–	*Aspergillus niger*	2.1 cm = 21 mm	Narasimha et al. (2013)
Schizophyllum commune	Ag	–	50–80	*Aspergillus niger*, *Candida albicans*	Not susceptible 1.9 mm	Chan and Mat Don (2012)

compound showed less effective activity than amphotericin B, with IC_{80} values of 5–7 mg ml^{-1} for *C. tropicalis* and *C. glabrata* (Yehia and Al-Sheikh 2017).

In an antifungal activity test carried out by Chan and Mat Don (2012), showed that *A. niger* was not susceptible to AgNPs. Nonetheless, the AgNPs demonstrated great antifungal impact against *C. albicans* and the outcome was comparable with discoveries by Sadhasivam et al. (2010). The colloidal AgNPs synthesized from mushroom extracts demonstrated maximum inhibition on the development of the fungal strain *Aspergillus niger* (Table 4.6). Maximum zone of inhibition of 2.1 cm was documented by Narasimha et al. (2013).

4.4.4 Anticancer

Malignancy or cancer is the most significant cause of mortality on the planet. Several cancers react to chemotherapy at first, but later create resistance. Accessible chemopreventive agents and chemotherapeutic operators cause bothersome symptoms; thus, building up a biocompatible and practical technique for the treatment of malignancy is vital. Cytotoxic specialists utilized for cancer treatment are costly and known to instigate certain reactions; Thus, the search for novel tumor therapies from natural products has been studied and exhibited strong activity against several cancer cell lines (Gurunathan et al. 2013).

Nanotechnology includes the utilization of structures, characterization, design, gadgets, production, and system on the nanometer scale. Regular difficulties relating to existing malignancy medicines are confinement of the treatment to tumor sites, drug resistance of tumors, and short medication flow times. Significant inconvenience, for example, cardiac issues and low white-platelet count. There are a few modes of delivery of NPs to tumors, for example, liposome-mediated drug delivery (doxorubicin and daunorubicin), biodegradable and biocompatible polymeric NPs [polycaprolactone (PCL) and poly (lactic-co-glycolic corrosive) (PLGA)], and dendrimers [(poly-l-lysine)- octa(3-aminopropyl) silsesquioxane] surface-modified with cyclo(RGDFK).

Currently, cancer kills roughly seven million individuals worldwide on a yearly basis. Cytotoxic impact corresponds conversely to the size of the bioactive compound AgNPs. Recent work likewise investigated the potential anticancer activity of biologically incorporated AgNPs in an in vitro tumor cell line (MCF-7) and demonstrated that there may be a successful option for the treatment of tumors (Mohanta et al. 2018; Suganya et al. 2016; Ismail et al. 2015; Yehia and Al-Sheikh 2014; Wu et al. 2012). The cytotoxicity effects demonstrated that the AgNP has potential anticancer activity in breast tumor cell lines, recommending that AgNPs may be a conceivable elective promoter for the treatment of human breast cancer (Mohanta et al. 2018).

Silver NPs showed a dose-dependent anti-proliferative impact against MCF-7 cells. Expanding the concentration of AgNPs initiated better inhibition of cell proliferation of 78%. Generally, AgNPs from *Pleurotus ostreatus* extracts were found to repress the proliferation of MCF-7 cells. The introduction of MCF-7 cells to AgNPs for 24 h lead to an increase in the variable from 5% to 78% at a concentration within the range 10–640 mg ml^{-1} (Yehia and Al-Sheikh 2014).

In addition, some research was directed at AgNP biosynthesis, utilizing *Pleurotus ostreatus* extracts and exploring the anticancer activity against human liver (HepG2) and breast (MCF-7) adenocarcinoma tumor cell lines. In conclusion, AgNPs arranged by *Pleurotus ostreatus* extracts demonstrated antitumor cytotoxicity against HepG2 and MCF-7 by means of caspase-subordinate apoptosis, related to the enhancement of p53 and the downregulation of Bcl-2. The anticancer activity of AgNPs on HepG2 and MCF-7 cancer cell lines was assessed using the 3-(4,5-dime thylthiazol-2-yl)-2,5-diphenyltetrazolium bromide method. AgNPs prepared by *Pleurotus ostreatus* edible mushroom extracts also showed antitumor cytotoxicity against HepG2 and MCF-7 via caspase-dependent apoptosis, associated with the activation of p53 and the downregulation of Bcl-2 (Ismail et al. 2015).

The antiproliferative action of AuNPs was assessed in vitro against the MCF-7 human breast cancer cell line (HTB-22) and NCI-N87 human stomach disease cell line (CRL-5822) at various concentrations (10, 50, and 100 ml). This uncovered an immediate dosage reaction relationship. At a concentration of 100 ml, the inhibitory impact was seen in MCF-7 (61.2%) and NCI-N87 (95.6%). The minimum inhibitory impact on MCF-7 (8.6%) and NCI-N87 (36.1%) was seen at a concentration of 10 ml. *G. lucidum* polysaccharide NPs indicated significant antitumor effectiveness, having both cytotoxic effects on tumor cells and development consequences for spleen cells, which make them promising candidates in the clinical setting. Generally, the cytotoxicity of AgNPs and (AuNPs) against cancerous cells tends to increase their concentration. However, AgNPs derived from mushrooms showed significant cytotoxicity against MDA-MB-231 cell lines at a comparatively low concentration (6 mg/ml) (Gurunathan et al. 2013).

The present investigation, by Ranjith Santhosh Kumar et al. (2017), reported that AuNPs synthesized and conjugated with doxorubicin can be synthesized by a basic technique using *G. lucidum* mushroom extract. All characterization procedures reveal that *G. lucidum* AuNPs are spherical in shape and 2–100 nm in size. The adequacy of the conjugation of mycosynthesized AuNPs with doxorubicin demonstrated the importance of the anticancer drug and its activity against the human breast cancer (MCF-7) doxorubicin-resistant cell line and as an anticancer drug has been resolved (Table 4.7).

Table 4.7 Summary of the anticancer properties of metal nanoparticle synthesis by mushrooms

Mushroom sp.	Polysaccharides	NPs	Cancer type	References
Ganoderma lucidum	Aqueous extract	Au	Human breast cancer cells (MCF-7)	Ranjith Santhosh Kumar et al. (2017)
Agaricus bisporus	Lectins	ND	Human breast cancer	Majumder (2017)
Ganoderma sessiliforme	Flavonoids, tannins, phenolic, compounds	Ag	Breast adenosarcoma cells (MCF-7 and MDA-MB231)	Mohanta et al. (2018)
Ganoderma lucidum	Pectin	Au	Human breast adenocarcinoma cells (MCF-7/MDA-MB-231)	Suganya et al. (2016)
Inonotus obliquus	Aqueous extract	Au	MCF-1 human breast cancer cell line and NCI-N87 human stomach cancer cell line	Lee et al. (2015)
Lentinus edodes	Lentinan	Se	Human cervix carcinoma cells (HeLa)	Jia et al. (2015)
Pleurotus ostreatus	Culture supernatant	Ag	MCF-7 cell line (breast carcinoma)	Yehia and Al-Sheikh (2014)
Pleurotus tuber-regium	Polysaccharide–protein complexes	Se	Human breast carcinoma (MCF-7)	Wu et al. (2012)
Pleurotus ostreatus	Aqueous extract	Ag	Human liver HepG2 and breast cancer (MCF-7) adenocarcinoma cell lines	Ismail et al. (2015)
Schizophyllum commune	Aqueous extract	Ag	Human epidermoid larynx carcinoma (HEP-2) cell	Arun et al. (2014)

4.4.5 Antidiabetic Application

Diabetes mellitus is a common illness that depicts a gathering of metabolic disorders characterized by hyperglycemia. It influences around 415 million individuals aged 20–79 years in the year 2015, and this number is anticipated to increase to 642 million by 2040 (Liu et al. 2018). In the current work, reported by Liu et al. 2018, as the essential characteristics of diabetes mellitus, hyperglycemia increases the oxidative stress through the overproduction of ROS, which prompts unevenness between free radicals and the antioxidant defense system of the cells. The creation of ROS in diabetes, which causes diabetic injury to liver, kidneys, and other organs, has been accounted for. In addition, superoxide dismutase, catalase, and glutathione peroxidase are viewed as the most imperative resistance systems against ROS with regard to oxidative harm. They also reported on the synthesis and antidiabetic activity of SeNPs in the presence of polysaccharides from *Catathelasma ventricosum* (CVPs). Moreover, the antidiabetic activity of CVPs-SeNPs was assessed by streptozocin-induced diabetic mice. The acquired outcomes demonstrated that, ideal blend states of CVPs-SeNPs were: ultrasound time 60 min with concentration at of Vc 0.04 M, response time 2 h, and pH 7.0. Under these conditions, the mean diameter of the incorporated CVPs-SeNPs was 49.73 nm. TEM of CVPs-SeNPs arranged under ideal conditions demonstrated individual and spherical nanostructure. CVPs-SeNPs (molecule size of around 50 nm) could be stable for roughly 3 months at 4 °C, but for only 1 month at 25 °C. The outcomes of serum profiles and antioxidant activity protein levels discovered that the CVPs-SeNPs had a potential antidiabetic impact. Also, CVPs-SeNPs indicated fundamentally higher antidiabetic activity ($p < 0.05$) than other selenium arrangements, for example, SeNPs, selenocysteine, sodium selenite (Liu et al. 2018).

4.4.6 Antiviral

Antiviral properties of Ag nanomaterials have been exhibited by the binding of AgNPs with glycoprotein, which are shown in the viral envelope and this method restrains the infectivity of HIV-1 infection. The size of these NPs ranges between 1 and 10 nm (Khan et al. 2017a, b). Narasimha (2013) reported that the viricidal properties of AgNPs synthesized from *Agaricus bisporus* have the capability to prevent viral strain bacteriophage. By expanding the focus of NPs from 20 ppm to 160 ppm the viricidal properties expanded with the sign of the plaque number on the medium. Concentration of NPs from 20 μl to 120 μl reduced plaque development, and at 140–160 ppm absolutely suppressed viral development in the host bacterial strain.

4.4.7 Other Nonmedical Applications

4.4.7.1 Photocatalytic Activity/Degradation of Textile Dye

Textile dye contains higher organic compounds that can enter into the soil and water, which can cause environmental pollution and eutrophication. Direct blue, with an empirical formula of $C_{40}H_{28}N_7NaO_{13}S_4$, is one example of a textile dye. It is a water-soluble anionic azo dye. The aggregation of water bodies (supplements and nutrients) and textile colors causes eutrophication, lessens the reoxygenation limit, and causes extreme harm to aquatic microorganisms. Currently, researchers are focusing on the photocatalytic degradation of dyes using metallic NPs. Sriramulu and Sumathi (2017) reported that the photocatalytic activity of AgNPs has been carried out by performing degradation of direct blue 71 (10 ppm) under UV light irradiation. The edible mushroom synthesis of ferritin NPs (10 ml of the mushroom extract was added to 90 ml of 1-mM Ag nitrate solution) showed greater degradation efficiency of approximately 97% in 60 min.

The use of silver, gold, and copper NPs for the degradation of water pollutants such as dyes and phenols were previously reported. Studies carried out by Narayanan et al. (2015) showed the synthesis of biologically immobilized AuNPs by utilizing the mushroom *Flammulina velutipes*, and investigated its heterogeneous catalytic potential in the decrease in methylene blue (MB) and 4-nitrophenol (4NP) in the presence of sodium borohydride ($NaBH_4$). These biomatrix-gold NPs were used as a heterogeneous catalyst in the reduction of organic pollutants such as methylene blue and 4-nitrophenol. The study on this immobilization of metal NPs can be extrapolated to other diverse groups of mushrooms, providing insight into the utilization of mushrooms as a potential candidate for bioremediation, and also as a bioorganic polymer for assembly/matrix of metal and/or metal oxides for various applications.

Sen et al. (2013b) reported the potential use of *Pleurotus florida*, cultivar Assam Florida, as a biofactory for AuNPs that were synthesized by reducing chloroauric acid with a glucan. The glucan acted as a reducing and stabilizing agent. The size scattering of AuNPs changed with the different concentrations of chloroauric acid and may be stable for up to 2 months in aqueous solution. The synthesized AuNPs-glucan bioconjugates showed brilliant catalytic properties in the lessening the harmful toxin 4-nitrophenol to 4-aminophenol. Utilization of these AuNPs-glucan bioconjugates in the chemoselective lessening of some different nitro-compounds such as isophthalic acids and isomers of nitrobenzoic acids may be applicable in the future. Henceforth, the extensive-scale production of NPs by utilizing this glucan and the use of the catalytic properties of AuNPs-glucan bioconjugates in industries are probable.

4.4.7.2 Bionanofertilizers

Suganya and Mahalingam (2017) reported the synthesis of extracellular zinc sulfide nanoparticles (ZnSNPs) by *Pleurotus florida* in spherical structures with particle size ranging from 10 to 20 nm. The seed germination test clearly reveals that ZnS NPs have a definite impact on seed germination and the early development of green gram under controlled conditions. In conclusion, the mushroom extract of *Pleurotus florida* is turned into a savvy and eco-friendly material for the green synthesis of ZnS NPs. Furthermore, the biologically synthesized ZnSNPs were appropriate as plant nutrients, which supports better germination and early development in green gram (*Vigna radiata*). Thus, this intercession opens up new a gateway for biofertilizer businesses to create "bionanofertilizers" for enhanced crop production and overseeing future requests for supply in the agricultural sector (Suganya and Mahalingam 2017).

4.5 Conclusions and Future Prospects

Mushrooms have been employed for many years as a source of food and a wide range of bioactive metabolites of high medicinal value. In addition, recent research has clearly shown the great potential of mushrooms for use as a biofactory for the production of metal nanoparticles. The metal NPs produced have shown promising application in the field of medicine. The main advantages of mushrooms over other biofactories are safety and the high yield. However, further studies are still needed to understand in depth the mechanisms of nanoparticle formation by mushrooms for both extracellular and intracellular pathways. The two main challenges for the development of drug-based nanoparticles using mushroom cells still require further research: first, the optimization of the production process on a large scale for the development of a validated industrial platform using mushroom cells; second, risk assessment for the development of nanoparticles when applied to medical formulations.

References

Abdel H, Nafady NA, Abdel-Rahim IR, Shaltout AM, Mohamed MA (2016) Biogenesis and optimisation of silver nanoparticles by the endophytic fungus *Cladosporium sphaerospermum*. Int J Nano Chem 2(1):11–19

Ahmad A, Mukherjee P, Senapati S, Mandal D, Khan MI, Kumar R, Sastry M (2003) Extracellular biosynthesis of silver nanoparticles using the fungus *Fusarium oxysporum*. Colloid Surf B 28:313–318

Al-Bahrani R, Raman J, Lakshmanan H, Hassan AA, Sabaratnam V (2017) Green synthesis of silver nanoparticles using tree oyster mushroom *Pleurotus ostreatus* and its inhibitory activity against pathogenic bacteria. Mater Lett 186:21–25

Aleksandra Z, Magdalena KO (2017) Fungal synthesis of size-defined nanoparticles. Adv Nat Sci Nanosci Nanotechnol 8:043001

Alghuthaymi MA, Almoammar H, Rai M, Said-Galiev E, Abd-Elsalam KA (2015) Myconanoparticles: synthesis and their role in phytopathogens management. Biotechnol Equip 29(2):221–236

Al-Hamadani AH, Abbass Kareem A (2017) Combination effect of edible mushroom-sliver nanoparticles and antibiotics against selected multidrug biofilm pathogens. Iraq Med J 1(3):68–74

Almonaci Hernández CA, Juarez-Moreno K, Castañeda-Juarez ME, Almanza-Reyes H, Pestryakov A, Bogdanchikova N (2017) Silver nanoparticles for the rapid healing of diabetic foot ulcers. Int J Med Nano Res 4:019

Amini N, Amin G, Jafari Azar Z (2017) Green synthesis of silver nanoparticles using *Avena sativa L.* extract. Nanomed Res J 2:57–63

Anthony KJP, Murugan M, Jeyaraj M, Rathinam NK, Sangiliyandi G (2014) Synthesis of silver nanoparticles using pine mushroom extract: A potential antimicrobial agent against *E. coli* and *B. subtilis*. Ind Eng Chem Res 20:2325–2331

Arun G, Eyini M, Gunasekaran P (2014) Green synthesis of silver nanoparticles using the mushroom fungus *Schizophyllum* commune and its biomedical applications. Biotechnol Bioprocess Eng 19:1083–1090

Arvizo RR, Bhattacharyya S, Kudgus R, Giri K, Bhattacharya R, Mukherjee P (2012) Intrinsic therapeutic applications of noble metal nanoparticles: past, present and future. Chem Soc Rev 41(7):2943–2970

Aziz N, Faraz M, Pandey R, Sakir M, Fatma T, Varma A, Barman I, Prasad R (2015) Facile algae-derived route to biogenic silver nanoparticles: synthesis, antibacterial and photocatalytic properties. Langmuir 31:11605–11612. https://doi.org/10.1021/acs.langmuir.5b03081

Aziz N, Pandey R, Barman I, Prasad R (2016) Leveraging the attributes of *Mucor hiemalis*-derived silver nanoparticles for a synergistic broad-spectrum antimicrobial platform. Front Microbiol 7:1984. https://doi.org/10.3389/fmicb.2016.01984

Balakumaran MD, Ramachandran R, Balashanmugam P, Mukeshkumar DJ, Kalaichelvan PT (2016) Mycosynthesis of silver and gold nanoparticles: optimization, characterization and antimicrobial activity against human pathogens. Microbiol Res 182:8–20

Balashanmugam P, Santhosh S, Giyaullah H, Balakumaran MD, Kalaichelvan PT (2007) Mycosynthesis, characterization and antibacterial activity of silver nanoparticles from *Microporus Xanthopus*: a macro mushroom. Int J Innov Res Sci Eng Technol 2(11):1–9

Bernardshaw S, Johnson E, Hetland G (2005) An extract of the mushroom *Agaricus blazei* Murill administers orally protects against systemic *Streptococcus pneumonia* infection in mice. Scand J Immunol 62:393–398

Bhat R, Deshpande R, Ganachari SV, Huh DS, Venkataraman A (2011) Photo-irradiated biosynthesis of silver nanoparticles using edible mushroom *Pleurotus florida* and their antibacterial activity studies. Bioinorg Chem Appl 2011:650979

Bhat R, Sharanabasava VG, Deshpande R, Shetti U, Sanjeev G, Venkataraman A (2013) Photo-bio-synthesis of irregular shaped functionalized gold nanoparticles using edible mushroom *Pleurotus florida* and its anticancer evaluation. J Photochem Photobio B: Biology 125:63–69

Bhattacharjee S, Debnath G, Roy AD, Saha AK, Das P (2017) Characterization of silver nanoparticles synthesized using an endophytic fungus, *Penicillium oxalicum* having potential antimicrobial activity. Adv Nat Sci 8:1–6

Birla SS, Gaikwad SC, Gade AK, Rai MK (2013) Rapid synthesis of silver nanoparticles from Fusarium oxysporum by optimizing physicocultural conditions. Scientific World Journal 2013:Article ID 796018

Blackwell M (2011) The fungi: 1, 2, 3 … 5.1 million species? Am J Bot 98:426–438

Borthakur M, Gogoi J, Joshi SR (2017) Macro and micro-fungi mediated synthesis of silver nanoparticles and its applications. ABDU-J Eng Technol 6:1–9

Bowman SM, Free SJ (2006) The structure and synthesis of the fungal cell wall. BioEssays 28(8):799–808

Can E (2011) Nanotechnological applications in aquaculture-sea food industries and adverse effects of nanoparticles on environment. J Mater Sci Eng 5:605–609

Castro-Longoria E, Vilchis-Nestor AR, Avalos-Borja M (2011) Biosynthesis of silver, gold and bimetallic nanoparticles using the filamentous fungus Neurospora crassa. Colloids Surf B Biointerfaces 83:42–48

Chan YS, Mat Don M (2012) Characterization of Ag nanoparticles produced by White-Rot Fungi and Its in vitro antimicrobial activities. Int Arab J Antimicrob Agents 2(3:3):1–7

Chan S, Mashita M. (2012). Instantaneous biosynthesis of silver nanoparticles by selected macro fungi. Aus J Basic Appl Sci 6(1): 86-88

Chen X, Yan J-K, Wu J-Y (2016) Characterization and antibacterial activity of silver nanoparticles prepared with a fungal exopolysaccharide in water. Food Hydrocolloids 53:69–74

Cyphert EL, Recum HA (2017) Emerging technologies for long-term antimicrobial device coatings: advantages and limitations. Exp Biol Med 242(8):788–798. https://doi.org/10.1177/1535370216688572

Dakal TC, Kumar A, Majumdar RS, Yadav V. (2016). Mechanistic basis of antimicrobial actions of silver nanoparticles. Front Microbiol. 7:1831. Front Microbiol. 2016; 7: 1831. https://doi.org/10.3389/fmicb.2016.01831

Dankovich TA, Gray DG (2011) Bactericidal paper impregnated with silver nanoparticle for point-of-use water treatment. Environ Sci Technol 45:1992–1998

Deyá C, Bellotti N (2017) Biosynthesized silver nanoparticles to control fungal infections in indoor environments. Adv Nat Sci Nanosci Nanotechnol 8:025005. (8pp)

Dhamodharan G, Mirunalini S (2013) A detail study of phytochemical screening, antioxidant potential and acute toxicity of *Agaricus bisporus* extract and its chitosan loaded nanoparticles. J Pharm Res 6:818–822

Dhanasekaran D, Latha S, Saha S, Thajuddin N, Panneerselvam A (2013) Extracellular biosynthesis, characterisation and in-vitro antibacterial potential of silver nanoparticles using *Agaricus bisporus*. J Exp Nanosci 8(4):579–588

Duran N, Marcarto PD, Alves OL, DeSouza GIH, Esposito E. (2005). Mechanistic aspects of biosynthesis of silver nanoparticles by several Fusarium oxysporum strains. J Nanobiotechnol, 3, 1–7. https://doi.org/10.1186/1477-3155-3-8

Elamawi RM, Al-Harbi RE, Hendi AA. 2018. Biosynthesis and chacterization of silver nanoparticles using Trichoderma longibrachiatum and their effect on phytopathogenic fungi. Egyptian J Biol Pest Cont. 28:28. https://doi.org/10.1186/s41938-018-0028-1

Elechiguerra JL, Burt JL, Morones JR, Camacho-Bragado A, Gao X, Lara HH (2005) Interaction of silver nanoparticles with HIV-1. J Nanobiotechnol 3:6

El-Sonbaty SM (2013) Fungus-mediated synthesis of silver nanoparticles and evaluation of antitumor activity. Cancer Nanotechnol 4:73–79

Eskandari-Nojedehi M, Jafarizadeh-Malmiri H, Rahbar-Shahrouz J (2016) Optimization of processing parameters in green synthesis of gold nanoparticles using microwave and edible mushroom (*Agaricus bisporus*) extract and evaluation of their antibacterial activity. Nanotechnology 5(6):530–537

Eskandari-Nojedehi M, Jafarizadeh-Malmiri H, Rahbar-Shahrouzi J (2017) Hydrothermal green synthesis of gold nanoparticles using mushroom (Agaricus bisporus) extract: physico-chemical characteristics and antifungal activity studies. Green Process Synth 7:38-47

Fedlheim DL, Foss CA (2001) Metal nanoparticles: synthesis, characterization, and applications. CRC Press, Taylor & Francis Group, Boca Raton, FL, USA

Ghareib M, Tahon MA, Saif MM, Abdallah WE (2016) Rapid extracellular biosynthesis of silver nanoparticles by *Cunninghamella phaeospora* culture supernatant. Iran J Pharm Res 15(4):915

Giannini C, Ladisa M, Altamura D, Siliqi D, Sibillano T, De Caro L (2016) X-ray diffraction: a powerful technique for the multiple-length-scale structural analysis of nanomaterials. CrystEngComm 6:1–22

Gudikandula K, Vadapally P, Singara Charya MA (2017) Biogenic synthesis of silver nanoparticles from white rot fungi: their characterization and antibacterial studies. Open Nano 2:64–78

Gupta S, Bector S (2013) Biosynthesis of extracellular and intracellular gold nanoparticles by *Aspergillus fumigatus* and *A. flavus*. Antonie Van Leeuwenhoek 103(5):1113–1123

Gupta VK, Nayak A, Agarwal S (2015) Bioadsorbents for remediation of heavy metals: current status and their future prospects. Environ Eng Res 20:1–18

Gurunathan S, Kalishwaralal K, Vaidyanathan R et al (2009) Purification and characterization of silver nanoparticles using *Escherichia coli*. Colloids Surf B Biointerfaces 74:328–335

Gurunathan S, Raman J, Malek SNA, John PA, Vikineswary S (2013) Green synthesis of silver nanoparticles using *Ganoderma neo-japonicum* Imazeki: a potential cytotoxic agent against breast cancer cells. Int J Nanomedicine 8:4399–4413

Gurunathan S, Park JH, Han JW, Kim J-H (2015) Comparative assessment of the apoptotic potential of silver nanoparticles synthesized by *Bacillus tequilensis* and *Calocybe indica* in MDA-MB-231 human breast cancer cells: targeting p53 for anticancer therapy. Int J Nanomed 10:4203–4223

Hoshyar N, Gray S, Han H, Bao G (2016) The effect of nanoparticle size on in vivo pharmacokinetics and cellular interaction. Nanomedicine 11(6):673–692

Hwang I-S, Hwang J-H, Choi H, Kim K-J, Lee DG (2012). Synergistic effects between silver nanoparticles and antibiotics and the mechanisms involved. J Med Microbiol 61: 1719–1726. https://doi.org/10.1099/jmm.0.047100-0

Ismail AFM, Ahmed MM, Salem AAM (2015) Biosynthesis of silver nanoparticles using mushroom extracts: induction of apoptosis in HepG2 and MCF-7 cells via caspases stimulation and regulation of BAX and Bcl-2 gene expressions. J Pharm Biomed Sci 5(1):1–9

Jia X, Liu Q, Zou S, Xu X, Zhang L (2015). Construction of selenium nanoparticles/β-glucan composition for enhancement of the antitumor activity. Carbohyd Polym 117: 434-442. https://doi.org/10.1016/j.carbpol.2014.09.088

Jennings MC, Minbiole KPC, Wuest WM (2015) Quaternary ammonium compounds: an antimicrobial mainstay and platform for innovation to address bacterial resistance. ACS Infect Dis 1(7):288–303

Jhansi K, Jayarambabu N, Reddy KP, Reddy NM, Suvarna RP, Rao KV, Kumar VR, Rajendar V (2017) Biosynthesis of MgO nanoparticles using mushroom extract: effect on peanut (*Arachis hypogaea L.*) seed germination. 3 Biotech 7:1–11

Jogaiah S, Kurjogi M, Abdelrahman M, Hanumanthappa N, Phan Tran L-S (2018) *Ganoderma* applanatum-mediated green synthesis of silver nanoparticles: structural characterization, and in vitro and in vivo biomedical and agrochemical properties. Arab J Chem. https://doi.org/10.1016/j.arabjc.2017.12.002

Jogaiah S, Kurjogi K, Abdel Rahman M, Hanumanthappa N, Tran L-M P (2017). Ganoderma applanatum-mediated green synthesis of silver nanoparticles: Structural characterization, and in vitro and in vivo biomedical and agrochemical properties. Arab J Chem. (In Press). https://doi.org/10.1016/j.arabjc.2017.12.002

Kanmani P, Lim S (2013) Synthesis and characterization of pullulan-mediated silver nanoparticles and its antimicrobial activities. Carbohydrate Polymers 97:421–428

Kannan M, Muthusamy P, Venkatachalam U, Rajarajeswaran J (2014) Mycosynthesis, characterization and antibacterial activity of silver nanoparticles (Ag-NPs) from fungus *Ganoderma lucidum*. Malaya J Biosci 1(3):134–142

Kashyap PL, Kumar S, Srivastava AK, Sharma AK (2013) Myconanotechnology in agriculture: a perspective. World J Microbiol Biotechnol 29:191–207

Khan NT, Jameel N (2016a) antifungal activity of silver nanoparticles produced from fungus, *Penicillium fellutanum* at different pH. J Microb Biochem Technol 8:440–443

Khan NT, Jameel J (2016b) Optimization of reaction parameters for silver nanoparticles synthesis from Fusarium oxysporum and determination of silver nanoparticles concentration. J Mater Sci Eng 5:283. https://doi.org/10.4172/2169-0022.1000283

Khan AU, Malik N, Khan M, Cho MH, Khan MM (2017a) Fungi-assisted silver nanoparticle synthesis and their applications. Bioprocess Biosyst Eng 41:1–20

Khan I, Saeed K, Khan I (2017b) Nanoparticles: properties, applications and toxicities. Arab J Chem. https://doi.org/10.1016/j.arabjc.2017.05.011

Kumar D (2011) Anti-inflammatory, analgesic, and antioxidant activities of methanolic wood extract of *Pterocarpus santalinus L.* J Pharmacol Pharmacother 2(3):200–202

Kumar A, Ghosh A (2016) Biosynthesis and characterization of silver nanoparticles with bacterial isolate from gangetic-alluvial soil. Int J Biochem Biotechnol 12:95–102

Kumari M, Mishra A, Pandey S, Singh SP, Chaudhry V, Mudiam MKR, Shukla S, Kakkar P, Nautiyal CS (2016) Physico-chemical condition optimization during biosynthesis lead to development of improved and catalytically efficient gold nano particles. Sci Rep 6:27575

Lateef A, Adeeyo AO (2015) Green synthesis and antibacterial activities of silver nanoparticles using extracellular laccase of *Lentinus edodes*. Not Sci Biol 7:405–411

Le MH, Do HD, Thi HHT, Dung LV, Nguyen HN, Thi HNT, Nguyen LD, Hoang CK, Le HC, Thi THL, Trinh HT, Ha PT (2016) The dual effect of curcumin nanoparticles encapsulated by 1-3/1-6 β-glucan from medicinal mushrooms *Hericium erinaceus* and *Ganoderma lucidum*. Adv Nat Sci Nanosci Nanotechnol 7:45019

Lee KD, Nagajyothi PC, Sreekanth TVM, Park S (2015) Eco-friendly synthesis of gold nanoparticles (AuNPs) using *Inonotus obliquus* and their antibacterial, antioxidant and cytotoxic activities. J Ind Eng Chem 26:67–72

Leopold JA (2015) Antioxidants and coronary artery disease: from pathophysiology to preventive therapy. Coron Artery Dis 26(2):176–183

Li Y, Leung P, Yao L, Song QW, Newton E (2006) Antimicrobial effect of surgical masks coated with nanoparticles. J Hosp Infect 629:58–63

Liao W, Yu Z, Lin Z, Lei Z, Ning Z, Regenstein JM, Yang J, Ren J (2015) Biofunctionalization of selenium nanoparticle with *Dictyophora indusiata* polysaccharide and its antiproliferative activity through death-receptor and mitochondria-mediated apoptotic pathways. Sci Rep 5:18629

Liu Y, Zeng S, Liu Y, Wu W, Shen Y, Zhang L, Li C, Chen H, Liu A, Shen L, Hu B, Wang C (2018). Synthesis and antidiabetic activity of selenium nanoparticles in the presence of polysaccharides from Catathelasma ventricosum. Int J Biol Macromol 114: 632–639. https://doi.org/10.1016/j.ijbiomac.2018.03.161

Madhanraj R, Eyini M, Balaji P (2017) Antioxidant assay of gold and silver nanoparticles from edible basidiomycetes mushroom fungi. Free Radic Antioxidants 7(2):137–142

Majumder P (2017) Nanoparticle-assisted herbal synergism an effective therapeutic approach for the targeted treatment of breast cancer: a novel prospective. Glob J Nanomed 2:555–595

Mandal D, Bolander ME, Mukhopadhyay D, Sarkar G, Mukherjee P (2006) The use of microorganisms for the formation of metal nanoparticles and their application. Appl Microbiol Biotechnol 69(5):485–492

Manzoor-ul-haq, Rathod V, Shivaraj P, Singh D, Krishnaveni R (2014) Isolation and screening of mushrooms for potent silver nanoparticles production from Bandipora District (Jammu and Kashmir) and their characterization. Inter J Curr Microb Appl Sci 3:704–714

Manzoor-ul-Haq RV, Shivraj N, Singh D, Yasin MAM (2015a) Silver nanoparticles from mushroom *Agaricus bisporus* and their activity against multi drug resistant strains of *Klebsiella* sps. *Pseudomonas* sp. and Acinetobacter sp. Int J Nat Prod Res 5(3):20–26

Manzoor-ul-Haq, Rathod V, Singh D, Singh KA, Ninganagouda S, Hiremath J (2015b) Dried mushroom Agaricus bisporus mediated synthesis of silver nanoparticles from Bandipora District (Jammu and Kashmir) and their efficacy against *Methicillin* Resistant *Staphylococcus aureus* (MRSA) strains. Nanosci Nanotechnol Int J 5(1):1–8

Maurya S, Bhardwaj AK, Gupta KK, Agarwal S, Kushwaha A, Vk C, Pathak RK, Gopal R, Uttam KN, Singh AK, Verma V, Singh MP (2016) Green synthesis of silver nanoparticles using *Pleurotus* and its bactericidal activity. Cell Mol Biol 62:3

Mazumdar H & Haloi N (2011). A study on biosynthesis of iron nanoparticles by Pleurotus sp. J. Microbiol Biotechnol Res 1(3): 39–49.

Mishra S, Dixit S, Soni S (2015) Methods of nanoparticles biosynthesis for medical and commercial applications. Bio-Nanopart Biosynth Sustain Biotechnol Implic:141–154. https://doi.org/10.1002/9781118677629

Mohanta YK, Nayak D, Biswas K, Kumar Singdevsachan S, Abd Allah EF, Hashem A, Alqarawi AA, Yadav D, Mohanta TK (2018) Silver nanoparticles synthesized using wild mushroom show potential antimicrobial activities against food borne pathogens. Molecules 23:655

Molnár Z, Bódai V, Szakacs G, Erdélyi B, Fogarassy Z, Sáfrán G, Varga T, Kónya Z, Tóth-Szeles E, Szűcs E, Lagzi I (2018) Green synthesis of gold nanoparticles by thermophilic filamentous fungi. Sci Rep 8(1):3943

Muhsin TM, Hachim AK (2014) Mycosynthesis and characterization of silver nanoparticles and their activity against some human pathogenic bacteria. World J Microbiol Biotechnol 30(7):2081–2090

Musa SF, Yeat TS, Kamal LZM, Tabana YM, Ahmed MA, El Ouweini A, Lim V, Keong LC, Sandai D (2017) *Pleurotus sajor-caju* can be used to synthesize silver nanoparticles with antifungal activity against *Candida albicans*. J Sci Food Agric 98(3):1197–1207

Narasimha G (2013) Viricidal properties of silver nanoparticles synthesized from white button mushrooms (*Agaricus bisporus*). Int J Nano Dimens 3(3):181–184

Narasimha G, Praveen B, Mallikarjuna K, Deva Prasad Raju B (2011a) Mushrooms (*Agaricus bisporus*) mediated biosynthesis of silver nanoparticles, characterization and their antimicrobial activity. Int J Nano Dimen 2:29–36

Narasimha G, Praveen B, Mallikarjuna K, Deva Prasad Raju B (2011b) Mushrooms (*Agaricus bisporus*) mediated biosynthesis of sliver nanoparticles, characterization and their antimicrobial activity. Int J Nano Dim 2(1):29–36

Narasimha G, Papaiah S, Praveen B, Sridevi A, Mallikarjuna K, Deva Prasad Raju B (2013) Fungicidal activity of silver nanoparticles synthesized by *Agaricus bisporus* (white button mushrooms). J Nanosci Nanotechnol 7(3):114–115

Narayanan KB, Park HH, Han SS (2015) Synthesis and characterization of biomatrixed-gold nanoparticles by the mushroom *Flammulina velutipes* and its heterogeneous catalytic potential. Chemosphere 141:169–175

Nath B, P, Niture SR, Jadhav SD, Boid SO (2015) Biosynthesis and characterization of silver nanoparticles produced by microorganisms isolated from *Agaricus bisporus*. Int J Curr Microbiol App Sci (Special Issue-2):330–342

Nithya R, Rangunathan R (2012) Synthesis of silver nanoparticles using probiotic microbe and its antibacterial effect against multidrug resistant bacteria. Afr J Biotechnol 11:11013–11021

Oksanen T, Pere J, Paavilainen L, Buchert J, Viikari L (2000) Treatment of recycled kraft pulps with *Trichoderma reesei* hemicellulases and cellulases. J Biotechnol 78:39–44

Oladipo OG, Awotoye OO, Olayinka A, Bezuidenhout CC, Maboeta MS (2018) Heavy metal tolerance traits of filamentous fungi isolated from gold and gemstone mining sites. Braz J Microbiol 49(1):29–37

Owaid MN, Al-Saeei SSS, Abed IA (2017a) Biosynthesis of gold nanoparticles using yellow oyster mushroom *Pleurotus cornucopiae* var. *citrinopileatus*. Environ Nanotechnol Monit Manage B 8:157–162

Owaid MN, Barish A, Ali SM (2017b) Cultivation of *Agaricus bisporus* (button mushroom) and its usages in the biosynthesis of nanoparticles. Open Agric 2:537–543

Owaid MN, Barish A, Shariati MA (2017c) Cultivation of *Agaricus bisporus* (cotton mushroom) and its usages in the biosynthesis of nanoparticles. De Gruyter Open 2:537–543

Owaid MN, Al-Saeedi SSS, Abed IA (2017d) Biosynthesis of gold nanoparticles using yellow oyster mushroom *Pleurotus cornucopiae var. citrinopileatus*. Environ Nanotech, Monitor Managet 8:157–162

Owaid MN, Saleem Al-Saeedi SS, Abed IA (2017e) Study on UV-visible for detection of biosynthesis of silver nanoparticles by oyster mushroom's extracts. J Water Environ Nanotechnol 2:66–70

Pantidos N, Horsfall LE (2014) Biological synthesis of metallic nanoparticles by bacteria, fungi and plants. J Nanomed Nanotech 5:1–10

Parikh RY, Ramanathan R, Coloe PJ, Bhargava SK, Patole MS, Shouche YS, Bansal V (2011) Genus-wide physicochemical evidence of extracellular crystalline silver nanoparticles biosynthesis by *Morganella* spp. PLoS One 6:21401

Paul S, Sasikumar CS, Singh AR (2015) Fabrication of silver nanoparticles synthesized from *Ganoderma lucidum* into the cotton fabric and its antimicrobial property. Int J Pharm Pharm Sci 7(8):53–56

Pellegrino D (2016) Antioxidants and cardiovascular risk factors. Diseases 4(1):–11

Phanjom P, Ahmed G (2017) Effect of different physicochemical conditions on the synthesis of silver nanoparticles using fungal cell filtrate of *Aspergillus oryzae* (MTCC No. 1846) and their antibacterial effect. Adv Nat Sci Nanosci Nanotechnol 8(4):045016

Philip D (2009) Biosynthesis of Au, Ag and Au–Ag nanoparticles using edible mushroom extract. Spectrochim Acta A Mol Biomol Spectrosc 73:374–381

Poudel M, Pokharel R, Sudip KC, Chandra Awal S, Pradhananga R (2017) Biosynthesis of silver nanoparticles using *Ganoderma Lucidum* and assessment of antioxidant and antibacterial activity. Int J Appl Sci Biotechnol 5(4):523–531

Prakash NS, Soni N (2011) Factors affecting the geometry of silver nanoparticles synthesis in *Chrysosporium tropicum* and *Fusarium oxysporum*. Am J Nanotechnol 2(1):112–121

Prasad R (2014) Synthesis of silver nanoparticles in photosynthetic plants. J Nanoparticles:963961. https://doi.org/10.1155/2014/963961

Prasad R (2016) Advances and applications through fungal nanobiotechnology. Springer, Cham. isbn:978-3-319-42989-2

Prasad R (2017) Fungal nanotechnology: applications in agriculture, industry, and medicine. Springer, Cham. isbn:978-3-319-68423-9

Prasad R, Kumar V, Prasad KS (2014) Nanotechnology in sustainable agriculture: present concerns and future aspects. Afr J Biotechnol 13(6):705–713

Prasad R, Pandey R, Barman I (2016) Engineering tailored nanoparticles with microbes: quo vadis. WIREs Nanomed Nanobiotechnol 8:316–330. https://doi.org/10.1002/wnan.1363

Prasad R, Bhattacharyya A, Nguyen QD (2017) Nanotechnology in sustainable agriculture: recent developments, challenges, and perspectives. Front Microbiol 8:1014. https://doi.org/10.3389/fmicb.2017.01014

Prasad R, Kumar V, Kumar M, Shanquan W (2018) Fungal nanobionics: principles and applications. Springer, Singapore. isbn:978-981-10-8666-3. https://www.springer.com/gb/book/9789811086656

Quester K, Avalos-Borja M, Castro-Longoria E (2016) Controllable biosynthesis of small silver nanoparticles using fungal extract. J Biomater Nanobiotechnol 7(02):118

Raziya S, Durga B, Rajamanathe SG, Govindh B, Annapura N (2016) Synthesis and characterization of CDS nanoparticles using Reishi mushroom. Int J Adv Tech Eng Sci 4:220–227

Ranjith Santhosh Kumar DS, Senthilkumar P, Surendran L, Sudhagar B (2017) *Ganoderma Lucidum*-oriental mushroom mediated synthesis of gold nanoparticles conjugated with doxorubicin and evaluation of its anticancer potential on human breast cancer Mcf-7/dox Cells. Int J Pharm Pharm Sci 9(9):267–274

Raziya S, Durga B, Rajamahanthe SG, Govindh B, Annapurna N (2016) Synthesis and characterization of CDS nanoparticles using reishi mushroom. Int J Adv Res Sci Eng Technol 4(6):220–227

Sadhasivam S, Shanmugam P, Yun K (2010) Biosynthesis of silver nanoparticles by Streptomyces hygroscopicus and antimicrobial activity against medically important pathogenic microorganisms. Colloids Surf B Biointerfaces 81(1):358–362

Saglam N, Yesilada O, Cabuk A, Sam M, Saglam S, Ilk S, Emul E, Celik PA, Gurel E (2016) Innovation of strategies and challenges for fungal nanobiotechnology, In advances and applications through fungal nanobiotechnology. Springer, Cham, pp 25–46

Sánchez C (2017) Reactive oxygen species and antioxidant properties from mushrooms. Synth Syst Biotechnol 2(1):13–22

Sanghi R, Verma P (2009) Biomimetic synthesis and characterization of protein capped silver nanoparticles. Bioresour Technol 100:501–504

Sarkar J, Roy SK, Laskar A, Chattopadhyay D, Acharya K (2013) Bioreduction of chloroaurate ions to gold nanoparticles by culture filtrate of *Pleurotus sapidus* Quel. Mater Lett 92:313–316

Saxena J, Sharma PK, Sharma MM, Singh A (2016) Process optimization for green synthesis of silver nanoparticles by *Sclerotinia sclerotiorum* MTCC 8785 and evaluation of its antibacterial properties. Springerplus 5(1):861

Schrofel A, Kratošova G, Săfarik I, Safarikova MS, Raška I, Shor LM (2014) Applications of biosynthesized metallic nanoparticles – a review. Acta Biomater 10:4023–4042

Sen IK, Mandal AK, Chakraborti S, Dey B, Chakraborty R, Islam SS (2013a) Green synthesis of silver nanoparticles using glucan from mushroom and study of antibacterial activity. Int Biol Macromol 62:439–449

Sen IK, Maity K, Islam SS (2013b) Green synthesis of gold nanoparticles using a glucan of an edible mushroom and study of catalytic activity. Carbohydr Polym 9:518–528

Senapati US, Sarkar D (2014) Characterization of biosynthesized zinc sulphide nanoparticles using edible mushroom *Pleurotus ostreatus*. Indian J Phys 88(6):557–562

Senapati US, Jha DK, Sarkar D (2015) Structural, optical, thermal and electrical properties of fungus guided biosynthesized zinc sulphide nanoparticles. Res J Chem Sci 5:33–40

Shulka VK, Yadav RS, Yadav P, Panday AC (2012) Green synthesis of nanosilver as a sensor for detection of hydrogen peroxide in water. J Hazard Mater 213:161–166

Siddiqi KS, Husen A, Rao RAK (2018) A review on biosynthesis of silver nanoparticles and their biocidal properties. J Nanobiotech 16(14):1–28

Singha K, Banerjee A, Pati BR, Mohapatra PKD (2017). Eco-diversity, productivity and distribution frequency of mushrooms in Gurguripal Eco-forest, Paschim Medinipur, West Bengal, India. Curr Res Environ Appl Mycol 7(1): 8–18.

Sintubin L, De Windt W, Dick J, Mast J, van der Ha D, Verstraete W, Boon N (2009) Lactic acid bacteria as reducing and capping agent for the fast and efficient production of silver nanoparticles. Appl Microbiol Biotechnol 84:741–749

Sonawane H, Bhosle S, Bapat G, Vikram G (2014) Pharmaceutical metabolites with potent bioactivity from mushrooms. J Pharm Res 8(7):969–972

Sriramulu M, Sumathi S (2017) Photocatalytic, antioxidant, antibacterial and anti-inflammatory activity of silver nanoparticles synthesised using forest and edible mushroom. Adv Nat Sci Nanosci Nanotechnol 8(4):045012

Starnes D, Jain A, Sahi S (2010) In planta engineering of gold nanoparticles of desirable geometries by modulating growth conditions: an environment-friendly approach. Environ Sci Technol 44:7110–7115

Sudhakar T, Nanda A, Babu SG, Janani S, Evans MD, Markose TK (2014) Synthesis of silver nanoparticles from edible mushroom and its antimicrobial activity against human pathogens. Int J Pharm Tech Res 6:1718–1723

Suganya P, Mahalingam PU (2017) Green synthesis and characterization of zinc sulphide nanoparticles from macro fungi *Pleurotus florida*. IOSR J Appl Chem 10(7):37–42

Suganya KSU, Govindaraju K, Kumar VG, Karthick V, Parthasarathy K (2016) Pectin mediated gold nanoparticles induces apoptosis in mammary adenocarcinoma cell lines. Int J Biol Macromol 93:1030–1040

Sujatha S, Tamilselvi S, Subha K, Panneerselvam A (2013) Studies on biosynthesis of silver nanoparticles using mushroom and its antibacterial activities. Int J Curr Microbiol App Sci 2(12):605–614

Tang B, Liu J, Fan L, Li D, Chen X, Zhou J, Li J (2018) Green preparation of gold nanoparticles with *Tremella fuciformis* for surface enhanced Raman scattering sensing. Appl Surf Sci 427:210–218

Tiquia-Arashiro S, Rodrigues DF (2016) Extremophiles: applications in nanotechnology. In: SpringerBriefs in microbiology. Springer, Berlin

Valverde ME, Hernández-Pérez T, Paredes-López, O (2015) Edible mushrooms: improving human health and promoting quality life. Int J Microbiol 2015(14): 376387.

Vijayaraghavan K, Ashokkumar T (2017) Plant-mediated biosynthesis of metallic nanoparticles: a review of literature, factors affecting synthesis, characterization techniques and applications. J Environ Chem Eng 5:4866–4883

Wang L, Hu C, Shao L. (2017). The antimicrobial activity of nanoparticles: present situation and prospects for the future. Int J Nanomedicine 12: 1227–1249. https://doi.org/10.2147/IJN.S121956

Wu H, Li X, Liu W, Chen T, Li Y, Zheng W, Man CWY, Wong MK, Wong KH (2012) Surface decoration of selenium nanoparticles by mushroom polysaccharides–protein complexes to achieve enhanced cellular uptake and antiproliferative activity. J Mater Chem 22:9602–9610

Xing ZC, Chae WP, Baek JY, Choi MJ, Jung Y, Kang IK (2010) In vitro assessment of antibacterial activity and cyto-compatibility of silver containing PHBV nanofibrous scaffolds for tissue engineering. Biomacromolecules 11:1248–1253

Xue B, He D, Gao S, Wang D, Yokoyama K, Wang L (2016) Biosynthesis of silver nanoparticles by the fungus *Arthroderma fulvum* and its antifungal activity against genera of *Candida, Aspergillus and Fusarium*. Int J Nanomedicine 11:1899-1906. https://doi.org/10.2147/IJN.S98339

Yadav KK, Singh JK, Gupta N, Kumar V (2017) A review of nanobioremediation technologies for environmental cleanup: a novel biological approach. J Mat Environ Sci 8:740–757

Yehia R and Al-Sheikh H. 2017. Biosynthesis and characterization of silver nanoparticles produced by Pleurotus ostreatus and their anticandidal and anticancer activities. World J Microbiol Biotechnol. 30(11): 2797–2803. https://doi.org/10.1007/s11274-014-1703-3

Yehia RS, Al-Sheikh (2014) Biosynthesis and characterization of silver nanoparticles produced by *Pleurotus ostreatus* and their anticandidal and anticancer activities. World J Microbiol Biotechnol 30:2797–2803

Zhang XF, Liu ZG, Shen W, Gurunathan S (2016) Silver nanoparticles: synthesis, characterization, properties, applications, and therapeutic approaches. Int J Mol Sci 17:1–34

Zhang J, Sun L, Zapata PA, Arias M, Atehortuac L, Webster TJ (2017) Anti-inflammatory bone protective effects of nano-protein extracts from mushroom species: *Ganoderma lucidum* and *Pleurotus ostreatus*. J Nanosci Nanotechnol 17:5884–5889

Zuber A, Purdey M, Schartner E, Forbes C, van der Hoek B, Giles D, Abell A, Monro T, Ebendorff-Heidepriem H (2016) Detection of gold nanoparticles with different sizes using absorption and fluorescence based method. Sensors Actuators 227:117–127

Chapter 5
Actinomycetes: Its Realm in Nanotechnology

T. Aswani, Sasi Reshmi, and T. V. Suchithra

Contents

5.1 Introduction

Nanotechnology is an emerging field in science which involves the manipulation of matter on atomic, molecular and super-molecular scale. It is a field of synthesis, characterization, exploration and application of nanosized (1–100 nm) materials (Anisa et al. 2003). Nanotechnologies are now widely considered to have the potential to bring benefits in areas such as bio-nanoparticle synthesis, water processing, drug development and the production of stronger and lighter materials. In the past 15 years, nanoscience research has been focused on developing easy, clean, simple, cost-effective and eco-friendly methods for nanoparticle synthesis. The use of microorganisms for the nanoparticle production filled this gap, and this method is called biosynthesis. It is a biochemical process in which biological agents reduce the dissolved metal ions into nano-metals. The biosynthesis is a green approach and is simple, single step and eco-friendly. The various biological agents like bacteria,

T. Aswani · S. Reshmi · T. V. Suchithra (✉)
School of Biotechnology, National Institute of Technology Calicut, Kozhikode, Kerala, India
e-mail: drsuchithratv@nitc.ac.in

R. Prasad (ed.), *Microbial Nanobionics*, Nanotechnology in the Life Sciences,
https://doi.org/10.1007/978-3-030-16383-9_5

fungi, plant tissues, etc. are used for metal nanoparticle biosynthesis (Prasad et al. 2016, 2018a, b; Prasad 2014).

Various physicochemical methods have explored for the development of nanoparticles, but most of these methods have the limitation in the synthesis of controlled sized and shaped nanoparticles. Hence, a biological process with the ability to do the same is going to be a stirring outlook (Kapil et al. 2015). The most explained approaches for metal nanoparticle synthesis are "bottom-up" approach and the "top-down" approach. In the bottom-up approach, larger complex materials with specific molecular structure are synthesized from small molecular level starting materials. In the top-down approach, nanoparticles are synthesized from larger complex materials and then assembled (Pattekari et al. 2011). Biosynthesis of nanoparticle comes under bottom-up approach where nanoparticles are formed by the oxidation and reduction of metals. Microbial and plant enzymes play the catalyst role in controlling this biosynthesis (Prabhu and Poulose 2012).

5.2 General Events in Nanobiosynthesis

Biosynthesis involves the use of enzymes produced from bacteria, fungi and phytochemicals obtained from plants (Fig. 5.1). For the production of metal nanoparticles, the cells are grown in culture media containing metal ions as inducers. The microbial cells are able to produce metal nanoparticles under optimum culture conditions including temperature, pH, mixing and kinetics. The metal nanoparticles harvested from the media are then processed to get it purified and powdered. The newly synthesized nano-metals are ready for use after proper physicochemical characterization.

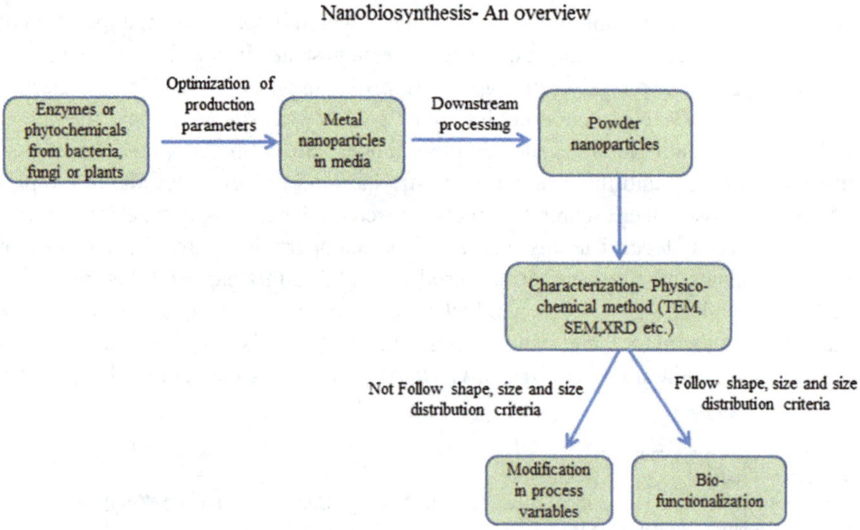

Fig. 5.1 An overview of nanobiosynthesis

5.3 Milestones in the Nanobiosynthesis

Many successful trials have been reported on the production of nanoparticles from microorganisms including bacteria, actinomycetes, fungi, algae, yeast and cyanobacteria and also from plants (Prasad et al. 2018a, b). A number of bacterial species were proved to produce metal nanoparticles in media. Bacterial species like *Salmonella* sp. and *Bacillus* sp. are widely used to synthesize silver nanoparticles. Similarly *Pseudomonas* sp. and *Geobacillus* sp. were known for the production of gold nanoparticles (Ghorbani 2013; Das et al. 2014; Rajasree and Suman 2012).

Looking on to fungi, various strains of *Fusarium* sp. were used for the production of silver and gold nanoparticles (Gaikwad et al. 2013; Thakker et al. 2013). Apart from this, *Penicillium* sp. and *Aspergillus* sp. are also playing as prominent candidates for nanoparticle synthesis. Other species like *P. waksmanii* and *A. sydowii* were undoubtedly proved for the production of silver and gold nanoparticles, respectively (Honary et al. 2013; Vala 2014). Some common marine macroalgae such as *Ulva* sp., *Sargassum* sp. and *Colpomenia* sp. are being used exclusively for the synthesis of silver nanoparticle (El-Rafie et al. 2013, Azizi et al. 2013). Parallel to microbial systems, various parts of plants are also used for the production of different metal nanoparticles. Leaf extract of olive plant and *Momordica charantia* for silver and gold nanoparticles were studied by Khalil et al. (2013) and Pandey et al. (2012) correspondingly. The uses of *Asparagus racemosus* for the synthesis of nano-platinum and nano-palladium were thoroughly studied (Raut et al. 2013). Similarly, *Cassia auriculata* plant part extract can be effectively used for the synthesis of zinc oxide nanoparticles (Vidya et al. 2013).

5.4 Actinomycetes: Bio-factories of Nanoparticles

Actinomycetes are Gram-positive fungi-like bacteria with high G-C content and have the suitability to synthesize nano-metal particles (Fig. 5.2). This production capability is mainly depending on the presence of reducing bioactive compounds.

A number of studies are available that show the potency of actinomycetes for nanoparticle synthesis, which includes the production of nanogold, nanosilver, etc. (Fig. 5.3). The findings of Alani et al. (2012), Balagurunathan et al. (2011) and Waghmare et al. (2011) suggest the uses of *Streptomyces* sp. for nanosilver, nanogold and nano-zinc synthesis.

Actinomycete-mediated nanoparticle synthesis is either intracellular or extracellular (Table 5.1). The nano-metal synthesis is mainly occurring extracellularly (Shah et al. 2012; Waghmare et al. 2014). If nanoparticles are synthesized intracellularly, either the metallic ion molecules will get reduced on the surface of mycelia along with cytoplasmic membrane, or they may trap the ions on the mycelial surface through electrostatic interactions between the positively charged ions and negatively charged carboxylate groups in enzymes that are usually found in the mycelial

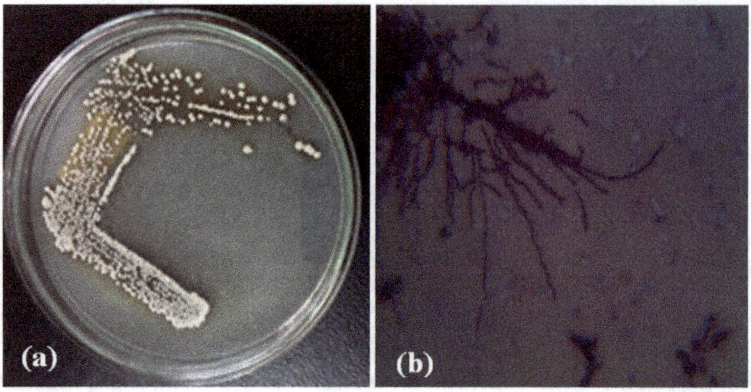

Fig. 5.2 (**a**) *Streptomyces* sp. on starch casein agar plate showing sporulating white color aerial mycelia; (**b**) under light microscope with 40× magnification

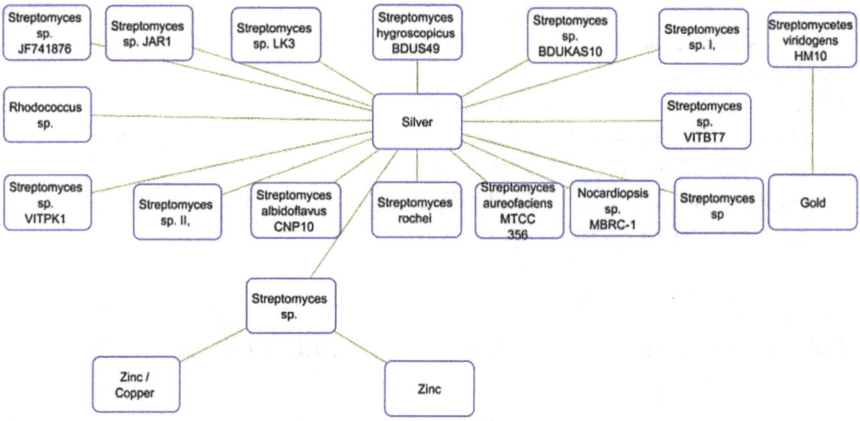

Fig. 5.3 Network model representation of bioactive nanoparticle from actinomycetes (by Cytoscape 3.6.1)

cell wall. In the case of silver nanoparticle production, silver ions reduced to silver nuclei, this reduction process is catalysed with the help of enzymes present on the cell walls. This process successively matures by further reduction reactions, and they gathered among the nuclei (Ahmad et al. 2003b; Sunitha et al. 2014; Sintubin et al. 2009; Prasad et al. 2016).

The possible mechanism of extracellular production involves reduction of nitrate to nitrite by nitrate reductase enzyme present in the nitrogen cycle (Fig. 5.4). This NADH-dependent nitrate reductase enzyme is a significant cause in the biologic synthesis of silver nanoparticles. Nitrate reductase is believed to be responsible for the bio-reduction of silver ions to metallic silver nanoparticles (Durán et al. 2005; Karthik et al. 2014; Prasad et al. 2016)

Table 5.1 Nanoparticles synthesized from diverse actinomycetes

Actinomycete source	Type of NPs	Location	Production medium and precursor	References
Silver nanoparticles				
Streptomyces sp.	Silver	Extracellular	MGYP medium and 1 mM $AgNO_3$	Alani et al. (2012)
Streptomyces sp. JAR1	Silver	Extracellular	ISP II medium and 1 mM $AgNO_3$	Chauhan et al. (2013)
Thermoactinomyces sp.	Silver	Extracellular	Starch casein agar and 10^{-1} mM $AgNO_3$	Deepa et al. (2013)
Nocardiopsis sp. MBRC-1	Silver	Extracellular	Starch casein agar and 10^{-1} mM $AgNO_3$	Manivasagan et al. (2013)
Actinomycetes sp.	Silver	Extracellular	Actinomycete broth and 1 mM $AgNO_3$	Narasimha et al. (2013)
Rhodococcus sp.	Silver	Extracellular	–	Otari et al. (2012)
Streptomyces albidoflavus CNP10	Silver	Extra-/ intracellular	Defined medium with 7 g/l KH_2PO_4, 2 g/l K_2HPO_4, 1 g/l $MgSO_4{\cdot}7H_2O$, 1 g/l $(NH_4)2SO_4$, 0.6 g/l yeast extract and 10 g/l dextrose and 1 mM $AgNO_3$	Prakasham et al. (2012)
Streptomyces hygroscopicus BDUS 49	Silver	Extracellular	–	Sadhasivam et al. (2010)
Streptomyces sp.VITPK1	Silver	Extracellular	Isp1 and 1 mM $AgNO_3$	Sanjenbam et al. (2014)
Streptomyces rochei	Silver	Extracellular	Bennett media and 0.1 mM $AgNO_3$	Selvakumar et al. (2012)
Streptomyces sp. BDUKAS10	Silver	Extracellular	Isp 4 and 1 mM $AgNO_3$	Sivalingham et al. (2012)
Streptomyces sp. VITBT7	Silver	Extracellular	Isp 1 and 1 mM $AgNO_3$	Subashini and Kannabiran (2013)
Streptomyces sp. I *Streptomyces* sp. II	Silver	Extracellular	–	Sukanya et al. (2013)
Streptomyces aureofaciens MTCC356	Silver	Extracellular	–	Sundarmoorthi et al. (2011)
Actinomycetes sp.	Silver	Extracellular	–	Sunitha et al. (2013)
Actinomycetes sp.	Silver	Extracellular	Actinomycete broth and 1 mM $AgNO_3$	Sunitha et al. (2014)
Streptomyces glaucus 71MD	Silver	Extracellular	Actinomycete broth	Tsibakhashvili et al. (2011)
Streptomyces aureofaciens MTCC356	Silver	Extracellular	Defined medium with 0.05% K_2HPO_4, 0.05% $MgSO_4$, 0.05% NaCl, 0.1% KNO_3, 0.001% $FeSO_4.7H_20$, 2% glucose, 0.03% yeast extract and 1 mM $AgNO_3$	Sundarmoorthi et al. (2011)

(continued)

Table 5.1 (continued)

Actinomycete source	Type of NPs	Location	Production medium and precursor	References
Streptomyces sp. JF741876	Silver	Extracellular	–	Vidyasagar et al. (2012)
Streptomyces sp. ERI-3	Silver	Extracellular	Mygp and synthetic media and 1 mM $AgNO_3$	Zonooz and Salouti (2011)
Streptomyces sp. LK3	Silver	Extracellular	SS medium	Karthik et al. (2014)
Gold nanoparticles				
Rhodococcus sp.	Gold	Extra-/intracellular	Mgyp	Ahmad et al. (2003a, b)
Streptomyces viridogens HM10	Gold	Extracellular	Yeast extra-malt extract broth and 1 mM $HAuCl_4$	Balagurunathan et al. (2011)
Thermoactinomyces spp. 44Th	Gold	Extracellular	Defined medium with 0.05% K_2HPO_4, 0.05% $MgSO_4$, 0.05% NaCl, 0.1% KNO_3, 0.001% $FeSO_4.7H_20$, 2% glucose, 0.03% yeast extract and 10^{-3} M $HAuCl_4$	Kalabegishvili et al. (2013)
Nocardia farcinica	Gold	Extracellular	–	Oza et al. (2012)
Thermomonospora sp.	Gold	Extracellular	Nutrient Broth and 1 mM $HAuCl_4$	Sastry et al. (2003)
Streptomyces hygroscopicus	Gold	Extracellular	Mgyp and 10^{-3} M $HAuCl_4$	Waghmare et al. (2014)
Other nanoparticles				
Streptomyces sp.	Zinc	Extracellular	Glycerol yeast extract agar and Zn	Rajamanickam et al. (2012)
Streptomyces sp.	Zinc/copper	Extracellular	Glycerol yeast extract agar and Zn/Cu	Usha et al. (2010)
Streptomyces sp. HBUM171191	Zinc/manganese	Extracellular	Supplemented-metal-nutrient (SMN) agar medium and Mg^{+2} and Mn^{+2}	Waghmare et al. (2011)

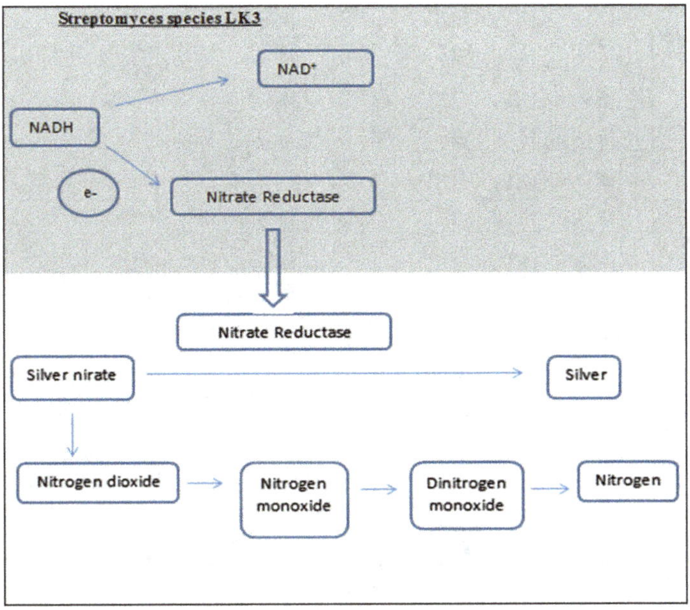

Fig. 5.4 Extracellular synthesis of silver nanoparticles in *Streptomyces* sp. LK3

The mechanism of gold nanoparticle synthesis is similar to nanosilver synthesis. This method comprises the reduction of gold ions from aqueous $AuCl^{4-}$. Bio-reduction of gold ions starts by electron transfer from NADH. The key enzyme in this process is NADH-dependent reductase, and it acts as an electron carrier. The gold ions obtained electrons are reduced to gold (Au^0) and then to gold nanoparticles.

5.5 Biomedical Applications of Actinomycete-Derived Nanoparticles

Nanomedicine is a growing research field with remarkable for the development of disease diagnosis and treatment in both human and animals. Bio-nanoparticles are used as fluorescent biological labels, gene and drug delivery agents and have various other applications including tissue engineering, medical imaging (MRI scanning) and phagokinetic studies. Researchers are now exploring these applications in fields of targeted drug delivery, cancer treatment, gene therapy, antimicrobial activity studies, development of biosensors and MRI. Metal nanoparticles synthesized by actinomycetes showed a wide range of antimicrobial activities against different types of bacteria and fungi including multidrug-resistant microorganisms.

The antimicrobial activity of silver nanoparticles produced by actinomycetes has found a variety of biomedical applications (Table 5.2). It has antibacterial

Table 5.2 Actinomycete-derived nanoparticles and their target microbes

Actinomycete source	Type of NPs	Target organism	References
Streptomyces sp. JAR1	Silver	*Enterococcus faecalis*, *Staphylococcus aureus*, *Escherichia coli*, *Salmonella typhimurium*, *Shigella* sp., *Fusarium* sp., *Scedosporium* sp. JAS1, *Ganoderma* sp. JAS4, *Aspergillus terreus* JAS1	Chauhan et al. (2013)
Streptomyces sp. LK3	Silver	*Rhipicephalus microplus*, *Haemaphysalis bispinosa*	Karthik et al. (2014)
Nocardiopsis sp. MBRC-1	Silver	*Escherichia coli*, *Bacillus subtilis*, *Enterococcus hirae*, *Pseudomonas aeruginosa*, *Shigella flexneri*, *Staphylococcus aureus*, *Aspergillus niger*, *Aspergillus brasiliensis*, *Aspergillus fumigatus*, *Candida albicans*	Manivasagan et al. (2013)
Actinomycetes	Silver	*Escherichia coli*, *Staphylococcus* sp., *Pseudomonas* sp., *Bacillus* sp.	Narasimha et al. (2013)
Streptomyces albidoflavus CNP10	Silver	*Bacillus subtilis*, *Micrococcus luteus*, *Escherichia coli*, *Klebsiella pneumoniae*	Prakasham et al. (2012)
Streptomyces hygroscopicus BDUS49	Silver	*Bacillus subtilis* KCTC 3014, *Enterococcus faecalis* KACC 13807, *Escherichia coli* KCTC 1682, *Salmonella typhimurium* KCCM 40253, *Candida albicans* KACC 30069, *Saccharomyces cerevisiae* KCTC 7906	Sadhasivam et al. (2010)
Streptomyces sp. VITPK1	Silver	*Candida albicans* MTCC 227, *Candida tropicalis* MTCC 184, *Candida krusei* MTCC 9215	Sanjenbam et al. (2014)
Streptomyces rochei	Silver	*Pseudomonas aeruginosa*, *Escherichia coli*, *Klebsiella pneumoniae*, *Enterobacter faecalis*, *Staphylococcus aureus*	Selvakumar et al. (2012)
Streptomyces sp.	Silver	*Staphylococcus aureus*, *Staphylococcus epidermidis*, *Escherichia coli*, *Salmonella typhi*, *Pseudomonas aeruginosa*, *Klebsiella pneumoniae*, *Proteus vulgaris*	Shirley et al. (2010)
Streptomyces sp. VITBT7	Silver	*Staphylococcus aureus* MTCC 739, *Klebsiella pneumonia*, *Pseudomonas aeruginosa* MTCC 424, *Bacillus cereus* MTCC 1168, *Escherichia coli* ATCC 25922, *Aspergillus niger* MTCC 1344, *Candida albicans* MTCC 227	Subashini and Kannabiran (2013)
Streptomyces sp. I, *Streptomyces* sp. II *Rhodococcus* sp.	Silver	*Staphylococcus aureus*, *Escherichia coli*, *Pseudomonas aeruginosa*, *Klebsiella pneumoniae*, *Proteus vulgaris*	Sukanya et al. (2013)
Streptomyces aureofaciens MTCC 356	Silver	*Staphylococcus aureus* ATCC 29737, *Bacillus subtilis* ATCC 6633, *Escherichia coli* ATCC 10536, *Pseudomonas aeruginosa* NCIM 2945	Sundarmoorthi et al. (2011)
Streptomyces viridogens HM10	Gold	*Staphylococcus aureus*, *Escherichia coli*	Balagurunathan et al. (2011)
Streptomyces sp.	Zinc	*Staphylococcus aureus*, *Escherichia coli*, *Salmonella* sp.	Rajamanickam et al. (2012)
Streptomyces sp.	Zinc and Copper	*Escherichia coli* ATCC 8739, *Staphylococcus aureus* ATCC 6538	Usha et al. (2010)

activity against a broad spectrum of bacterial species including *Enterococcus faecalis*, *Staphylococcus aureus*, *Escherichia coli*, *S. aureus*, *Klebsiella pneumoniae*, *Pseudomonas aeruginosa*, *Proteus mirabilis*, *Bacillus subtilis*, *Proteus vulgaris*, *Salmonella typhimurium* and *Micrococcus luteus* (Subashini and Kannabiran 2013; Chauhan et al. 2013; Sukanya et al. 2013; Selvakumar et al. 2012; Sunitha et al. 2013, 2014; Manivasagan et al. 2013; Prakasham et al. 2012; Sadhasivam et al. 2010; Sundarmoorthi et al. 2011). Apart from antibacterial activities, these silver nanoparticles show antifungal, antidermatophytic and antiparasitic activities. Antifungal activity mainly covers the bioactivity against genus *Candida* (*C. albicans*, *C. tropicalis* and *C. krusei*), *Aspergillus* (*A. fumigatus*, *A. flavus*, *A. brasiliensis*, *A. terreus* and *A. niger*) and *Saccharomyces cerevisiae.* Antidermatophytic activity mainly targets dermatophytes like *Trichophyton* (*T. tonsurans* and *T. rubrum*), *Ganoderma* sp. and *Scedosporium* sp. They have strong antiparasitic activity against *Haemaphysalis bispinosa* and *Rhipicephalus microplus* (Karthik et al. 2014; Manivasagan et al. 2013; Sadhasivam et al. 2010; Sanjenbam et al. 2014; Vidyasagar et al. 2012).

5.6 Mechanism of Antimicrobial Action of Nanoparticles

The antimicrobial activity of actinomycete-based nanosilver particles against multidrug-resistant bacterial strains has been thoroughly investigated. Some studies report the synergistic effect of actinomycete-derived nanosilver with commercially available antibiotics. This synergism increases the rate of inhibition against Gram-positive bacteria, Gram-negative bacteria and fungi. The bioactivity of these silver nanoparticles has a correlation with the size and shape of nanoparticles. Triangular-shaped silver nanoparticles exhibit strongest bactericidal activity even in their lowest concentration against *E. coli* when related with rod- and spherical-shaped nanoparticles. The smaller nanoparticles have better toxicity on bacterial pathogens; this may be due to the better diffusion capacity of smaller particle compared to larger ones which proves relativeness of toxicity with the size of nanoparticles (Chauhan et al. 2013; Manivasagan et al. 2013; Shirley et al. 2010; Pal et al. 2007; Sharma et al. 2009; Mohan et al. 2007; Panacek et al. 2006).

A very little information is available on the mechanism of antibacterial activities of metal nanoparticles. The three suggested mechanisms of antibacterial action are mainly based on the activity of silver nanoparticle (Fig. 5.5).

In the first mechanism, the metal nanoparticles bind with the cell membrane, and the binding causes disturbance to the membrane affecting its power functions, such as permeability and respiration. Silver nanoparticles are believed to cause depletion of intracellular ATP by rupturing plasma membrane or by blocking respiration in association with oxygen and sulfhydryl groups on the cell wall to form R-S-S-R bonds. This leads to cell death by the release of cell contents out of the cell due to membrane damage. The interaction of silver nanoparticles with cell membrane depends on the availability of the surface. Smaller nanoparticles will have stronger bactericidal effect by binding with the large surface area of the bacterial cell membrane

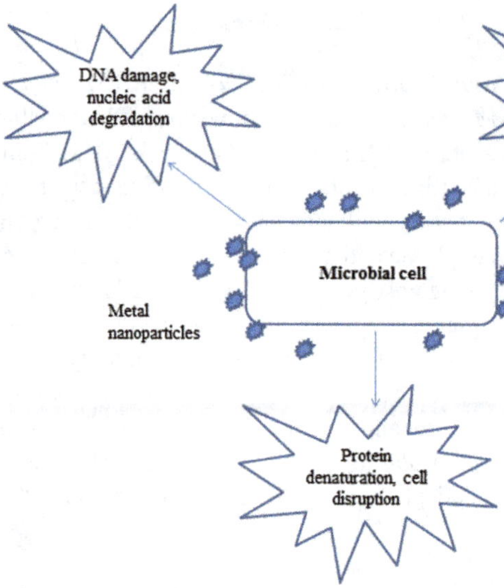

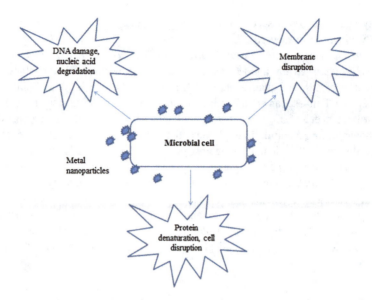

Fig. 5.5 Antimicrobial activity of metal nanoparticles

(Pal et al. 2007; Shang et al. 2014; Sivalingham et al. 2012; Chauhan et al. 2013; Manivasagan et al. 2013; Aziz et al. 2014, 2015, 2016, 2019; Prasad et al. 2016). The second mechanism involves the penetration of silver nanoparticles into the bacterial cell membrane, interacts with phosphorus-containing and sulphur-containing compounds, such as DNA, and causes damages inside it (Chauhan et al. 2013; Prasad and Swamy 2013). The third mechanism explains the release of silver ions by silver nanoparticles which may contribute to the bactericidal activity. It is believed that after interacting with silver nanoparticles, the cellular proteins become inactivated and DNA loses its replication ability. The higher concentration of silver nanoparticles has the ability to interact with DNA, RNA and other cytoplasmic components (Manivasagan et al. 2013).

5.7 Conclusion

Actinomycetes are a diverse group of bacteria and extensively used by scientists for different types of researches. They are able to produce various kinds of bioactive compounds which have great therapeutic values. Recently it was discovered that actinomycetes are capable of producing nanoparticles which have antimicrobial activities, and that was a milestone in the area of therapeutics. Thereby nanotechnology offers an excellent way of utilizing the abilities of actinomycetes for the eco-friendly production of nanoparticles, and it also aims to enhance their utility in preventing microbial infections.

References

Ahmad A, Senapati S, Khan MI, Kumar R, Ramani R, Srinivas V, Sastry M (2003a) Extracellular biosynthesis of monodisperse gold nano-particles by a novel extremophilic actinomycete *Thermomonospora* sp. Langmuir 19:3550–3553

Ahmad A, Senapati S, Khan MI, Kumar R, Ramani R, Srinivas V, Sastry M (2003b) Intracellular synthesis of gold nanoparticles by a novel alkalotolerant actinomycete, *Rhodococcus* species. Nanotechnology 14:824

Alani F, Moo-Young M, Anderson W (2012) Biosynthesis of silver nanoparticles by a new strain of Streptomyces sp. compared with *Aspergillus fumigatus*. World J Microbiol Biotechnol 28(3):1081–1086

Anisa M, Daar AS, Singer PA (2003) Mind the gap': science and ethics in nanotechnology. Nanotechnology 14(3):R9

Aziz N, Fatma T, Varma A, Prasad R (2014) Biogenic synthesis of silver nanoparticles using *Scenedesmus abundans* and evaluation of their antibacterial activity. J Nanopart: 689419. https://doi.org/10.1155/2014/689419

Aziz N, Faraz M, Pandey R, Sakir M, Fatma T, Varma A, Barman I, Prasad R (2015) Facile algae-derived route to biogenic silver nanoparticles: synthesis, antibacterial and photocatalytic properties. Langmuir 31:11605–11612. https://doi.org/10.1021/acs.langmuir.5b03081

Aziz N, Pandey R, Barman I, Prasad R (2016) Leveraging the attributes of *Mucor hiemalis*-derived silver nanoparticles for a synergistic broad-spectrum antimicrobial platform. Front Microbiol 7:1984. https://doi.org/10.3389/fmicb.2016.01984

Aziz N, Faraz M, Sherwani MA, Fatma T, Prasad R (2019) Illuminating the anticancerous efficacy of a new fungal chassis for silver nanoparticle synthesis. Front Chem 7:65. https://doi.org/10.3389/fchem.2019.00065

Azizi S, Namvar F, Mahdavi M, Ahmad MB, Mohamad R (2013) Biosynthesis of silver nanoparticles using brown marine macroalga, *Sargassum muticum* aqueous extract. Mater 6:5942–5950

Balagurunathan R, Radhakrishnan M, Rajendran RB, Velmurugan D (2011) Biosynthesis of gold nanoparticles by actinomycete *Streptomyces viridogens* strain HM10. Indian J Biochem Biophys 48:331–335

Chauhan R, Kumar A, Abraham J (2013) A biological approach to the synthesis of silver nanoparticles with Streptomyces sp. JAR1 and its antimicrobial activity. Sci Pharm 81:607–621

Das VL, Thomas R, Varghese RT, Soniya EV, Mathew J, Radhakrishnan EK (2014) Extracellular synthesis of silver nanoparticles by the *Bacillus* strain CS 11 isolated from industrialized area. 3Biotech 4:121–126

Deepa S, Kanimozhi K, Panneerselvam A (2013) Antimicrobial activity of extracellularly synthesized silver nanoparticles from marine derived actinomycetes. Int J Curr Microbiol Appl Sci 2(9):223–230

Durán N, Marcato PD, Alves OL, De Souza GIH, Esposito E (2005) Mechanistic aspects of biosynthesis of silver nanoparticles by several *Fusarium oxysporum* strains. J Nanobiotechnol 3:8

El-Rafie HM, El-Rafie MH, Zahran MK (2013) Green synthesis of silver nanoparticles using polysaccharides extracted from marine macro algae. Carbohydr Polym 96(2):403–410

Gaikwad S, Birla S, Ingle A, Gade A, Marcato P, Rai M, Duran N (2013) Screening of different Fusarium species to select potential species for the synthesis of silver nanoparticles. J Braz Chem Soc 24(2):1974–1982

Ghorbani HR (2013) Biosynthesis of silver nanoparticles using *Salmonella typhimurium*. J Nanostruct Chem 3:29

Honary S, Gharaei-Fathabad E, Barabadi H, Naghibi F (2013) Fungus-mediated synthesis of gold nanoparticles: a novel biological approach to nanoparticle synthesis. J Nanosci Nanotechnol 13(2):1427–1430

Kalabegishvili T, Kirkesali E, Ginturi E, Rcheulishvili A, Murusidze I, Pataraya D, Gurielidze M, Bagdavadze N, Kuchava N, Gvarjaladze D, Lomidze L (2013) Synthesis of gold nanoparticles by new strains of thermophilic actinomycetes. Nano Stud 7:255–260

Kapil P, Choudhary P, Samant L, Mukherjee S, Vaidya S, Chowdhary A (2015) Biosynthesis of nanoparticles: a review. Int J Pharm Sci Rev Res 30(1):219–226

Karthik L, Kumar G, Vishnu Kirthi A, Rahuman AA, Bhaskara Rao KV (2014) Streptomyces sp. LK3 mediated synthesis of silver nanoparticles and its biomedical application. Bioprocess Biosyst Eng 37:261–267

Khalil MMH, Ismail EH, El-Baghdady KZ, Mohamed D (2013) Green synthesis of silver nanoparticles using olive leaf extract and its antibacterial activity. Arab J Chem. https://doi.org/10.1016/j.arabjc.2013.04.007s

Manivasagan P, Venkatesan J, Senthilkumar K, Sivakumar K, Kim S (2013) Biosynthesis, antimicrobial and cytotoxic effect of silver nanoparticles using a novel Nocardiopsis sp. MBRC-1. Bio Med Res Int: 287638, 9 pages

Mohan YM, Lee K, Premkumar T, Geckeler KE (2007) Hydrogel net-works as nanoreactors: a novel approach to silver nanoparticles for antibacterial applications. Polymer 48:158–164

Narasimha G, Janardhan A, Alzohairy M, Khadri H, Mallikarjuna K (2013) Extracellular synthesis, characterization and antibacterial activity of silver nanoparticles by actinomycetes isolative. Int J Nano Dimens 4:77–83

Otari SV, Patil RM, Nadaf NH, Ghosh SJ, Pawar SH (2012) Green biosynthesis of silver nanoparticles from an actinobacteria *Rhodococcus* sp. Mater Lett 72:92–94

Oza G, Pandey S, Gupta A, Kesarkar R, Sharon M (2012) Biosynthetic reduction of gold ions to gold nanoparticles by *Nocardia farcinica*. J Microbiol Biotechnol Res 4:511–515

Pal S, Tak YK, Song JM (2007) Does the antibacterial activity of silver nanoparticles depend on the shape of the nanoparticle? A study of the gram-negative bacterium *Escherichia coli*. Appl Environ Microbiol 27:1712–1720

Panacek A, Kvitek L, Prucek R, Kolar M, Vecerova R, Pizurova N, Sharma VK, Nevecna T (2006) Silver colloid nanoparticles: synthesis, characterization, and their antibacterial activity. J Phys Chem 110:16248–16253

Pandey S, Oza G, Mewada A, Sharon M (2012) Green synthesis of highly stable gold nanoparticles using Momordica charantia as nano fabricator. Arch App Sci Res 4(2):1135–1141

Pattekari P, Z heng Z, Zhang X, Levchenko T, Torchilinb V, Lvov Y (2011) Top-down and bottom-up approaches in production of aqueous nanocolloids of low solubility drug paclitaxel. Phys Chem Chem Phys 13:9014–9019

Prabhu S, Poulose EK (2012) Silver nanoparticles: mechanism of anti-microbial action, synthesis, medical applications, and toxicity effects. Int Nano Lett 2:32

Prakasham RS, Buddana SK, Yannam SK, Guntuku GS (2012) Characterization of silver nanoparticles synthesized by using marine isolate *Streptomyces albidoflavus*. J Microbiol Biotechnol 22:614–621

Prasad R (2014) Synthesis of silver nanoparticles in photosynthetic plants. J Nanopart: 963961. https://doi.org/10.1155/2014/963961

Prasad R, Swamy VS (2013) Antibacterial activity of silver nanoparticles synthesized by bark extract of *Syzygium cumini*. Journal of Nanoparticles. https://doi.org/10.1155/2013/431218

Prasad R, Pandey R, Barman I (2016) Engineering tailored nanoparticles with microbes: quo vadis. WIREs Nanomed Nanobiotechnol 8:316–330. https://doi.org/10.1002/wnan.1363

Prasad R, Jha A, Prasad K (2018a) Exploring the Realms of Nature for Nanosynthesis. Springer International Publishing (ISBN 978-3-319-99570-0). https://www.springer.com/978-3-319-99570-0

Prasad R, Kumar V, Kumar M, Shanquan W (2018b) Fungal Nanobionics: Principles and Applications. Springer Singapore (ISBN 978-981-10-8666-3). https://www.springer.com/gb/book/9789811086656

Rajamanickam U, Mylsamy P, Viswanathan S, Muthusamy P (2012) Biosynthesis of zinc nanoparticles using actinomycetes for antibacterial food packaging. In: International Conference on Nutrition and Food Sciences IPCBEE vol 39 IACSIT

Rajasree SRR, Suman TY (2012) Extracellular biosynthesis of gold nanoparticles using a gram negative bacterium *Pseudomonas fluorescens*. Asian Pac J Trop Dis 2:S796–S799

Raut RW, Haroon ASM, Malghe US, Nikam BT, Kashid SB (2013) Rapid biosynthesis of platinum and palladium metal nanoparticles using root extract of *Asparagus racemosus* Linn. Adv Mater Lett 4(8):650–654

Sadhasivam S, Shanmugam P, Yun K (2010) Biosynthesis of silver nanoparticles by *Streptomyces hygroscopicus* and antimicrobial activity against medically important pathogenic microorganisms. Colloids Surf B: Biointerfaces 81:358–362

Sanjenbam P, Gopal JV, Kannabiran K (2014) Anticandidal activity of silver nanoparticles synthesized using Streptomyces sp. VITPK1. J De Mycologie Médicale. Available from: https://doi.org/10.1016/j.mycmed.2014.03.004

Sastry M, Ahmed A, Khan MI, Kumar R (2003) Biosynthesis of metal nanoparticles using fungi and actinomycetes. Curr Nanosci 85(2):162–170

Selvakumar P, Viveka S, Prakash S, Jasminebeaula S, Uloganathan R (2012) Antimicrobial activity of extracellularly synthesized silver nanoparticles from marine derived *Streptomyces rochei.* Int J Pharm Biol Sci 3:188–197

Shang L, Nienhaus K, Nienhaus GU (2014) Engineered nanoparticles interacting with cells: size matters. J Nano Biotechnol 12:5

Shah R, Oza G, Pandey S, Sharon M (2012) Biogenic fabrication of gold nanoparticles using Halomonas Salina. J Microbiol Biotechnol Res 2(4):485–492

Sharma VK, Yngard RA, Lin Y (2009) Silver nanoparticles: green synthesis and their antimicrobial activities. Adv Colloid Interf Sci 145:83–96

Shirley AD, Dayanand A, Sreedhar B, Dastager SG (2010) Antimicrobial activity of silver nanoparticles synthesized from novel Streptomyces species. Dig J Nanomater Biostruc 5:447–451

Sintubin L, De Windt W, Dick J, Mast J, van der Ha D, Verstraete W, Boon N (2009) Lactic acid bacteria as reducing and capping agent for the fast and efficient production of silver nanoparticles. Appl Microbiol Biotechnol 84:741–761

Sivalingham P, Antony JJ, Siva D, Achiraman S, Anbarasu K (2012) Mangrove Streptomyces sp. BDUKAS10 as nanofactory for fabrication of bactericidal silver nanoparticles. Colloids Surf B: Biointerfaces 98:12–17

Subashini J, Kannabiran K (2013) Antimicrobial activity of Streptomyces sp. VITBT7 and its synthesized silver nanoparticles against medically important fungal and bacterial pathogens. Der Pharm Lett 5:192–200

Sukanya MK, Saju KA, Praseetha PK, Sakthivel G (2013) Potential of biologically reduced silver nanoparticles from Actinomycete cultures. J Nanosci:1–8

Sundarmoorthi E, Devarasu S, Vengadesh Prabhu K (2011) Antimicrobial and wound healing activity of silver nanoparticles synthesized from *Streptomyces aureofaciens*. Int J Pharm Res Dev 12:69–75

Sunitha A, Rimal IRS, Geo S, Sornalekshmi S, Rose A, Praseetha PK (2013) Evaluation of antimicrobial activity of biosynthesized iron and silver nanoparticles using the fungi *Fusarium oxysporum* and *Actinomycetes* sp. on human pathogens. Nano Biomed Eng 5:39–45

Sunitha A, Geo S, Sukanya S, Praseetha PK, Dhanya RP (2014) Biosynthesis of silver nanoparticles from actinomycetes for therapeutic applications. Int J Nano Dimens 5:155–162

Thakker JN, Dalwadi P, Dhandhukia PC (2013) Biosynthesis of gold nanoparticles using *Fusarium oxysporum* f.sp. cubense JT1, a plant pathogenic fungus. ISRN Biotechnol, 5 pages. https://doi.org/10.5402/2013/515091

Tsibakhashvili NY, Kirkesali EI, Pataraya DT, Gurielidze MA, Kalabegishvili TL, Gvarjaladze DN, Tsertsvadze GI, Frontasyeva MV, Zinicovscaia II, Wakstein MS, Khakhanov SN, Shvindina NV, Shklover VY (2011) Microbial synthesis of silver nanoparticles by *Streptomyces glaucus* and *Spirulina platensis*. Adv Sci Lett 4:3408–3417

Usha R, Prabu E, Palaniswamy M, Venil CK, Rajendran R (2010) Synthesis of metal oxide nanoparticles by Streptomyces sp for development of antimicrobial textiles. Glob J Biotechnol Biochem 5:153–160

Vala AK (2014) Exploration on green synthesis of gold nanoparticles by a marine-derived fungus *Aspergillus sydowii*. Environ Prog Sustain Energy. https://doi.org/10.1002/ep.11949

Vidya C, Hiremath S, Chandraprabha MN, Antonyraj MAL, Gopala IV, Jain A, Bansal K (2013) Green synthesis of ZnO nanoparticles by *Calotropis Gigantea*. Int J Curr Eng Technol 1:118–120
Vidyasagar GM, Shankaravva B, Begum R, Imrose, Raibagkar RL (2012) Antimicrobial activity of silver nanoparticles synthesized by Streptomyces species JF714876. Int J Pharm Sci Nanotechnol 5:1638–1642
Waghmare SS, Deshmukh AM, Kulkarni W, Oswaldo LA (2011) Biosynthesis and characterization of manganese and zinc nanoparticles. Univ J Environ Res Technol 1:64–69
Waghmare SS, Deshmukh AM, Sadowski Z (2014) Biosynthesis, optimization, purification and characterization of gold nanoparticles. Afr J Microbiol Res 8:138–146
Zonooz NF, Salouti M (2011) Extracellular biosynthesis of silver nanoparticles using cell filtrate of Streptomyces sp. ERI-3. Scientia Iranica 18(6):1631–1635

Chapter 6
Impact of Nanomaterials on the Microbial System

Rishabh Anand Omar, Shagufta Afreen, Neetu Talreja, Divya Chauhan, Mohammad Ashfaq, and Werayut Srituravanich

Contents

R. A. Omar
Centre for Environmental Science and Engineering, Indian Institute of Technology Kanpur, Kanpur, India

S. Afreen
CAS Key Laboratory of Bio-based materials, Qingdao Institute of Bioenergy and Bioprocess Technology, Chinese Academy of Sciences, Qingdao, China

N. Talreja
Department of Bio-nanotechnology, Gachon University, Incheon, South Korea

D. Chauhan
Department of Chemistry, Punjab University, Chandigarh, India

M. Ashfaq (✉)
School of Life Science, BS Abdur Rahaman Institute of Science and Technology, Chennai, India

Department of Mechanical Engineering, Faculty of Engineering, Chulalongkorn University, Bangkok, Thailand

R. Prasad (ed.), *Microbial Nanobionics*, Nanotechnology in the Life Sciences,
https://doi.org/10.1007/978-3-030-16383-9_6

6.1 Introduction

Presently, infectious diseases are one of the leading causes of morbidity and mortality globally. The morbidity and mortality have decreased in the last decade owing to the use of antimicrobial agents. This drug resistance in communities and hospitals may be caused by both Gram-negative and Gram-positive bacteria, which may have developed resistance to antibiotics, thereby resulting in severe infections (Khan et al. 2011). These antimicrobial agents lead to the development of resistance against micro-organisms; thereby rendering most drugs ineffective in clinical practice. The development of newer antibiotics or the modification of existing drugs may resolve microbial resistance against drugs. On the other hand, newer antimicrobial agents may be entrapped rapidly and killed; therefore, microbes or pathogens may not able to develop resistance (Lu et al. 2009).

The most important challenge in infectious diseases and the emergence of bacterial strains is to demand newer antibiotics for prolonged effectiveness in addition to longer term solutions to resolve this most challenging and expected issue. In this context, nanotechnology resolves such issues by using antimicrobial nanomaterials against micro-organisms that may not be able to develop resistance. Various studies have suggested that metal-based nanomaterials, such as silver (Ag), copper (Cu), zinc (Zn), and gold (Au) have antimicrobial properties that are able to control infectious diseases efficiently (Kumar et al. 2011a, b). These antimicrobial nanomaterials have various advantages: economic viability, insignificant toxicity, and overcoming resistance in comparison with commercially available antibiotics (Lara et al. 2011). In addition, several nano-sized materials are used as a carrier for the delivery of antibiotic drugs in an effective manner. These nanocarriers also have several advantages, such as enhanced accumulation and pharmacokinetics, and reduced side effects of antimicrobial drugs (Salem et al. 2015). On the other hand, nanomaterials are retained for much longer in the body than antibiotics, which may be beneficial to achieve therapeutic efficacy. Moreover, the biocompatibility of the nanomaterials, especially for long-term exposure, must be considered with their therapeutic efficacy. This chapter focuses on the different types of nanomaterials used for controlling microbial infectious diseases with an emphasis on drug resistance (Avella et al. 2005; Ashkarran et al. 2012).

W. Srituravanich
Department of Mechanical Engineering, Faculty of Engineering, Chulalongkorn University, Bangkok, Thailand

Biomedical Engineering Research Center, Faculty of Engineering, Chulalongkorn University, Bangkok, Thailand

6.2 Emerging Roles of Nanomaterials for Infection Control

Presently, engineered nanomaterials are continuously increasing on the basis of functionality and structure in both organic and inorganic nanomaterials, permitting changes in their electrical, mechanical, catalytic, and optical activity (Geiser et al. 2005). The synthesis of nanomaterials is mainly based on the production of small nanoparticles (NPs) from bulky materials. Nanomaterials exhibit antimicrobial activity that enhances effectiveness against micro-organisms in addition to the safe administration of antibiotic drugs. Several antimicrobial agents are used in clinical practice (Cha et al. 2013; Applerot et al. 2012), as these nanomaterials may not have any direct or adverse effects. However, long-term exposure of the materials to cells may cause some cytotoxicity, which remains a concern (Devi and Joshi 2012). Moreover, these antimicrobial nanomaterials are involved in various biological pathways of the micro-organism; therefore, it is difficult to develop resistance against nanomaterials. These nanomaterials have advantages over commercially available antibiotics such as economic viability, a long shelf-life, safety for long-term storage, and prolonged effectiveness. Moreover, some of the nanomaterials require high-temperature treatment for sterilization and at high temperatures commercially available antibiotic drugs are deactivated (Allahverdiyev et al. 2014). On the other hand, nanomaterials-based antibiotics delivery has various benefits, such as controlled release, uniform distribution of drugs to targeted sites, high solubility, enhanced patient compliance, and improved cellular internalization. Therefore, nanomaterials used in various applications, for example, drug delivery, agriculture, environmental remediation, energy, pharmaceutical removal, immunization, sensors, and wound dressing (Prasad 2014; Prasad et al. 2014, 2016, 2017a, b, c, d, 2018).

6.3 Antimicrobial Nanomaterials

The classification of nanomaterials includes four main types on the basis of dimension: (1) zero dimension (0D), the 0D nanomaterials, mainly quantum dots, hollow spheres, nano lenses, core-shell quantum dots, and heterogenous particles, are synthesized by using several physical and chemical processes (Bakry et al. 2007); (2) one dimension (1D), the 1D nanomaterials, mainly nanotubes, are of continuously increasing interest because of their applicability toward end applications, i.e., energy, nanoelectronics, nanodevice, and nanocomposite materials; (3) two dimensions (2D), the 2D nanomaterials such as nanosheets, nanoplates, nanowalls, and nanodisks with a unique shape have gained significant attention from researchers because of their low-dimensional characteristics and wider applicability in the development of nanoreactors, nanocontainers, photocatalysts, and template synthesis for other nanomaterials (Sun et al. 2018; Tao et al. 2017); and (4) three dimensions (3D), the 3D nanomaterials such as cylindrical micelles, vesicles, polymeric clusters, and spherical micelles that have a large specific surface area. The 3D

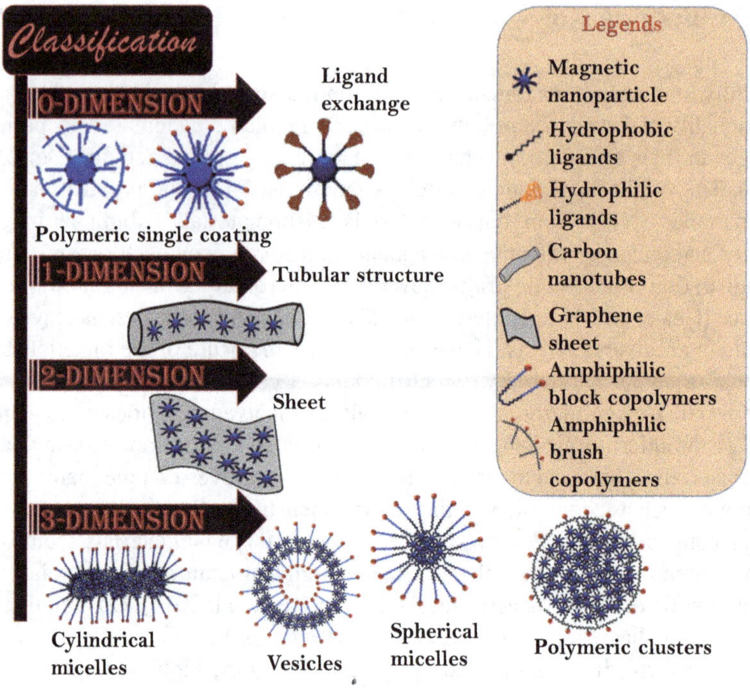

Fig. 6.1 Schematic representation of the different nanomaterials on the basis of classification

nanomaterials extensively used in various applications are mainly catalysts, batteries, and electrode materials. The 3D nanomaterials have high reactive sites for adsorption that enhances the transport of molecules (Gajjar et al. 2009; Pan et al. 2010; Qiu et al. 2012; Egger et al. 2009; Feris et al. 2009). Figure 6.1 shows the schematic representation of the different nanomaterials on the basis of dimension. In this chapter, we briefly discuss metal- and carbon-based nanomaterials within the context of antimicrobial activity (Baker-Austin et al. 2006).

6.3.1 Metal Nanoparticles

Various synthesis processes such as photochemical, electrochemical, green synthesis, chemical batch deposition, chemical, and radiolytic have been used for the synthesis of metal NPs. Generally, synthesis of metal NPs from chemical processes is mainly based on the reduction of metal ions by using reducing agents (Krishna et al. 2015). Moreover, radioactivity and X-rays are processes also effectively used in the synthesis of metal NPs. The gamma radiation energy is deposited within the solution, forming radiolytic radicals. These radiolytic radicals are extremely reactive and undergo redox reaction with appropriate molecules. The suitable scavenger is used for oxidizing radicals for the formation of a reducing environment. The solution also contains metal ion precursors that convert into zerovalent metals. The radiolytic process is easier for controlling the size and shape of the NPs than chemical reduction processes (Baek and An 2011; Bolla et al. 2011).

Several metal NPs such as Cu, Ag, Zn, Au, Al, Fe, and their oxides were extensively used in antimicrobial activity or against infectious diseases that easily control infection against pathogens (Chauhan et al. 2016; Mustafa et al. 2011; Friedman et al. 2011; Fabrega et al. 2009; Santo et al. 2008, Guzman et al. 2012; Jones et al. 2008; You et al. 2011). The antimicrobial activity of these metal NPs was mainly because of their small or nano-sized structures. Metal NPs have different chemical characteristics compared with bulk materials because the nano-sized structures have more reactive sites, thereby easily reacting with micro-organisms (Huang et al. 2008). The nano-sized metal NPs have high reactivity and cytotoxicity compared with bulky materials. Moreover, unique characteristics such as large surface area, shape, and reactive sites are also important for providing a suitable alternative to antibiotic drugs (Lellouche et al. 2012; Aziz et al. 2016). However, the main mechanism of metal NPs against micro-organisms is still not clear. Therefore, various studies have been performed so far to understand the exact mechanism of metal NPs against micro-organisms. On the other hand, most of the studies suggested that metal and its oxides have antimicrobial activity via the generation of reactive oxygen species (ROS) because of their shape and release of metal ions. These metal ions (only trace amounts) are also required for various biological pathways, including DNA transcription, DNA replication, and central metabolism (Leid et al. 2011; Swamy and Prasad 2012; Prasad and Swamy 2013). The excess levels of these metal ions are damaging the cells of living organisms. Conversely, analysis of the antibacterial activity on the basis of both Gram-negative and Gram-positive bacteria is necessary because many metal NPs have strong activity against Gram-positive bacteria and are less effective against Gram-negative bacteria and vice versa (Huh and Kwon 2011; Allahverdiyev et al. 2014; Aziz et al. 2015, 2016). Therefore, various facets should be considered when carrying out a comparative study of metal and bacterial species: (1) bulky materials are not used at a nanoscale level; (2) various antibacterial processes have varying results; (3) the peptidoglycan layer plays an important role against the activity of NPs; and (4) pH, the shape of the NPs, and carrier may change the activity. In general, a thick peptidoglycan layer of Gram-positive bacteria compared with Gram-negative bacteria prohibits the penetration of metal NPs into the bacterial cells (Bouwmeester et al. 2009; Thill et al. 2006; Suri et al. 2007). Therefore, the antimicrobial efficiency of the NPs may be related to the interaction of bacterial cell walls in the case of Gram-positive bacterial strains, whereas the penetration of the NPs depends on the size, surface charge, and functionality of the nanomaterials in Gram-negative bacteria.

6.3.2 Carbon-Based Nanomaterials

Carbon-based nanomaterials such as carbon nanotubes (CNTs), carbon nanofibers (CNFs), graphene, fullerenes, and their derivatives have been used in various applications, mainly drug delivery, agriculture, medicine, electronics, mechanics, optics, and environmental remediation. CNTs have become the most extensively used carbon-based nanomaterials (Sankararamakrishnan et al. 2016; Sankararamakrishnan and Chauhan 2014; Fendler 2001, Talreja and Kumar 2018; Kumar et al. 2019). CNTs are commonly synthesized by a chemical vapor deposition process. They have a cylindrical sp2-carbon structure and tunable physical properties such as diameter, length, surface functionality, chirality, and single- or multiple-walled structure. CNTs have great mechanical strength; thereby, they are utilized as reinforcing materials for various polymeric composites. CNTs are also used as bioimaging owing to their unique optical characteristics. Cherukuri et al. observed CNTs in phagocytic cells in mice by using near-infrared fluorescence. Liu et al. also observed CNTs in several tissues after intravenous delivery into mice by using Raman spectroscopy. In addition, CNTs are used in drug delivery systems because they easily interact with several biomolecules such as protein and DNA. Several studies have suggested that CNTs easily interact with DNA molecules, which indicates the potential ability of CNTs for gene delivery (Guzman et al. 2012; Poveda and Gupta 2016; Afreen et al. 2018).

Metal NPs (Cu, Zn, Fe, Ag, and Ni)-dispersed CNFs are relatively new carbon based nanomaterials. CNFs are widely used in various end applications such as water treatment, plant protection, drug delivery, agriculture, medicine, and wound dressing (Bhadauriya et al. 2018; Ashfaq et al. 2018; Kumar et al. 2018; Talreja et al. 2016; Talreja et al. 2014; Saraswat et al. 2012; Kumar et al. 2011a, b). CNFs have the remarkable ability to adsorb pollutants from water, and gases because of their large surface area. CNFs are also used in various sensor applications, including phytohormones, heavy metals, and biomolecules. Ashfaq et al. developed Cu-CNF (nano-antibiotics) and encapsulated with polyvinyl alcohol and cellulose acetate phthalate-based composite for a nano-antibiotics delivery system. The study suggested that Cu acted as an antibacterial agent and the wound dressing materials produced have potential ability for safe use in wound healing (Ashfaq et al. 2014). Ashfaq et al. (2016) developed asymmetrically dispersed bi-metallic CNFs for antibiotics materials. The study focused on the initial and prolonged effectiveness of the materials; the data suggested that the robust release of Zn and controlled release of Cu metal ions enhanced the biocompatibility and the effectiveness of the materials.

Graphene is the newest carbon-based nanomaterial and possesses similar optical, thermal, and electrical properties. The two-dimensional structure of the graphene makes it a more suitable candidate for electronic characteristics (Chen et al. 2006). Graphene and its oxide are produced using the chemical vapor deposition process and oxidation graphitic powder under acidic conditions. Various biomedical applications, including cell-labeling agents, drug delivery, and biomaterials (scaffold) have been synthesized by using or reinforcing graphene/graphene oxide (Huh and Kwon 2011; Donaldson et al. 2004).

Fullerenes are the third allotropic form of carbon and consist of 20 hexagonal and 12 pentagonal rings. Fullerenes are antioxidants, reacting readily and generating ROS at a high rate, and, subsequently, cell damage or death. Fullerenes are also used in catalysts, water treatment, biohazard protection, portable power, and medicine (Feng et al. 2014; Fang et al. 2007).

The carbon-based nanomaterials are widely used as antibacterial agents or to control infections. The interaction of carbon-based nanomaterials with pathogens may be dependent on the surface functional group, composition, nature of microbes, and environments (where carbon-based nanomaterials interact with cells or biomolecules). Usually, antibacterial activity of the carbon-based nanomaterials involves a combination of both a physical and a chemical process (Hajipour et al. 2012). In the physical process, carbon-based nanomaterials may cause morphological changes or structural damage to the cellular membrane of the micro-organism, subsequently leading to cell death, whereas in the chemical process, carbon-based nanomaterials interact with the surface of micro-organisms, which may lead to the generation of reactive oxygen stress, subsequently resulting in the death of the micro-organism. Moreover, significant production and their use increase the risk of exposure for humans; therefore, the biocompatibility of these carbon-based nanomaterials remains a concern. On the other hand, carbon-based nanomaterials are efficiently used in bioimaging, drug delivery, antibacterial agents, diagnosis of diseases, and therapy (Huh and Kwon 2011).

Several researchers have focused their studies on the translocation of carbon-based nanomaterials within living things, including humans, animals, and agriculture. Carbon-based nanomaterials readily enter living things by using various routes, including digestion, injection, inhalation, and translocation within the organs and tissues by using the circulatory system. Several studies have suggested that carbon-based nanomaterials employ antagonistic effects on the physiological function of the cardiovascular system, immune system, and respiratory system (Ramos et al. 2018). However, the cytotoxicity effects of the carbon-based nanomaterials depend on its bioavailability, which is determined by the interaction of carbon-based nanomaterials with cells or biomolecules. Therefore, the interaction of nanomaterials with cells or biomolecules may be necessary to understand the biocompatibility or biosafety concerns in living organisms (Huh and Kwon 2011; Simon-Deckers et al. 2009).

Despite several positive aspects of carbon-based nanomaterials in biomedical applications, safety is remains a concern. However, some studies have suggested that the enhanced toxicity of CNTs is due to their agglomeration, cellular uptake, and induced oxidative stress. On the other hand, the biocompatibility of carbon-based nanomaterials depends on the presence or release of metal contents from them (Weiss et al. 2006). The cytotoxicity of the carbon-based nanomaterials varies depending on the different cell lines. Several studies have also suggested that the cytotoxicity of the carbon-based nanomaterials might be minimized by attaching a functional group to the surface of the carbon-based nanomaterials and by using a surfactant to avoid agglomeration of the nanomaterials. Ashfaq et al. (2013) suggested that CNFs have relatively less toxicity than other carbon-based nanomaterials including CNTs. Ashfaq et al. also suggested that the cytotoxicity of the carbon-based nanomaterials might be due to the presence of metal NPs (Ashfaq et al. 2013, 2016, 2017a, b).

In general, these nanomaterials and their biological applications, especially antimicrobial applications, may be recognized, which clearly indicates the promising future of nanomaterials in health applications. On the other hand, these nanomaterials, either alone or incorporated into different materials, may enhance the efficiency of the materials for various applications (Yadav and Kumar 2008). These nanomaterials also have various advantages over commercially available antibiotic drugs, as discussed earlier in this chapter.

6.4 Role of Nanomaterials Against Micro-Organisms

The antimicrobial activity of the drugs mainly depends on the chemical composition, which inhibits or suppresses the growth of pathogens, or kills the pathogen without affecting the normal cells. Recently, most of the commercially available antimicrobial drugs modified with natural compounds such as β-lactams and also pure natural products such as aminoglycosides have been frequently used clinically. However, the wider use of these antibiotic drugs is one of the reasons for developing resistance to antimicrobial drugs (Jarboe et al. 2008; Brown et al. 1988). Antimicrobial drug resistance is a serious issue nowadays. In this context, the synthesis of nanomaterials has emerged as a novel alternative strategy for controlling bacterial infection. Several antimicrobial nanomaterials and nano-sized carriers are widely used for controlling infection against pathogens. The nanomaterials enhance antibacterial activity in comparison with the commercially available antibiotic drugs owing to a larger surface area (Huh and Kwon 2011; Jain 2008; Aziz et al. 2014, 2015, 2016).

The antimicrobial activity of Ag is well-known and has been used in many biological applications such as dental work, catheters, wound dressing, and control of infections. Currently, Ag is used in wound dressing materials because of the emergence of microbial drug resistance and relative limited effectiveness of the commercially available drugs (Impellitteri et al. 2009; Liu et al. 2009). Ag-based NPs are the most efficient antimicrobial agents among all metals and oxide-based NPs efficiently kill all micro-organisms such as viruses, fungi, and bacteria (Juan et al. 2010; Echegoyen and Nerín 2013). The AgNPs targeting the cell division and respiratory chain that leads to the generation of ROS, and eventually to the death of the cell (Morones et al. 2005). Moreover, the release of Ag ions facilitates antibacterial activity. The antimicrobial activity of Ag is dependent on shape and size. Additionally, combination therapy, such as NPs with antibiotics, may facilitate antimicrobial activity owing to the synergetic effects (Choi and Hu 2008; Ajitha et al. 2014; Aziz et al. 2016).

Zinc is also well-known to have antibacterial activity, stability under harsh conditions, and insignificant toxicity against human or animal cells. ZnNPs have advantages over AgNPs such as biocompatibility, cost-effectiveness, UV blocking ability, and white appearance (Heinlaan et al. 2008). The antimicrobial mechanism of ZnNPs is still under study. Researchers believe that ZnNPs destruct the protein and lipids of bacterial cellular membrane. Moreover, photocatalytic production of hydrogen peroxide and release of Zn ions are also responsible for antimicrobial activity (Sinha et al. 2011). Similar to Ag and Zn, Cu is also extensively used as an antimicrobial agent. Cu is a co-factor in many enzymatic reactions in living microorganisms. The high level of Cu ions may produce ROS, which disrupts the synthesis of DNA and amino acid (Ashfaq et al. 2016, 2017a, b; Ruparelia et al. 2008). Titanium dioxide (TiO_2)-based nanomaterials are extensively used as semiconductors and photocatalytic antimicrobial agents. The antimicrobial activity of TiO_2 is based on the thickness of the cellular membrane of the pathogen (Musee et al. 2011; Baek and An 2011). The mode of action of TiO_2-based nanomaterials is mainly the generation of ROS, primarily peroxide and free hydroxyl radicals. The hydroxyl radicals target the cell wall of the pathogen, which leads to cell death (Chen et al. 2008; Maness et al. 1999).

Like metal NPs, carbon-based nanomaterials such as CNTs, CNFs, graphene, and fullerene are also used as antimicrobial agents. Several studies have suggested that CNTs have antimicrobial activity against Gram-positive and Gram-negative bacteria, rupturing the cellular membranes of the bacteria and generating ROS, consequently leading to cell death. Moreover, aggregation, stabilization, and bioavailability of the CNTs are considered for effective control of infection (Donsì et al. 2010). Additionally, several studies have suggested that the metal NPs incorporating CNTs might improve the antibacterial efficiency (Hajipour et al. 2012). Transitional metal NPs-dispersed CNFs have superior antibacterial effects in comparison with the CNTs. The CNFs have advantages over CNTs or graphene or fullerenes because CNFs have metal NPs and the release of metal NPs is slow. The slow release of metal NPs from CNFs makes them suitable candidates for various biological applications. In general, Ag has excellent antimicrobial properties among all metal- and oxide-based NPs (Hammel et al. 2004). However, toxicity to human or animal cells and higher costs have limited their applicability. On the other hand, of all carbon-based nanomaterials, CNFs have excellent antibacterial activity without having any adverse effects on human or animal cells. The antibacterial activity is mainly due to the metal NPs present in the carbon-based nanomaterials (Jiang et al. 2009).

In general, the antimicrobial activity of nanomaterials mainly depends on two factors: (1) physico-chemical properties such as size, shape, and surface area of the nanomaterials, and (2) bacteria types such as Gram-positive or Gram-negative.

6.5 Mechanism of Nanomaterials for Controlling Infection

The antimicrobial agents used in treatment control the infection against the pathogen. Presently, around more than 100 antimicrobial agents are available on the market. However, the mechanism of these antimicrobial agents differs because of their chemical composition, structure, and their affinity for certain target sites within the pathogens. Most of the antimicrobial agents, such as cephalosporin, bacitracin, penicillin, and vancomycin, target the cell wall, which inhibits the synthesis of the cell wall. Cationic peptides such as colistin also inhibit the functioning of the cell wall. Membranes are important barriers; a small disruption or leakage from the cellular membrane usually leads to cell death. On the other hand, some antimicrobial agents, such as rifampin, quinolones, and metronidazole, inhibit the synthesis of nucleic acid. Other antibiotics, e.g., trimethoprim and sulfonamides, target the folic acid pathways. Such antimicrobial agents disrupt these pathways. The folic acid pathways are important for the synthesis of DNA (Landini et al. 2010).

On the other hand, nanomaterials inhibit or kill wide ranges of pathogens depending upon the size, surface charges, and chemical composition. The nanomaterials interact with cellular membrane of the micro-organism, which changes their permeability, and transport activity; therefore, particles penetrate within the cells, leading subsequently to disruption of the cellular membrane, resulting in cell death (Aziz et al. 2019). Moreover, interaction of metal NPs with sulfur-containing proteins present in the cell wall of the pathogen may cause cellular disruption due to electrostatic force, bringing about changes in the integrity of the lipid bilayer and cellular permeability. The increase in cellular permeability leads to the leakage of cells. Additionally, the cell wall thickness of the micro-organism is an important factor for the interaction of the nanomaterials, as Gram-negative bacteria are more susceptible to damaging the cellular membrane in comparison with Gram-positive bacteria, because of the difference in the peptidoglycan layer (Mollasalehi and Yazdanparast 2013). A schematic illustration of the mode of action of antimicrobial agents and nanomaterials against pathogens is shown in Fig. 6.2.

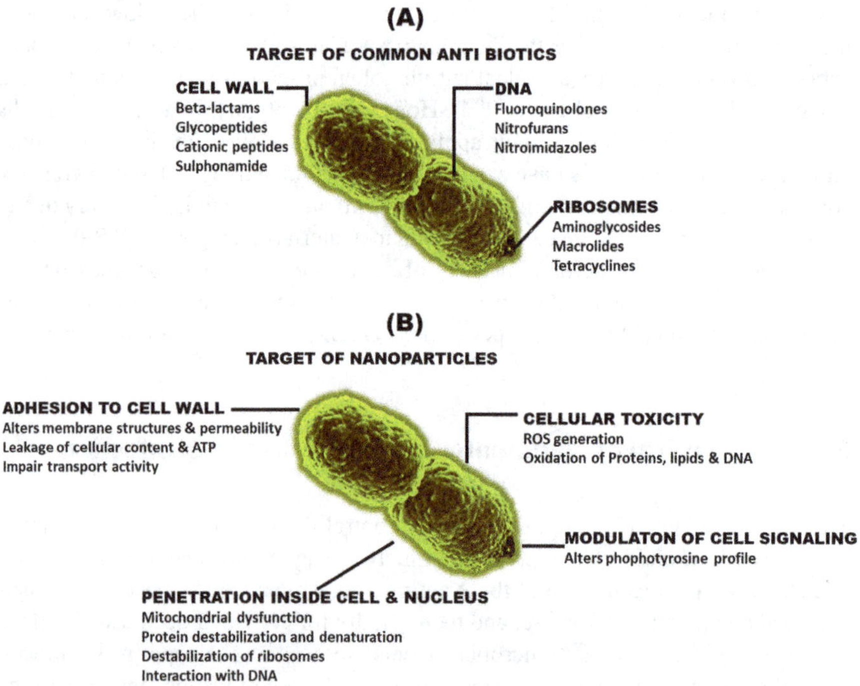

Fig. 6.2 Schematic illustration of the mode of action of antimicrobial agents and nanomaterials against pathogens

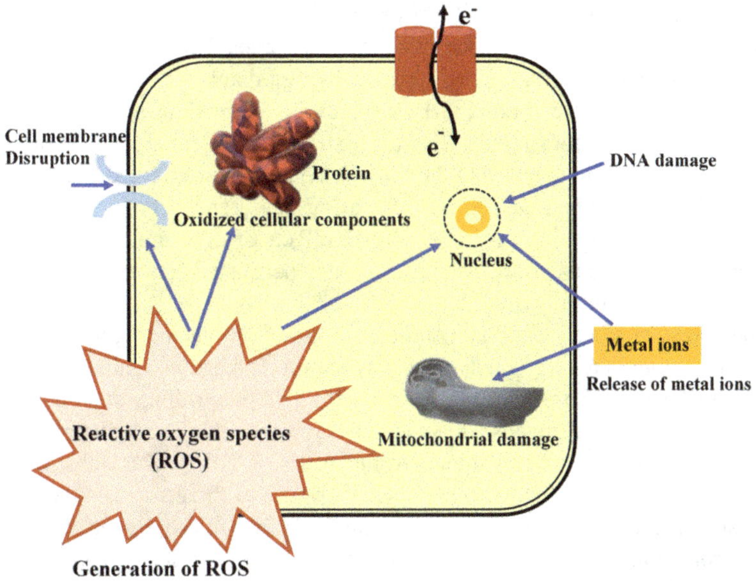

Fig. 6.3 Schematic illustration of the mode of action of nanomaterials against pathogens

Another aspect of the mode of action of nanomaterials is based on the generation of ROS such as hydrogen peroxide, hydroxyl radicals, singlet oxygen, and superoxide ions. The ROS may be undergoing mitochondrial oxidative phosphorylation, as phosphorylation and dephosphorylation are essential for the cellular activity and growth of micro-organisms. Therefore, the phosphotyrosine profile may be useful in understanding the interaction of nanomaterials with both Gram-positive and Gram-negative bacteria (Pramanik et al. 2012). Figure 6.3 shows the schematic illustration of the mode of action of nanomaterials against pathogens.

In general, the antimicrobial mechanism against pathogen as follows: generation of ROS via photocatalysis; (2) disruption of the cellular membrane of the pathogen; (3) disruption of electron transport; (4) inhibition of enzymes; and (5) inhibition of the synthesis of DNA.

6.6 Nanomaterials Against the Environment and Ecosystems

Significant progress has been made in the development of nanomaterials because of their applicability in various fields such as cosmetics, food packaging, drug delivery, therapeutics, biosensors, diagnostics, agriculture, and medicine (Prasad et al. 2014, 2017a, b, c, d, 2018). The nanomaterials also have antibacterial and odor-fighting abilities owing to their similar size compared with biological macromolecules, and therefore are extensively used in wound dressing, detergents, and antimicrobial coating. The nanotechnology-based consumer product are continuously bringing to the market around three or four products every week, because of the wide applicability of the nanomaterials. Therefore, the living organism's exposure to nanomaterials will also increase continuously. Despite the tremendous success and various positive aspects of nanomaterials, long-term exposure-related toxicity remains a concern. This toxicity to humans and the environment is one of the most critical issues associated with nanomaterials. Moreover, various researchers continue to investigate to attempt to understand the possible adverse effects of exposure of humans and the environment to nanomaterials. Additionally, any substance could be toxic at a high level (Neethirajan and Jayas 2011).

The wide use of nanomaterials in the biological field, including medical science and various commercial products, that may lead to leakage in addition to the accumulation of nanomaterials in the environment. The controlling of infection from pathogens by nanomaterials is advantageous. However, these nanomaterials may be released from metal ions into the environment (Nasr 2015). The leakage of metal ions or NPs into the environment is one of the grave threats to beneficial micro-organisms in eco-systems and in public health. Various micro-organisms do benefit the environment, ecosystem, subsequently, humans. These micro-organisms play an important role in bioremediation, nitrogen fixation, and element cycling for plant growth. Therefore, researchers continue to examine the adverse effects of nanomaterials on the environment, ecosystem, and human health (Oberdorster 2004; Prasad and Aranda 2018).

6.7 Conclusion and Future Prospects

Antibiotics have been used since ancient times for controlling infection against pathogens. The developing resistance to antimicrobial drugs is one of the most serious issues in fighting infectious diseases. To overcome this issue, researchers have developed antibiotics combined with natural compounds to achieve synergetic antibacterial effects. However, this success is limited and although microbial resistance has been overcome temporarily, the situation remains similar. In this context, nanomaterials are emerging tools for controlling infections without the development of resistance. Nanomaterials have been used as effective antimicrobial agents for various biological applications, mainly wound dressing, medical devices, nanolotion, and nanogels, which control against infections with different modes of action. These nanomaterial-based products have the potential ability to overcome drug resistance. Several nanomaterials have also been used as carriers for antibiotic drugs, with promising efficiency against pathogens. Interestingly, combined treatment of nanomaterials with drugs or two different nanomaterials may facilitate antimicrobial activity and overcome drug resistance. The extensive use of nanomaterials may release their ions into the environment, which may kill some useful micro-organisms. However, very few studies have so far addressed the complication of long-term exposure to nanomaterials of the environment, ecosystem, and human health. Most of the literature suggests that nanomaterials might be effective against pathogens, safer, economically viable, and targeted therapy for controlling infectious diseases.

Acknowledgement The authors acknowledge support from Chulalongkorn University Pathumwan, Bangkok, 10330, Thailand, through Chulalongkorn Academic Advancement into its Second Century Project (Small Medical Device).

References

Afreen S, Omar RA, Talreja N, Chauhan D, Ashfaq M (2018) Carbon-Based Nanostructured Materials for Energy and Environmental Remediation Applications. In: Prasad R, Aranda E (eds) Approaches in Bioremediation. Nanotechnology in the Life Sciences. Springer, Cham

Ajitha B, Reddy YAK, Reddy PS (2014) Biosynthesis of silver nanoparticles using *Plectranthus amboinicus* leaf extract and its antimicrobial activity. Spectrochim Acta A Mol Biomol Spectrosc 128:257–262

Allahverdiyev AM, Abamor ES, Bagirova M, Rafailovich M (2014) Antimicrobial effects of TiO_2 and Ag_2O nanoparticles against drug resistant bacteria and leishmania parasites. Future Microbiol 6:933–940

Applerot G, Lellouche J, Perkas N, Nitzan Y, Gedanken A, Banin E (2012) ZnO nanoparticle-coated surfaces inhibit bacterial biofilm formation and increase antibiotic susceptibility. RSC Adv 2(6):2314–2321

Ashfaq M, Singh S, Sharma A, Verma N (2013) Cytotoxic evaluation of the hierarchal web of carbon micro-nanofibers. Ind Eng Chem Res 52:4672–4682

Ashfaq M, Khan S, Verma N (2014) Synthesis of PVA-CAP-based biomaterial in situ dispersed with Cu nanoparticles and carbon micro-nanofibers for antibiotic drug delivery applications. Biochem Eng J 90:79–89

Ashfaq M, Verma N, Khan S (2016) Copper/Zinc bimetal nanoparticles-dispersed carbon Nanofibers: a novel potential antibiotics. Mater Sci Eng C 59:938–947

Ashfaq M, Verma N, Khan S (2017a) Carbon nanofibers as a micronutrient carrier in plants: efficient translocation and controlled release of Cu nanoparticles. Environ Sci Nano 4:138–148

Ashfaq M, Verma N, Khan S (2017b) Highly effective Cu/Zn-carbon micro/nanofiber-polymer nanocomposite-based wound dressing biomaterial against the *P. aeruginosa* multi-and extensively drug-resistant strains. Mater Sci Eng C 77:630–641

Ashfaq M, Verma N, Khan S (2018) Novel polymeric composite grafted with metal nanoparticle-dispersed CNFs as a chemiresistive non-destructive fruit sensor material. Mater Chem Phys 217:216–227

Ashkarran AA, Ghavami M, Aghaverdi H, Stroeve P, Mahmoudi M (2012) Bacterial effects and protein corona evaluations: crucial ignored factors in the prediction of bio-efficacy of various forms of silver nanoparticles. Chem Res Toxicol 25(6):1231–1242

Avella M, De Vlieger JJ, Errico ME, Fischer S, Vacca P, Volpe MG (2005) Biodegradable starch/clay nanocomposite films for food packaging applications. Food Chem 93:467–474

Aziz N, Fatma T, Varma A, Prasad R (2014) Biogenic synthesis of silver nanoparticles using *Scenedesmus abundans* and evaluation of their antibacterial activity. J Nanopar: 689419. http://sci-hub.tw/10.1155/2014/689419

Aziz N, Faraz M, Pandey R, Sakir M, Fatma T, Varma A, Barman I, Prasad R (2015) Facile algae-derived route to biogenic silver nanoparticles: synthesis, antibacterial and photocatalytic properties. Langmuir 31:11605–11612. http://sci-hub.tw/10.1021/acs.langmuir.5b03081

Aziz N, Pandey R, Barman I, Prasad R (2016) Leveraging the attributes of *Mucor hiemalis*-derived silver nanoparticles for a synergistic broad-spectrum antimicrobial platform. Front Microbiol 7:1984. http://sci-hub.tw/10.3389/fmicb.2016.01984

Aziz N, Faraz M, Sherwani MA, Fatma T, Prasad R (2019) Illuminating the anticancerous efficacy of a new fungal chassis for silver nanoparticle synthesis. Front Chem 7:65. https://doi.org/10.3389/fchem.2019.00065

Baek YW, An YJ (2011) Microbial toxicity of metal oxide nanoparticles (CuO, NiO, ZnO, and Sb_2O_3) to *Escherichia coli, Bacillus subtilis*, and *Streptococcus aureus*. Sci Total Environ 409(8):1603–1608

Baker-Austin C, Wright MS, Stepanauskas R, McArthur JV (2006) Co-selection of antibiotic and metal resistance. Trends Microbiol 14(4):176–182

Bakry R, Vallant RM, Najam-ul-Haq M, Rainer M, Szabo Z, Huck CW, Bonn GK (2007) Medicinal applications of fullerenes. Int J Nanomedicine 2(4):639–649

Bhadauriya P, Mamtani H, Ashfaq M, Raghav A, Teotia AK, Kumar A, Verma N (2018) Synthesis of yeast-immobilized and copper nanoparticle-dispersed carbon nanofiber-based diabetic wound dressing material: simultaneous control of glucose and bacterial infections. ACS Appl Bio Mater 1(2):246–258

Bolla JM, Alibert-Franco S, Handzlik J, Chevalier J, Mahamoud A, Boyer G, Kieć-Kononowicz K, Pagès JM (2011) Strategies for bypassing the membrane barrier in multidrug resistant gram-negative bacteria. FEBS Lett 585(11):1682–1690

Bouwmeester H, Dekkers S, Noordam MY, Hagens WI, Bulder AS, Heer C, Voorde SECG, Wijnhoven SWP, Marvin HJP, Sips AJAM (2009) Review of health safety aspects of nanotechnologies in food production. Regul Toxicol Pharmacol 53:52–62

Brown MR, Allison DG, Gilbert P (1988) Resistance of bacterial biofilms to antibiotics a growth-rate related effect? J Antimicrob Chemother 22(6):777–780

Cha C, Shin SR, Annabi N, Dokmeci MR, Khademhosseini A (2013) Carbon-based nanomaterials: multifunctional materials for biomedical engineering. ACS Nano 7(4):2891–2897

Chauhan D, Afreen S, Mishra S, Sankararamakrishnan N (2016) Synthesis, characterization and application of Zinc augmented aminated PAN nanofibers towards decontamination of chemical and biological contaminants. J Ind Eng Chem 55:50–64

Chen H, Weiss J, Shahidi F (2006) Nanotechnology in nutraceuticals and functional foods. Food Technol 03.06:30–36

Chen WJ, Tsai PJ, Chen YC (2008) Functional Fe_3O_4/TiO_2 core/shell magnetic nanoparticles as photo killing agents for pathogenic bacteria. Small 4(4):485–491

Choi O, Hu Z (2008) Size dependent and reactive oxygen species related nanosilver toxicity to nitrifying bacteria. Environ Sci Technol 42(12):4583–4588

Devi LS, Joshi SR (2012) Antimicrobial and synergistic effects of silver nanoparticles synthesized using soil fungi of high altitudes of eastern Himalaya. Mycobiology 40(1):27–34

Donaldson K, Stone V, Tran CL, Kreyling W, Borm PJ (2004) Nanotoxicology. Occup Environ Med 619:727–728

Donsì F, Annunziata M, Sessa M, Ferrari G (2010) Nanoencapsulation of essential oils to enhance their antimicrobial activity in foods. J Biotechnol 44:1908–1914

Echegoyen Y, Nerín C (2013) Nanoparticle release from nano-silver antimicrobial food containers. Food Chem Toxicol 62:16–22

Egger S, Lehmann RP, Height MJ, Loessner MJ, Schuppler M (2009) Antimicrobial properties of a novel silver-silica nanocomposite material. Appl Environ Microbiol 75:2973–2976

Fabrega J, Fawcett SR, Renshaw JC, Lead JR (2009) Silver nanoparticle impact on bacterial growth: effect of pH, concentration, and organic matter. Environ Sci Technol 43(19):7285–7290

Fang J, Lyon DY, Wiesner MR, Dong J, Alvarez PJ (2007) Effect of a fullerene water suspension on bacterial phospholipids and membrane phase behavior. Environ Sci Technol 41(7):2636–2642

Fendler JH (2001) Colloid chemical approach to nanotechnology. Korean J Chem Eng 18:1–13

Feng L, Xie N, Zhong J (2014) Carbon nanofibers and their composites: a review of synthesizing, properties and applications. Materials 7(5):3919–3945

Feris K, Otto C, Tinker J, Wingett D, Punnoose A, Thurber A, Kongara M, Sabetian M, Quinn B, Hanna C, Pink D (2009) Electrostatic interactions affect nanoparticle-mediated toxicity to gram-negative bacterium *Pseudomonas aeruginosa* PAO1. Langmuir 26(6):4429–4436

Friedman A, Blecher K, Sanchez D, Tuckman-Vernon C, Gialanella P, Friedman JM, Martinez LR, Nosanchuk JD (2011) Susceptibility of gram-positive and-negative bacteria to novel nitric oxide-releasing nanoparticle technology. Virulence 2(3):217–221

Gajjar P, Pettee B, Britt DW, Huang W, Johnson WP, Anderson AJ (2009) Antimicrobial activities of commercial nanoparticles against an environmental soil microbe, *Pseudomonas putida* KT2440. J Biol Eng 3(9):1–13

Geiser M, Rothen-Rutishauser B, Kapp N, Schürch S, Kreyling W, Schulz Y, Semmler M, Im Hof V, Heyder V, Gehr P (2005) Ultrafine particles cross cellular membranes by nonphagocytic mechanisms in lungs and in cultured cells. Environ Health Perspect 113:1555–1560

Guzman M, Dille J, Godet S (2012) Synthesis and antibacterial activity of silver nanoparticles against gram-positive and gram-negative bacteria. Nanomedicine 8(1):37–45

Hajipour MJ, Fromm KM, Ashkarran AA, de Aberasturi DJ, de Larramendi IR, Rojo T, Serpooshan V, Parak WJ, Mahmoudi M (2012) Antibacterial properties of nanoparticles. Trends Biotechnol 30(10):499–511

Hammel E, Tang X, Trampert M, Schmitt T, Mauthner K, Eder A, Pötschke P (2004) Carbon nanofibers for composite applications. Carbon 42(5–6):1153–1158

Heinlaan M, Ivask A, Blinova I, Dubourguier HC, Kahru A (2008) Toxicity of nanosized and bulk ZnO, CuO and TiO_2 to bacteria *Vibrio fischeri* and crustaceans *Daphnia magna* and *Thamnocephalus platyurus*. Chemosphere 71(7):1308–1316

Huang Z, Zheng X, Yan D, Yin G, Liao X, Kang Y, Yao Y, Huang D, Hao B (2008) Toxicological effect of ZnO nanoparticles based on bacteria. Langmuir 24(8):4140–4144

Huh AJ, Kwon YJ (2011) Nanoantibiotics: a new paradigm for treating infectious diseases using nanomaterials in the antibiotics resistant era. J Control Release 156(2):128–145

Impellitteri CA, Tolaymat TM, Scheckel KG (2009) The speciation of silver nanoparticles in antimicrobial fabric before and after exposure to a hypochlorite/detergent solution. J Environ Qual 38:1528–1530

Jain KK (2008) The handbook of nanomedicine. Springer, Humana Press, Totowa, NJ

Jarboe LR, Hyduke DR, Tran LM, Chou KJ, Liao JC (2008) Determination of the *Escherichia coli* S-nitrosoglutathione response network using integrated biochemical and systems analysis. J Biol Chem 283(8):5148–5157

Jiang W, Mashayekhi H, Xing B (2009) Bacterial toxicity comparison between nano-and micro-scaled oxide particles. Environ Pollut 157(5):1619–1625

Jones N, Ray B, Ranjit KT, Manna AC (2008) Antibacterial activity of ZnO nanoparticle suspensions on a broad spectrum of microorganisms. FEMS Microbiol Lett 279:71–76

Juan L, Zhimin Z, Anchun M, Lei L, Jingchao Z (2010) Deposition of silver nanoparticles on titanium surface for antibacterial effect. Int J Nanomedicine 5:261–267

Khan SS, Mukherjee A, Chandrasekaran N (2011) Studies on interaction of colloidal silver nanoparticles (SNPs) with five different bacterial species. Colloids Surf B Biointerfaces 87(1):129–138

Krishna G, Kumar SS, Pranitha V, Alha M, Charaya S (2015) Biogenic synthesis of silver nanoparticles and their synergistic effect with antibiotics: a study against *Salmonella* sp. Int J Pharm Pharm Sci 7(11):84–88

Kumar A, Pandey AK, Singh SS, Shanker R, Dhawan A (2011a) Cellular uptake and mutagenic potential of metal oxide nanoparticles in bacterial cells. Chemosphere 83(8):1124–1132

Kumar V, Talreja N, Deva D, Sankararamakrishnan N, Sharma A, Verma N (2011b) Development of bi-metal doped micro and nano multi-functional polymeric adsorbent for the removal of fluoride and arsenic in waste-water. Desalination 282:27–38

Kumar R, Ashfaq M, Verma N (2018) Novel PVA/Starch-encapsulated Cu/Zn bimetal nanoparticle carrying carbon nanofibers as a biodegradable and anti-reactive oxidative nanofertilizer. J Mater Sci 53(10):7150–7164

Kumar D, Talreja N (2019) Nickel nanoparticles-doped rhodamine grafted carbon nanofibers as colorimetric probe: Naked eye detection and highly sensitive measurement of aqueous Cr3+ and Pb2+. Korean J Chem Eng 36:126–135

Landini P, Antoniani D, Burgess JG, Nijland R (2010) Molecular mechanisms of compounds affecting bacterial biofilm formation and dispersal. Appl Microbiol Biotechnol 86(3):813–823

Lara HH, Garza-Treviño EN, Ixtepan-Turrent L, Singh DK (2011) Silver nanoparticles are broad-spectrum bactericidal and virucidal compounds. J Nanobiotechnol 9(30):1–8

Leid JG, Ditto AJ, Knapp A, Shah PN, Wright BD, Blust R, Christensen L, Clemons CB, Wilber JP, Young GW, Kang AG (2011) In vitro antimicrobial studies of silver carbene complexes: activity of free and nanoparticle carbene formulations against clinical isolates of pathogenic bacteria. J Antimicrob Chemother 67(1):138–148

Lellouche J, Friedman A, Lahmi R, Gedanken A, Banin E (2012) Antibiofilm surface functionalization of catheters by magnesium fluoride nanoparticles. Int J Nanomedicine 7:1175–1188

Liu L, Xu K, Wang H, Tan PKJ, Fan W, Venkatraman SS, Li L, Yang Y (2009) Self-assembled cationic peptide nanoparticles as an efficient antimicrobial agent. Nat Nanotechnol 4:457–463

Lu C, Brauer MJ, Botstein D (2009) Slow growth induces heat-shock resistance in normal and respiratory-deficient yeast. Mol Biol Cell 20(3):891–903

Maness PC, Smolinski S, Blake DM, Huang Z, Wolfrum EJ, Jacoby WA (1999) Bactericidal activity of photocatalytic TiO_2 reaction: toward an understanding of its killing mechanism. Appl Environ Microbiol 65(9):4094–4098

Mollasalehi H, Yazdanparast R (2013) An improved non-crosslinking gold nanoprobe-NASBA based on 16S rRNA for rapid discriminative bio- sensing of major salmonellosis pathogens. Biosens Bioelectron 47(231):236

Morones JR, Elechiguerra JL, Camacho A, Holt K, Kouri JB, Ramírez JT, Yacaman MJ (2005) The bactericidal effect of silver nanoparticles. Nanotechnology 16(10):2346–2356

Musee N, Thwala M, Nota N (2011) The antibacterial effects of engineered nanomaterials: implications for wastewater treatment plants. J Environ Monit 13(5):1164–1183

Mustafa S, Khan HM, Shukla I, Shujatullah F, Shahid M, Ashfaq M, Azam A (2011) Effect of ZnO Nanoparticles on ESBL producing *Escherichia coli* and *Klebsiella* sp. East J Med 16:253–257

Nasr NF (2015) Applications of nanotechnology in food microbiology. Int J Curr Microbiol App Sci 4(4):846–853

Neethirajan S, Jayas SD (2011) Nanotechnology for the food and bioprocessing industries. Food Bioprocess Technol 4:39–47

Oberdorster E (2004) Manufactured nanomaterials fullerenes C60 induce oxidative stress in the brain of juvenile largemouth bass. Environ Health Perspect 112:1058–1062

Pan X, Redding JE, Wiley PA, Wen L, McConnell JS, Zhang B (2010) Mutagenicity evaluation of metal oxide nanoparticles by the bacterial reverse mutation assay. Chemosphere 79(1):113–116

Poveda RL, Gupta N (2016) Carbon nanofibers: structure and fabrication. In: Carbon nanofiber reinforced polymer composites. Springer, Cham, pp 11–26

Pramanik A, Laha D, Bhattacharya D, Pramanik P, Karmakar P (2012) A novel study of antibacterial activity of copper iodide nanoparticle mediated by DNA and membrane damage. Colloids Surf B: Biointerfaces 96:50–55

Prasad R (2014) Synthesis of silver nanoparticles in photosynthetic plants. J Nanopart: 963961. http://sci-hub.tw/10.1155/2014/963961

Prasad R, Aranda E (2018) Approaches in Bioremediation: The New Era of Environmental Microbiology and Nanobiotechnology. Springer International Publishing (978-3-030-02369-0). https://doi.org/10.1007/978-3-030-02369-0

Prasad R, Swamy VS (2013) Antibacterial activity of silver nanoparticles synthesized by bark extract of *Syzygium cumini*. J Nanopart. http://sci-hub.tw/10.1155/2013/431218

Prasad R, Kumar V, Prasad KS (2014) Nanotechnology in sustainable agriculture: present concerns and future aspects. Afr J Biotechnol 13(6):705–713

Prasad R, Pandey R, Barman I (2016) Engineering tailored nanoparticles with microbes: quo vadis. WIREs Nanomed Nanobiotechnol 8:316–330. http://sci-hub.tw/10.1002/wnan.1363

Prasad R, Bhattacharyya A, Nguyen QD (2017a) Nanotechnology in sustainable agriculture: recent developments, challenges, and perspectives. Front Microbiol 8:1014. http://sci-hub.tw/10.3389/fmicb.2017.01014

Prasad R, Pandey R, Varma A, Barman I (2017b) Polymer based nanoparticles for drug delivery systems and cancer therapeutics. In: Kharkwal H, Janaswamy S (eds) Natural polymers for drug delivery. CAB International, Wallingford, pp 53–70

Prasad R, Kumar M, Kumar V (2017c) Nanotechnology: An Agriculture Paradigm. Springer Singapore (ISBN: 978-981-10-4573-8). http://www.springer.com/us/book/9789811045721

Prasad R, Kumar V, Kumar M (2017d) Nanotechnology: Food and Environmental Paradigm. Springer Singapore (ISBN 978-981-10-4678-0). http://www.springer.com/us/book/9789811046773

Prasad R, Jha A, Prasad K (2018) Exploring the realms of nature for nanosynthesis. Springer, Cham. https://www.springer.com/978-3-319-99570-0

Qiu Z, Yu Y, Chen Z, Jin M, Yang D, Zhao Z, Wang J, Shen Z, Wang X, Qian D, Huang A (2012) Nanoalumina promotes the horizontal transfer of multiresistance genes mediated by plasmids across genera. Proc Natl Acad Sci 109(13):4944–4949

Ramos MADS, Da Silva PB, Spósito L, De Toledo LG, Bonifácio BV, Rodero CF, Dos Santos KC, Chorilli M, Bauab TM (2018) Nanotechnology-based drug delivery systems for control of microbial biofilms: a review. Int J Nanomedicine 13:1179–1213

Ruparelia JP, Chatterjee AK, Duttagupta SP, Mukherji S (2008) Strain specificity in antimicrobial activity of silver and copper nanoparticles. Acta Biomater 4(3):707–716

Salem W, Leitner DR, Zingl FG, Schratter G, Prassl R, Goessler W, Reidl J, Schild S (2015) Antibacterial activity of silver and zinc nanoparticles against *Vibrio cholerae* and enterotoxic *Escherichia coli*. Int J Med Microbiol 305(1):85–95

Sankararamakrishnan N, Chauhan D (2014) Studies on the use of novel nano composite (CNT/Chitosan/Fe(0)) towards arsenate removal. J Environ Res Develop 8:594–599

Sankararamakrishnan N, Chauhan D, Dwivedi J (2016) Synthesis of functionalized carbon nanotubes by floating catalytic chemical vapor deposition method and their sorption behavior toward arsenic. Chem Eng J 284:599–608

Santo CE, Taudte N, Nies DH, Grass G (2008) Contribution of copper ion resistance to survival of *Escherichia coli* on metallic copper surfaces. Appl Environ Microbiol 74(4):977–986

Saraswat R, Talreja N, Deva D, Sankararamakrisnan N, Sharma A, Verma N (2012) Development of novel in-situ nickel-doped, phenolic resin-based micro-nanoactivated carbon adsorbents for the removal of vitamin B-12. Chem Eng J 197:250–260

Simon-Deckers A, Loo S, Mayne-L'hermite M, Herlin-Boime N, Menguy N, Reynaud C, Gouget B, Carriere M (2009) Size-, composition-and shape-dependent toxicological impact of metal oxide nanoparticles and carbon nanotubes toward bacteria. Environ Sci Technol 43(21):8423–8429

Sinha R, Karan R, Sinha A, Khare SK (2011) Interaction and nanotoxic effect of ZnO and Ag nanoparticles on mesophilic and halophilic bacterial cells. Bioresour Technol 102(2):1516–1520

Sun YZ, Talreja N, Tao CH, Texter J, Muhler M, Strunk J, Chen J (2018) Catalysis of carbon dioxide photoreduction on nanosheets: fundamentals and challenges. Angew Chem Int Ed 57:7610–7627

Suri SS, Fenniri H, Singh B (2007) Nanotechnology-based drug delivery systems. J Occup Med Toxicol 2(1):16

Swamy VS, Prasad R (2012) Green synthesis of silver nanoparticles from the leaf extract of *Santalum album* and its antimicrobial activity. J Optoelectron Biomed Mater. 4(3):53–59

Talreja N, Verma N, Kumar D (2014) Removal of hexavalent chromium from water using Fe-grown carbon nanofibers containing porous carbon microbeads. J Water Process Eng 3:34–45

Talreja N, Verma N, Kumar D (2016) Carbon bead-supported ethylene diamine functionalized carbon nanofibers: an excellent adsorbent for salicylic acid. CLEAN-Soil Air Water 44(11):1461–1470

Talreja N, Kumar D (2018) Engineered nanoparticles' toxicity: environmental aspects. In: Nanotechnology in Prof. Chaudhery Mustansar Hussain, Ajay Kumar Mishra, Environmental Science, 2 Volumes,Wiley-VCH

Tao CH, Gao YN, Talreja N, Guo F, Texter J, Yan C, Sun YZ (2017) Two-dimensional nanosheets for electrocatalysis in energy generation and conversion. J Mater Chem A 5:7257–7284

Thill A, Zeyons O, Spalla O, Chauvat F, Rose J, Auffan M, Flank AM (2006) Cytotoxicity of CeO_2 nanoparticles for *Escherichia coli*. Physico-chemical insight of the cytotoxicity mechanism. Environ Sci Technol 40(19):6151–6156

Weiss JP, Takhistov P, McClements DJ (2006) Functional materials in food nanotechnology. J Food Sci 71:107–116

Yadav BC, Kumar R (2008) Structure, properties and applications of fullerenes. Int J Nanotechnol Appl 2(1):15–24

You J, Zhang Y, Hu Z (2011) Bacteria and bacteriophage inactivation by silver and zinc oxide nanoparticles. Colloids Surf B: Biointerfaces 85(2):161–167

Chapter 7
Microbial Production of Nanoparticles: Mechanisms and Applications

Madan L. Verma, Sneh Sharma, Karuna Dhiman, and Asim K. Jana

Contents

M. L. Verma (✉)
Centre for Chemistry and Biotechnology, Deakin University, Geelong, VIC, Australia

Department of Biotechnology, Dr YS Parmar University of Horticulture and Forestry, Himachal Pradesh, India

S. Sharma · K. Dhiman
Department of Biotechnology, Dr YS Parmar University of Horticulture and Forestry, Himachal Pradesh, India

A. K. Jana
Department of Biotechnology, Dr BR Ambedkar National Institute of Technology, Jalandhar, India

R. Prasad (ed.), *Microbial Nanobionics*, Nanotechnology in the Life Sciences,
https://doi.org/10.1007/978-3-030-16383-9_7

7.1 Introduction

Microbial bioprocessing for the sustainable production of nanomaterials holds great promise. It is a clean, nontoxic, eco-friendly and cost-effective process that is emerging as a safe biogenic route in the field of nanobiotechnology. It is pertinent that physico-chemical methods for nanomaterial production are capital-intensive, inefficient with regard to materials and energy balance, and may require/produce toxic chemicals. Microbial production of nanomaterials is a type of bottom-up approach where formations of nanomaterials occur by the oxidation or reduction of a metal, and the agents mainly responsible for such processes are the different enzymes secreted by microbial systems. Versatile nanoparticles (NPs) possess unique and biocompatible properties that have encouraged scientists to explore biogenic routes of NP production from different types of microbes, in particular bacteria and fungi (Prasad et al. 2016; Aziz et al. 2016, 2019; Elgorban et al. 2016; Vago et al. 2015; Neveen-Mohamed 2014), and algae (Ebrahiminezhad et al. 2016; Kumar et al. 2014; Aziz et al. 2014, 2015).

Nanoparticles have unique physico-chemical features such as a definite size, a surface area to mass ratio, chemical stability with high reactivity and functionalized structure with desired biocompatibilities (Kong et al. 2017). Microbial NPs have gained attention because of the simplicity of its mode of action, ease of surface modifications, a plethora of applications, such as data storage, antimicrobial, sensing, sustainable agriculture, environment, especially in biotechnology as a nanocarrier for enzyme immobilizations and for drug delivery (Golchin et al. 2018; Prasad 2016, 2017; Prasad et al. 2016, 2018a, b; Verma 2018; Verma et al. 2016; Verma and Barrow 2015; Dykman and Khlebtsov 2017; Verma 2017a, b, c; Kumar et al. 2014; Verma et al. 2013a, b, c, d; Verma et al. 2012). Functionalized microbial NPs can effectively penetrate across obstacles through small capillaries into individual cells, thereby presenting huge probabilities for specific drug delivery to the disease site. Microbial NPs have tremendous potential to deliver multiple drug molecules, recombinant proteins, vaccines, or nucleotides to their target sites effectively (Pelaz et al. 2017).

This chapter provides an overview of microorganism-mediated biogenic synthesis of extracellular/intracellular NPs under ambient conditions. Recent applications of computational techniques to understand the role of capping agents binding to microbial NPs are discussed. The possibility of scaling up the study with respect to fungal gold (Au) NPs is discussed in particular, to envision the scope of large-scale production in an industrial setting. Various applications of microbial NPs, including antimicrobial, specific delivery (bioactives, drugs), and sensing, are also critically discussed.

7.2 Biosynthesis of Extracellular/Intracellular Nanoparticles from Microbes: Mechanism and Capping Agents

Recently, many research studies have been published on the synthesis of NPs through various microorganisms. Amongst various microorganisms, the most notable microorganisms employed for the synthesis of various NPs such as Au, silver (Ag), and iron (Fe), etc., are the bacteria and fungi. Microorganisms are considered cost-effective and eco-friendly nanofactories for the production of NPs. Because of their intrinsic potential, they produce NPs, which are intra- and/or extracellular in nature (Asmathunisha and Kathiresan 2013; Prasad et al. 2016).

7.2.1 Mechanism

Formation of gold nanoparticles (AuNPs) can occur in either the intracellular or the extracellular space. Extracellular AuNP formation is commonly reported for fungi when Au^{3+} ions are trapped and reduced by proteins in the cell wall. The average size of synthesized AuNPs is approximately 15 nm. The reduction of Au^{3+} ions occurs through the cell membrane and cytoplasmic region (Das et al. 2012a). It is suggested that these regions might be responsible for reduction of Au^{3+} to Au^0 because of the presence of electron dense particles in these regions, i.e., cell wall and cytoplasmic regions (Das et al. 2012b).

There are two main precursors of AuNPs in the biosynthetic process: $HAuCl_4$, which dissociates to Au^{3+} ions (Khan et al. 2013), and AuCl, which dissociates to Au^+ (Zeng et al. 2010). However, it is not clear whether the diffusion of the Au^{3+} ions through the membrane occurs via active bioaccumulation or passive biosorption. It may be due to the toxicity of Au^{3+} ions, which increases the porosity of the cellular membrane. The enzymatic reduction mechanism of Au^{3+} is the same for intracellular and extracellular AuNPs (Gupta and Bector 2013). It has been observed that NADH-dependent reductases are involved in the bioreduction process while working on AuNP biosynthesis by the soluble protein extract of the fungus *Fusarium oxysporum* (Mukherjee et al. 2002). However, the role of the specific protein(s) involved in Au reduction has not yet been identified.

Some fungi, namely *Candida albicans*, produce phytochelatins, an oligopeptide chain of glutathione, cysteine, and glycine that is involved in the biosynthesis of AuNPs. Phytochelatins are formed in the pathogenic fungus with the aid of the transpeptidation reaction of dipeptides from a glutathione molecule.

7.2.2 Capping Agents

Microorganisms use extracellular proteins as capping agents to minimize AuNP aggregation and thus stabilize the nanocrystal because small NPs are unstable. As fungi secrete a variety of enzymes and proteins, there are specific organic molecules that act as capping agents. Three capping proteins with a molecular weight of about 100, 25, and 19 kDa from AuNPs synthesized by *Fusarium oxysporum* were identified as plasma membrane ATPase, 3-glucan binding protein, and glyceraldehyde-3-phosphate dehydrogenase (Zhang et al. 2011).

Capping agents are used to minimize NP aggregation, thus stabilizing the nanocrystal, and resulting in the production of NPs with a narrow size and shape distribution that may further be applicable for biomedical and industrial purposes. Many surfactants have also been reported and used as capping agents to alter the desired shape and size of the NPs, but these are difficult to remove and do not easily degrade. Thus, the commercial surfactants are hazardous to the environment (Liu et al. 2005; Gittins et al. 2000). Keeping in mind the limitation possessed by these chemicals, there is an urgent need to use environmentally friendly capping agents and design green biochemicals on a commercial scale for NP synthesis. There are various molecules that could act as capping agents, but there are some green capping agents with their potential role.

7.2.2.1 Biomolecules

Microbes secrete various biomolecules; these molecules of microbial origin improve the homogeneous preparation of NPs by adhering to green chemistry rules. Amino acids act as efficient reducing and capping agents to synthesize NPs. Different types of amino acids were used as capping agents and the same were employed for the synthesis of AuNPs (4–7 nm in size) using tetra auric acid. Of the 20 amino acids, L-histidine was adopted, which was found to reduce tetra auric acid ($AuCl_4$) to AuNPs. The concentration of L-histidine was found to affect the size of the NPs and their aggregates. Moreover, the amino and carboxy groups present in the amino acids caused the reduction of $AuCl_4$ and coating of the NP surface (Maruyama et al. 2015).

7.2.2.2 Polysaccharides

Polysaccharides act as capping agents in NP synthesis as they are low-cost, hydrophilic, stable, safe, biodegradable, and nontoxic. In NP synthesis, water is used as a solvent, in place of toxic solvents (Duan et al. 2015; Akhlaghi et al. 2013). A polysaccharide such as dextran, a polymer of glucose molecules, is hydrophilic, biocompatible, nontoxic, and used for coating many metal NPs (Virkutyte and Varma 2011). The components of natural honey act as a source of both reducing and protecting agents to synthesize spherical AuNPs of 15 nm in size in water. Fructose present in the honey is supposed to act as a reducing agent, whereas proteins are responsible for the stabilization of the NPs (Philip 2009). AgNPs are synthesized within the size range of 2–14 nm using aminocellulose as a reducing and capping agent. The amino cellulose stabilizes aqueous colloidal solutions of NPs and shows significant antibacterial action against all bacterial isolates (Cheng et al. 2013).

7.3 Understanding the Role of Capping Agents or Biomolecules Binding to Nanoparticles via Computational Techniques

With the upcoming new field parameters such as supercomputing and computational fields, we can compile the interaction of molecules. Different techniques are being employed to study interactions, namely, density functional theory (DFT), molecular dynamics simulations, docking, etc. The interaction of biomolecules using a phase display approach identifies the physical link between peptide and substrate interactions. These interactions provide the controlled placement and assembly of molecules, thereby broadening the scope of NP synthesis (Whaley et al. 2000).

Amino acid residues present in the biomolecules show facet-specific binding for the formation of Au nanoplates and helps in the fabrication of nanostructures (Shao et al. 2004). This mechanism has been further explored using molecular dynamics simulation with the application of an intermolecular potential CHARMM-METAL. The adsorption strength correlates with the degree of coordination of polarizable atoms, i.e., O, N, and C to different epitaxial sites. It has been observed that the size and geometry of NPs determine the adsorption energy and show significant attraction to the metal surface (Feng et al. 2011). Facet-specific interaction of biomolecules with inorganic materials was also investigated using the DFT method. It is further demonstrated that the specific surface recognition of an amino acid side chain occurs because of the combination of various processes such as electron exchange, dispersion, and partial charge transfer showed great binding affinity (Ramakrishnan et al. 2015). The results showed that electrostatic interactions are responsible for the binding of biomolecules, i.e., amino acid residues. The short-term elevation of reaction temperature provides fast and high adsorption affinity of oligonucleotides on the surface of AuNPs and it was observed that the binding ability depends on the length of the oligonucleotide and its nucleotide composition (Epanchintseva et al. 2017).

7.4 Case Study of Scaling Up with Respect to Fungal Biosynthesis of Gold Nanoparticles

Microbes are found to be small nanofactories, and microorganism synthesis of NPs has united biotechnology, biological science, and technology in a brand-new field of nanobiotechnology (Fariq et al. 2017; Abdel-Aziz et al. 2018). Microbes are used everywhere in the world for the biological synthesis of NPs because they grow fast, are simple to cultivate, and can grow at numerous temperatures, pH values, and pressure (Rai and Duran 2011). Microbes use their intrinsic potential to synthesize NPs of inorganic material by an intracellular and extracellular reduction

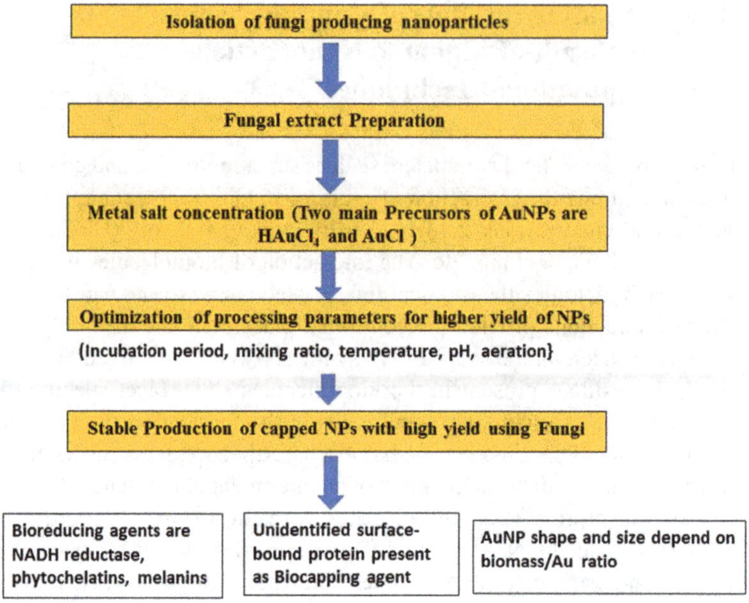

Fig. 7.1 Biosynthesis of gold nanoparticles (NPs) using fungi

mechanism (Fig. 7.1). Some enzymatic activities of microbes turn metal ions, which are trapped by microorganisms, into the elemental form (Li et al. 2011). Heavy metal ions can be reduced by bacteria to produce NPs. Large-scale sustainable production and less frequent use of toxic chemicals are some of the advantages of bacteria-based NP synthesis, but less control over size, shape, and distribution of NPs and laborious culturing processes are some of the disadvantages. Fungi also possess various intracellular and extracellular enzymes capable of producing monodispersed NPs with well-defined geometries and sizes (Fig. 7.2). Stable and easy biological NP synthesis can be achieved by mycosynthesis. Because of the relatively larger biomass, the yield of NPs is high in fungi compared with bacteria. Fungi exhibit a great capacity to bind metal salts to their cell wall, which leads to a higher uptake of metal and provides greater tolerance of metals, eventually resulting in massive NP productions (Fig. 7.3). Three possible mechanisms have been proposed to explain the mycosynthesis of metal NPs: nitrate reductase action; electron shuttle quinones; or both. Fungi have been reported to produce NPs with diverse sizes and shapes. The studies on fungi can be easily extrapolated to others. Fungal production of NPs, is achieved at the extracellular and at the intracellular level.

Industrial production of homologous and heterologous proteins is achieved using a high concentration of the fungal secretome. For example, the entomopathogenic fungus *Beauveria bassiana* has been reported for the expression of a functionally active class I fungal hydrophobin (Kirkland and Keyhani 2011). The tripeptide glutathione, which is a well-known reducing agent, is involved in metal reduction, and in yeasts and fungi it participates in cadmium sulfide (CdS) biosynthesis. Recombinant

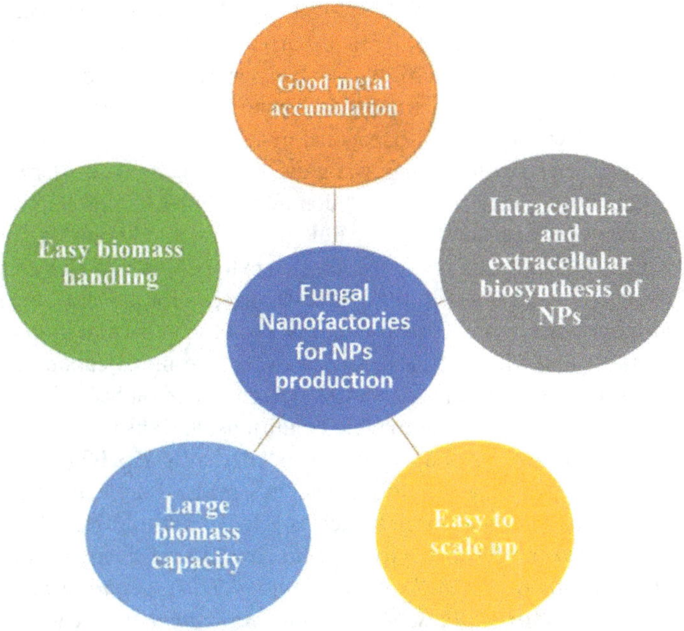

Fig. 7.2 Fungal are the ideal nanofactories for NP production

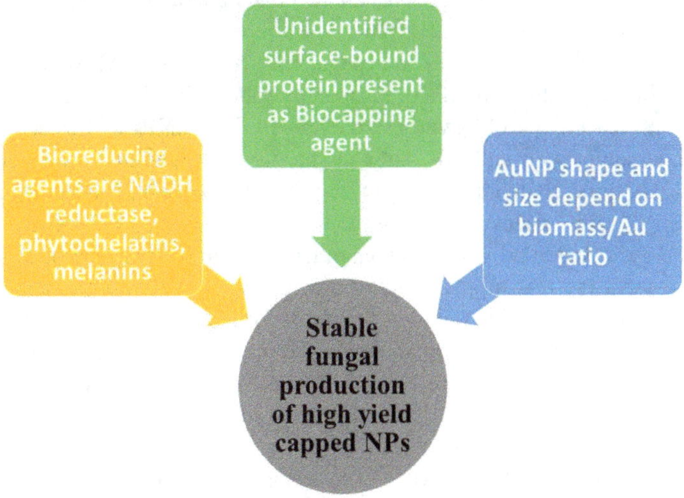

Fig. 7.3 Factors responsible for the stable fungal production of capped NPs

expression of glutathione in *E. coli* for CdSNP production has been reported by Chen et al. (2009). Use of *Verticillium* sp. for the synthesis of AuNPs was the first case where eukaryotic organisms were used for NP synthesis (Mukherjee et al. 2001). Extracellular and intracellular synthesis of NPs was reported on the cytoplasmic membrane and on the surface of the fungal mycelia. Intracellular generation of AuNPs has been demonstrated by typical purple color formation of the mycelia mass. Transmission electron microscopy (TEM) analysis showed that particles of well-defined geometry such as triangular, hexagonal or spherical shapes, were formed on the cell wall and quasi-hexagonal morphology was formed on the cytoplasmic membrane. *Verticillium* fungi can grow and replicate even after exposure to metal ions; therefore, fungi can be used commercially for the production of NPs. Fungal secretome has been studied very little until now, and knowledge of it is still at an early stage. In the case of fungi, the role that extracellular proteins and enzymes play in Au reduction and AuNP capping is advantageous for the large and relatively unexplored fungal secretome. Fungal biomass has a high concentration of cationic biosorption sites; thus, they have been used to remove metal cations from water (Das 2010). Biosorption on fungal biomass is higher than on bacteria, particularly at low pH. For example, under nonviable conditions, various Gram-negative bacteria can immobilize Au^{3+} at about 0.35 mM g^{-1} dry cells at pH 3 (Tsuruta 2004). At pH 2.5, *Aspergillus* sp. can immobilize about 1 mM g^{-1} dry cells (Kuyucak and Volesky 1988). Various fungal species, such as *Verticillium* sp., *Verticillium luteoalbum*, *Colletotrichum* sp., *Fusarium oxysporum*, *Trichothecium* sp., *Fusarium semitectum*, *Alternata niger*, *Helminthosporium solani*, *Trichoderma viride*, *Rhizopus oryzae*, *Aureobasidium pullulans*, *Penicillium brevicompactum*, *Cylindrocladium floridanum*, *Mucor hiemalis*, *Candida albicans*, etc., have been exploited for the biosynthesis of NPs (Zhang et al. 2011; Kumar et al. 2011; Das et al. 2009; Xie et al. 2007; Gericke and Pinches 2006; Shankar et al. 2003; Mukherjee et al. 2002; Mukherjee et al. 2001; Aziz et al. 2016).

Mukherjee et al. (2002) incubated fungal extract with 10^{-3} M $AuCl_4$ in the dark and were able to produce AuNPs of various morphologies. Bhambure et al. (2009) used *Aspergillus niger* to biosynthesize extracellular AuNPs and treatment of the fungal supernatant with aqueous Au^+ ions produced NPs with an average particle size of 12.79 nm. Castro-Longoria et al. (2011) also demonstrated that fungus *Neurospora crassa* strains N150 can synthesize Ag, Au, and bimetallic NPs. *N. crassa* was indicated to be a potential nanofactory for metallic NP synthesis. Several advantages of using this fungus are that it is a nonpathogenic organism, has a fast growth rate, induces rapid reduction of metallic ions, stabilizes NPs, and carries out facile and economical biomass handling. In another study, on the assumption that all the Au in a solution is reduced to form NPs, the authors suggested that the approximate mole concentration of the synthesized NPs might be given (Link et al. 1999). Du et al. (2011) estimated the concentration of synthesized AuNPs measuring 45 nm within the range 10^{-9} to10^{-10} M in accordance with TEM analysis and the density of bulk face-centered cubic Au.

Das et al. (2012a, b) employed the protein extract of *Rhizopus oryzae* for the biogenic production of NPs by using reducing chloroauric acid ($HAuCl_4$). A plant pathogenic fungus named *Fusarium oxysporum* produced extracellular AuNPs (Thakker et al. 2013). Another plant pathogenic fungus named *Nigrospora oryzae* produced AuNPs with anthelmintic efficacy (Kar et al. 2014). Vago et al. (2015) demonstrated one-step biosynthesis of AuNPs by mesophilic filamentous fungi.

Magdi and Bhushan (2015) produced extracellular AuNPs by *Penicillium chrysogenum* using Au chloride ion solution ($HAuCl_4 \bullet 3H_2O$). NPs were characterized to determine the composition, shape, structure, and particle size. Dhanasekar et al. (2015) demonstrated the use of cell-free filtrate of filamentous fungus *Alternaria* sp. for the synthesis of isotropic and anisotropic AuNPs for the first time.

Pei et al. (2017) used new fungus *Mariannaea* sp. HJ cells (cell-AuNPs) and cell-free extracts (extract-AuNPs) for biosynthesis of AuNPs. Bioprocessing of AuNP synthesis was optimized for initial Au ion concentrations and pH. The authors reported that initial Au ion concentrations of 2 mM under a neutral pH of 7 were optimized for both cells and their extract. Cell-AuNPs of various shapes, including spherical, hexagonal, and irregular shapes were produced, with an average size of 37.4 nm, whereas the extracts-AuNPs with an average size of 11.7 nm were almost spherical- and pseudo-spherical-shaped.

It can be inferred from the recent studies discussed above that fungi are ideal microorganisms that can have the potential for use in an industrial setting for large-scale production.

7.5 Applications of Microbial Nanoparticles

Microbial NPs are being used in different sectors ranging from biomedical to the food industry. NPs possess a plethora of applications and act as antimicrobial agents, anticancer agents, drug delivery agents, and sensing agents, etc.

7.5.1 Antimicrobial Agents

Applications in the field of medicine include formulations of many potential antimicrobial agents that are effective against human pathogens, including multidrug-resistant bacteria (Ingle et al. 2008). Multidrug-resistant strains of bacteria have become a serious public health problem (Wright 2005). The emerging resistance of bacteria and the high cost of advanced antimicrobial drugs have encouraged scientists to search for effective, economically viable, and broadly applicable drugs (Jones et al. 2004). Therefore, the development of novel antimicrobial compounds or modification of the available ones to combat resistant pathogens is urgently needed. AgNPs produced by microorganisms are good candidates for a new generation of antimicrobial materials (Rai et al. 2009). This study focused on the biogenic synthesis of metal NPs by acidophilic actinobacteria strain HGG16n and evaluation of their antimicrobial activity. The physico-chemical characteristics of the biosynthesized AgNPs were also determined.

Microbial infections represent serious threats to human health. In addition, extensive usage of antimicrobial agents such as antibiotics render the antimicrobial-resistant microbes ineffective (Raghunath and Perumal 2017). In such scenarios, applications of microbial NPs show strong antimicrobial cells and are safe to humans (Sirelkhatim et al. 2015; Syed et al. 2010; Ren et al. 2009; Kim et al. 2007) NPs may act against microbial infections by inducing oxidative stress in addition to non-oxidative mechanisms in the targeted microbes that act against them, providing a substitute for antibiotics and subsequently multiple drug resistance (Das et al. 2017; Wang et al. 2017).

Microbial production of AgNPs was achieved by using *Streptacidiphilus durhamensis*. NPs showed highest antimicrobial activity against pathogenic Gram-positive and Gram-negative bacteria, such as *Pseudomonas aeruginosa*, *Staphylococcus aureus*, *Proteus mirabilis*, *Escherichia coli*, *Klebsiella pneumoniae*, and *Bacillus subtilis* (Buszewski et al. 2018). Copper, tungsten carbide, and Ag showed strong antimicrobial effects against Gram-positive (*S. aureus*) and Gram-negative (*P. aeruginosa*) bacteria (Bankier et al. 2018).

7.5.2 Anticancer Agents

Biogenic synthesis of Ag NPs was achieved using microalgal secretory carbohydrates (Ebrahiminezhad et al. 2016). The synthesized NPs showed anticarcinogenic properties as novel anticancer and antimicrobial agents.

Nanoparticle-bound enzymes with peroxidase activity have dual application; first, the detection (selective quantitation and colorimetric analysis) of cancer cells, and second, cancer therapy by activating oxidative stress. Both the detection and therapeutic processes are selective to cancer cells, indicating the high specificity and robustness of the hybrid NP conjugate, proving it to be a promising candidate for clinical cancer diagnostics and treatment and their targeted drug delivery approach (Nasrabadi et al. 2016).

Nanoparticle-immobilized serratiopeptidase conjugate was developed to improve therapeutic benefit (Venkatpurwar and Pokharkar 2010). Conjugate was characterized by using UV-visible spectroscopy, TEM, X-ray diffractometry, and Fourier transform infrared spectroscopy. In vitro enzymatic activity and in vivo anti-inflammatory activity of the synthesized serratiopeptidase capped gold NPs complex confirmed the retention of biological activity. The tri-functional role of serratiopeptidase was reported, such as reduction, stabilization, and therapeutic activity, which demonstrated the viable nanocarrier for oral administration with improved therapeutic benefit (Venkatpurwar and Pokharkar 2010).

Recently, the green synthesis of biogenic NPs from microbial sources has become an emerging field owing to their safer, eco-friendly, simple, fast, energy-efficient, low-cost, and less toxic nature. Interestingly, NPs play a key role in the diagnosis of tumors at the initial stage by allowing cellular visualization (Barabadi et al. 2017).

Nanoparticles are the ideal nanocarrier for the delivery of anticancer agents, showing anticarcinogenic properties. Recently, the anticancer activity of biogenic AuNPs synthesized by using marine bacteria *Enterococcus* sp. was demonstrated against cancerous cell lines derived from lung (A549) and liver (HepG2) cells (Rajeshkumar 2016).

7.5.3 Drug Delivery Agents

Microbial NPs possess unique properties and biocompatibility, and are considered to be the most efficient nanocarriers for drug delivery (Xin et al. 2017; Pelaz et al. 2017). Nanomaterials possess unique properties because of the nanosized and quantum effects (Kong et al. 2017).

Nanoparticles have been employed to target cancer cells (Baskar et al. 2018). AuNPs were covalently bound to the fungal enzyme asparaginase. Molecular characterization of the NP-bound biocatalyst was done using Fourier transform infrared spectroscopy and proton nuclear magnetic resonance spectroscopy techniques. The cytotoxicity studies of lung and ovarian cancerous cells demonstrated very promising results for lung cancer cells compared with ovarian cancer cells. A therapeutic enzyme named superoxide dismutase was immobilized onto NPs that demonstrated antioxidant properties against free radicals (Golchin et al. 2018). NP-immobilized serratiopeptidase enzyme was employed for oral drug delivery and demonstrated efficient anti-inflammatory activity (Venkatpurwar and Pokharkar 2010).

Tao et al. (2015) investigated silica-coated NPs, which exhibited both oxidase- and peroxidase-mimicking activities and imparted reactive oxygen species (ROS) reactions. Antibacterial properties were demonstrated against both Gram-negative and Gram-positive bacteria. Biocompatibility of NPs, with ease of their biological and chemical nature, mimic the function of some enzymes, including super oxide dismutase (SOD), esterase, peroxidase, glucose oxidase, for various therapeutic applications such as tissue regeneration (Golchin et al. 2018).

Nanoparticle-based targeted drug deliveries have considerable ability to overcome the limitations of traditional therapeutics (Xin et al. 2017; Daraee et al. 2016). For example, many drugs are manifestly stuck owing to their inability to cross the blood–brain barrier. The ability of NPs to deliver across this barrier is enormously promising because NPs can cross several biological barriers for sustained delivery of therapeutic agents for difficult-to-treat diseases such as brain tumors (Bosio et al. 2016; Nazir et al. 2014; Hainfeld et al. 2013).

Serratiopeptidase (STP), a proteolytic endopeptidase bioenzyme is recognized as one of the most important therapeutic enzymes, having anti-inflammatory activity (Salamone and Wodzinski 1997). Traditionally, therapeutic enzyme delivery has been limited because of their poor uptake and vulnerability to degradation inside the gastrointestinal tract. For efficient drug delivery, today, NPs such as the AuNP complex have immense potential from the therapeutic perspective of biomedicine formulation. In this, the prerequisite is the nanocarrier, which plays an important role in the bioavailability of the pharmaceutical active compound and efficiently improves absorption across the gastrointestinal mucosa (Dykman and Khlebtsov 2017). Maji et al. (2015) developed a new nanostructured hybrid as a mimetic enzyme for in vitro detection and therapeutic treatment of cancer cells. In vitro studies demonstrated enhanced cytotoxicity to HeLa cells. However, it was safe for normal cells; the treatment caused no damage, proving the selective killing effect of the NPs on cancer cells.

It is inferred from the recent studies discussed above of the application of NPs that a nanocarrier-based approach such as NP-immobilized enzymes represents an important modality within therapeutic and diagnostic biomedical applications, including cancer, cardiovascular diseases, and neurological diseases.

7.5.4 Sensing Agents

Owing to the indiscriminate use of pesticides, the agricultural sector is highly contaminated by the excessive misuse of pesticides (Ghormade et al. 2011). In this area, nanosensors can detect on-site soil monitoring of real-time concentrations of pesticide residues and confirm the health of the soil for agricultural usage (Periasamy et al. 2009; Prasad et al. 2014, 2017). Nanosensors play a vital role in the real-time monitoring of pesticide concentrations in the agricultural sector. Zhang et al. (2015) developed nanosensors for the detection of organophosphorus and non-organophosphorus pesticides with a higher sensitivity limit within the range 0.5 μM to 1 μM. Da Silva et al. (2014) developed nanosensors using an atomic force microscopy tip for detection of the acetolactate synthase-inhibitor herbicides metsulfuron-methyl and imazaquin. Nanosensors were developed by coating an atomic force microscopy tip with acetyl co-enzyme A carboxylase (Amarante et al. 2014; Franca et al. 2011). Molecular modeling techniques were employed to measure the interaction at the molecular level. Gan et al. (2010) developed nanosensors for the detection of the organophosphorus pesticide at the limit of 50 pgl^{-1}. Ramanathan et al. (2009) developed nanosensors with the detection limit of 34 μM for paraoxon.

The food industry is facing challenges with the increasing incidence of food contamination by pathogenic microorganisms (Verma 2017b). Nanosensors have become the prerequisite of quality assurance in food industry. They play a crucial role in the early detection of contaminants in food samples that can avoid future loss to the food industry. NP-immobilized enzymes are being used to check the quality of food samples (Perez-Lopez and Merkoci 2011). Nanosensors were developed for biomolecule detection in food samples by immobilizing enzymes to check sugar molecules such as glucose, lactose, and fructose in food (Ozdemir et al. 2010). Nanosensors are being used in many applications in the food industry. For example, Pal et al. (2014) developed nanosensors for choline detection in milk samples within the range 0.5 μM to 2 mM; Devi et al. (2012) for the detection of the contamination of fish meat with xanthine residues at a limit of 0.1 μM; Miranda et al. (2011) for the early detection of microbial contamination; and Li et al. (2011) for aflatoxin B_1 detection at a limit of 1.6 nM.

7.6 Conclusion and Future Prospects

Biosynthesis of NPs using microorganisms offers multifarious advantages. However, to economize NP production on a commercial scale, factors such as the time-consuming production process and the cost-intensive downstreaming process need to be addressed so that cost-effective methods of production can be developed.

This write-up concludes that the production of microbial NPs is the most sought-after bioprocessing technology, which has the possibility for scaling up NP production. Microbial production of NPs demonstrates a sustainable approach for the large-scale production of NPs.

The chapter concludes with discussions on the current limitations and prospects of the biogenic production of NPs.

References

Abdel-Aziz SM, Prasad R, Hamed AA, Abdelraof M (2018) Fungal nanoparticles: A novel tool for a green biotechnology? In: Fungal Nanobionics (ed. Prasad R, Kumar V, Kumar M, and Shanquan W), Springer Nature Singapore Pte Ltd. 61–87

Akhlaghi SP, Peng B, Yao Z, Tam KC (2013) Sustainable nano materials derived from polysaccharides and amphiphilic compounds. Soft Matter 9(33):7905–7918

Amarante AM, Oliveira GS, Bueno CC, Cunha RA, Lerich JCM, Freitas LCG, Franca EF, Oliveira ON Jr, Leite FL (2014) Modeling the coverage of an AFM tip by enzymes and its application in nanobiosensors. J Mol Graph Model 53:100–104

Asmathunisha N, Kathiresan K (2013) A review on biosynthesis of nanoparticles by marine organisms. Colloids Surf B Biointerfaces 103:283–287

Aziz N, Fatma T, Varma A, Prasad R (2014) Biogenic synthesis of silver nanoparticles using *Scenedesmus abundans* and evaluation of their antibacterial activity. J Nanopart: 689419. https://doi.org/10.1155/2014/689419

Aziz N, Faraz M, Pandey R, Sakir M, Fatma T, Varma A, Barman I, Prasad R (2015) Facile algae-derived route to biogenic silver nanoparticles: synthesis, antibacterial and photocatalytic properties. Langmuir 31:11605–11612. https://doi.org/10.1021/acs.langmuir.5b03081

Aziz N, Pandey R, Barman I, Prasad R (2016) Leveraging the attributes of *Mucor hiemalis*-derived silver nanoparticles for a synergistic broad-spectrum antimicrobial platform. Front Microbiol 7:1984. https://doi.org/10.3389/fmicb.2016.01984

Aziz N, Faraz M, Sherwani MA, Fatma T, Prasad R (2019) Illuminating the anticancerous efficacy of a new fungal chassis for silver nanoparticle synthesis. Frontiers in Chemistry 7:65. https://doi.org/10.3389/fchem.2019.00065

Bankier C, Cheong Y, Mahalingam S, Edirisinghe M, Ren G, Cloutman-Green E, Ciric L (2018) A comparison of methods to assess the antimicrobial activity of nanoparticle combinations on bacterial cells. PLoS One 13(2):e0192093

Barabadi H, Ovais M, Shinwari ZK, Saravanan M (2017) Anti-cancer green bionanomaterials: present status and future prospects. Green Chem Lett Rev 10:285–314

Baskar G, Garrick BG, Lalitha K, Chamundeeswari M (2018) Gold nanoparticle mediated delivery of fungal asparaginase against cancer cells. J Drug Delivery Sci Technol 44:498–504

Bhambure R, Bule M, Shaligram N, Kamat M, Singhal R (2009) Extracellular biosynthesis of gold nanoparticles using *Aspergillus niger*-its characterization and stability. Chem Eng Technol 32:1036–1041

Bosio VE, German A, Yanina N, Martinez ND, Guillermo R (2016) Nanodevices for the immobilization of therapeutic enzymes. Crit Rev Biotechnol 36(3):447–464

Buszewski B, Railean-Plugaru V, Pomastowski P, Rafińska K, Szultka-Mlynska M, Golinska P, Wypij M, Laskowski D, Dahm H (2018) Antimicrobial activity of biosilver nanoparticles produced by a novel *Streptacidiphilus durhamensis* strain. J Microbiol Immunol Infect 51:45–54

Castro-Longoria E, Vilchis-Nestor AR, Avalos-Borja M (2011) Biosynthesis of silver, gold and bimetallic nanoparticles using the filamentous fungus *Neurospora crassa*. Colloids Surf B Biointerfaces 83:42–48

Chen YL, Tuan HY, Tien CW, Lo WH, Liang HC, Hu YC (2009) Augmented biosynthesis of cadmium sulfide nanoparticles by genetically engineered *Escherichia coli*. Biotechnol Prog 25:1260–1266

Cheng F, Betts JW, Kelly SM, Schaller J, Heinze T (2013) Synthesis and antibacterial effects of aqueous colloidal solutions of silver nanoparticles using amino cellulose as a combined reducing and capping reagents. Green Chem 15(4):989–998

Da Silva AC, Deda DK, Bueno CC, Moraes AS, Da Roz AL, Yamaji FM, Prado RA, Viviani V, Oliveira ON Jr, Leite FL (2014) Nanobiosensors exploiting specific interactions between an enzyme and herbicides in atomic force spectroscopy. J Nanosci Nanotechnol 14(9):6678–6684

Daraee H, Eatemadi A, Abbasi E, Fekri AS, Kouhi M, Akbarzadeh A (2016) Application of gold nanoparticles in biomedical and drug delivery. Artif Cells Nanomed Biotechnol 44:410–422

Das N (2010) Recovery of precious metals through biosorption–a review. Hydrometallurgy 103:180–189

Das SK, Das AR, Guha AK (2009) Gold nanoparticles: microbial synthesis and application in water hygiene management. Langmuir 25:8192–8199

Das SK, Dickinson C, Laffir F, Brougham DF, Marsili E (2012a) Synthesis, characterisation and catalytic activity of gold nanoparticles, biosynthesised with *Rhizopus oryzae* protein extract. Green Chem 14:1322–1344

Das SK, Liang J, Schmidt M, Laffir F, Marsili E (2012b) Biomineralization mechanism of gold by zygomycete fungi *Rhizopus oryzae*. ACS Nano 6:6165–6173

Das B, Dash SK, Mandal D, Ghosh T, Chattopadhyay S, Tripathy S, Das S, Dey SK, Das D, Roy S (2017) Green synthesized silver nanoparticles destroy multidrug resistant bacteria via reactive oxygen species mediated membrane damage. Arab J Chem 10(6):862–876

Devi R, Yadav S, Pundir CS (2012) Amperometric determination of xanthine in fish meat by zinc oxide nanoparticle/chitosan/multiwalled carbon nanotube/polyaniline composite film bound xanthine oxidase. Analyst 137:754–759

Dhanasekar NN, Rahul GR, Narayanan KB, Raman G, Sakthivel N (2015) Green chemistry approach for the synthesis of gold nanoparticles using the fungus *Alternaria* sp. J Microbiol Biotechnol 25(7):1129–1135

Du L, Xian L, Feng JX (2011) Rapid extra-intracellular biosynthesis of gold nanoparticles by the fungus *Penicillium* sp. J Nanopart Res 13:921–930

Duan H, Wang D, Li Y (2015) Green chemistry for nanoparticle synthesis. Chem Soc Rev 44(16):5778–5792

Dykman LA, Khlebtsov NG (2017) Immunological properties of gold nanoparticles. Chem Sci 8:1719–1735

Ebrahiminezhad A, Bagheri M, Taghizadeh SM, Berenjian A, Ghasemi Y (2016) Biomimetic synthesis of silver nanoparticles using microalgal secretory carbohydrates as a novel anticancer and antimicrobial. Adv Nat Sci Nanosci Nanotechnol 7:015018

Elgorban AM, Al-Rahmah AN, Sayed SR, Hirad A, Mostafa AAF, Bahkali AH (2016) Antimicrobial activity and green synthesis of silver nanoparticles using *Trichoderma viride*. Biotechnol Equip 30:299–304

Epanchintseva A, Vorobjev P, Pyshnyi D, Pyshnaya I (2017) Fast and strong adsorption of native oligonucleotides on citrate quoted gold nanoparticles. Langmuir 34(1):164–172

Fariq A, Khan T, Yasmin A (2017) Microbial synthesis of nanoparticles and their potential applications in biomedicine. J Appl Biomed 15(4):241–248

Feng J, Pandey RB, Berry RJ, Farmer BL, Naik RR, Heinz H (2011) Adsorption mechanism of single amino acid and surfactant molecules to Au 111 surfaces in aqueous solution: design rules for metal binding molecules. Soft Matter 7(5):2113–2120

Franca EF, Leite FL, Cunha RA, Oliveira ON Jr, Freitas LCG (2011) Designing an enzyme-based nanobiosensor using molecular modeling techniques. Phys Chem Chem Phys 13:8894–8899

Gan N, Yang X, Xie D, Wu Y, Wen WA (2010) Disposable organophosphorus pesticides enzyme biosensor based on magnetic composite nanoparticles modified screen printed carbon electrode. Sensors 10:625–638

Gericke M, Pinches A (2006) Microbial production of gold nanoparticles. Gold Bull 239:22–28

Ghormade V, Deshpande MV, Paknikar KM (2011) Perspectives for nano-biotechnology enabled protection and nutrition of plants. Biotechnol Adv 29:792–803

Gittins DI, Bethell D, Schiffrin DJ, Nichols RJ (2000) A nanometre-scale electronic switch consisting of a metal cluster and redox-addressable groups. Nature 408(6808):67–69

Golchin K, Golchin J, Ghaderi S, Alidadiani N, Eslamkhah S, Eslamkhah M, Davaran S, Akbarzadeh A (2018) Gold nanoparticles applications: from artificial enzyme till drug delivery. Artif Cells Nanomed Biotechnol 46(2):250–254

Gupta S, Bector S (2013) Biosynthesis of extracellular and intracellular gold nanoparticles by *Aspergillus fumigatus* and *A. flavus*. Anton Leeuw 103:1113–1123

Hainfeld JF, Smilowitz HM, O'Connor MJ, Dilmanian FA, Slatkin DN (2013) Gold nanoparticle imaging and radiotherapy of brain tumors in mice. Nanomedicine (Lond) 8:1601–1609

Ingle AP, Gade AK, Pierrat S, Sonnichsen C, Rai MK (2008) Mycosynthesis of silver nanoparticles using the fungus *Fusarium acuminatum* and its activity against some human pathogenic bacteria. Curr Nanosci 4:41–144

Jones ME, Karlowsky JA, Draghi DC, Thornsberry C, Sahm DF, Bradley JS (2004) Rates of antimicrobial resistance among common bacterial pathogens causing respiratory, blond, urine, and skin and soft tissue infections in pediatric patients. Eur J Clin Microbiol Infect Dis 23:445–455

Kar PK, Murmu S, Saha S, Tandon V, Acharya K (2014) Anthelmintic efficacy of gold nanoparticles derived from a phytopathogenic fungus, *Nigrospora oryzae*. PLoS One 9(1):e84693

Khan MM, Kalathil S, Han TH, Lee J, Cho MH (2013) Positively charged gold nanoparticles synthesized by electrochemically active biofilm-a biogenic approach. Nanosci Nanotechnol 13:6079–6085

Kim JS, Kuk E, Yu KN, Kim JH, Park SJ, Lee HJ, Kim SH, Park YK, Park YH, Hwang CY, Kim YK, Lee YS, Jeong DH, Cho MH (2007) Antimicrobial effects of silver nanoparticles. Nanomed: Nanotechnol Biol Med 3:95–101

Kirkland BH, Keyhani NO (2011) Expression and purification of a functionally active class I fungal hydrophobin from the entomopathogenic fungus *Beauveria bassiana* in *Escherichia coli*. J Ind Microbiol Biotechnol 38:327–335

Kong FY, Zhang JW, Li RF, Wang ZX, Wang WJ, Wang W (2017) Unique roles of gold nanoparticles in drug delivery, targeting and imaging applications. Molecules 22:1445–1451

Kumar SK, Amutha R, Arumugam P, Berchmans S (2011) Synthesis of gold nanoparticles: an ecofriendly approach using *Hansenula anomala*. ACS Appl Mater Interfaces 3:1418–1425

Kumar S, Malarkodi C, Paulkumar K, Vanaja M, Gnanajobitha G, Annadurai G (2014) Algae-mediated green fabrication of silver nanoparticles and examination of its antifungal activity against clinical pathogens. Int J Met 2014:692643

Kuyucak N, Volesky B (1988) Biosorbents for recovery of metals from industrial solutions. Biotechnol Lett 10:137–142

Li SC, Chen JH, Cao H, Yao DS, Liu DL (2011) Amperometric biosensor for aflatoxin B_1 based on aflatoxin oxidase immobilized on multiwalled carbon nanotubes. Food Control 22(1):43–49

Link S, Wang ZL, El-Sayed MA (1999) Alloy formation of gold silver nanoparticles and the dependence of the plasmon absorption on their composition. J Phys Chem B 103:3529–3533

Liu FK, Ko FH, Huang PW, Wu CH, Chu TC (2005) Studying the size/shape separation and optical properties of silver nanoparticles by capillary electrophoresis. J Chromatogr A 1062(1):139–145

Magdi HM, Bhushan B (2015) Extracellular biosynthesis and characterization of gold nanoparticles using the fungus *Penicillium chrysogenum*. Microsyst Technol 21(10):2279–2285

Maji SK, Mandal AK, Nguyen KT, Borah P, Zhao Y (2015) Cancer cell detection and therapeutics using peroxidase-active nanohybrid of gold nanoparticle-loaded mesoporous silica-coated graphene. ACS Appl Mater Interfaces 7(18):9807–9816

Maruyama T, Fujimoto Y, Markawa T (2015) Synthesis of gold nanoparticles using various amino acids. J Colloid Interface Sci 447:254–257

Miranda OR, Li X, Garcia-Gonzalez L, Zhu ZJ, Yan B, Bunz UHF, Rotello VM (2011) Colorimetric bacteria sensing using a supramolecular enzyme-nanoparticle biosensor. J Am Chem Soc 133:9650–9653

Mukherjee P, Ahmad A, Mandal D, Senapati S, Sainkar SR, Khan MI et al (2001) Bioreduction of $AuCl_4$ ions by the fungus, *Verticillium* sp. and surface trapping of the gold nanoparticles formed. Angew Chem Int Ed Engl 40:3585–3588

Mukherjee P, Senapati S, Mandal D, Ahmad A, Khan MI, Kumar R, Sastry M (2002) Extracellular synthesis of gold nanoparticles by the fungus *Fusarium oxysporum*. Chembiochem 3:461–463

Nasrabadi HT, Abbasi E, Davaran S, Kouhi M, Akbarzadeh A (2016) Bimetallic nanoparticles: preparation, properties, and biomedical applications. Artif Cells Nanomed Biotechnol 44(1):376–380

Nazir S, Hussain T, Ayub A, Rashid U, MacRobert AJ (2014) Nanomaterials in combating cancer: therapeutic applications and developments. Nanomedicine 10:19–34

Neveen-Mohamed K (2014) Biogenic silver nanoparticles by *Aspergillus terreus* as a powerful nanoweapon against *Aspergillus fumigates*. Afr J Microbiol Res 7:5645–5651

Ozdemir C, Yeni F, Odaci D, Timur S (2010) Electrochemical glucose biosensing by pyranose oxidase immobilized in gold nanoparticle-polyaniline/AgCl/gelatin nanocomposite matrix. Food Chem 119:380–385

Pal S, Sharma MK, Danielsson B, Willander M, Chatterjee R, Bhand S (2014) A miniaturized nanobiosensor for choline analysis. Biosens Bioelectron 54:558–564

Pei X, Qu Y, Shen W, Li H, Zhang X, Li S, Zhang Z, Li X (2017) Green synthesis of gold nanoparticles using fungus *Mariannaea* sp. HJ and their catalysis in reduction of 4-nitrophenol. Environ Sci Pollut Res 24(27):21649–21659

Pelaz B, Alexiou C, Alvarez-Puebla RA, Alves F, Andrews AM, Ashraf S et al (2017) Diverse applications of nanomedicine. ACS Nano 11:2313–2381

Perez-Lopez B, Merkoci A (2011) Nanomaterials based biosensors for food analysis applications. Trends Food Sci Techol 22:625–639

Periasamy AP, Umasankar Y, Chen SM (2009) Nanomaterials-acetylcholinesterase enzyme matrices for organophosphorus pesticides electrochemical sensors: a review. Sens (Basel, Switzerland) 9(6):4034–4055

Philip D (2009) Biosynthesis of Au, Ag and Au-Ag nanoparticles using edible mushroom extract. Spectrochim Acta A Mol Biomol Spectrosc 73(2):374–381

Prasad R (2016) Advances and applications through fungal nanobiotechnology. Springer, Cham

Prasad R (2017) Fungal nanotechnology: applications in agriculture, industry, and medicine. Springer

Prasad R, Kumar V, Prasad KS (2014) Nanotechnology in sustainable agriculture: present concerns and future aspects. Afr J Biotechnol 13(6):705–713

Prasad R, Pandey R, Barman I (2016) Engineering tailored nanoparticles with microbes: quo vadis. WIREs Nanomed Nanobiotechnol 8:316–330. https://doi.org/10.1002/wnan.1363

Prasad R, Bhattacharyya A, Nguyen QD (2017) Nanotechnology in sustainable agriculture: recent developments, challenges, and perspectives. Front Microbiol 8:1014. https://doi.org/10.3389/fmicb.2017.01014

Prasad R, Kumar V, Kumar M, Shanquan W (2018a) Fungal Nanobionics: principles and applications. Singapore, Springer. https://www.springer.com/gb/book/9789811086656

Prasad R, Jha A, Prasad K (2018b) Exploring the Realms of Nature for Nanosynthesis. Springer International Publishing (ISBN 978-3-319-99570-0). https://www.springer.com/978-3-319-99570-0

Raghunath A, Perumal E (2017) Metal oxide nanoparticles as antimicrobial agents: a promise for the future. Int J Antimicrob Agents 49(2):137–152

Rai M, Duran N (2011) Metal nanoparticles in microbiology. Springer Verlag, Berlin/Heidelberg. https://doi.org/10.1007/978-3-642-18312-6_1

Rai M, Yadav A, Gade A (2009) Silver nanoparticles as a new generation of antimicrobials. Biotechnol Adv 27:76–83

Rajeshkumar S (2016) Anticancer activity of eco-friendly gold nanoparticles against lung and liver cancer cells. J Genet Eng Biotechnol 14:195–202

Ramakrishnan SK, Martin M, Cloitre T, Firlej L, Gergely C (2015) Design rules for metal binding biomolecules: understanding of amino acid adsorption on platinum crystallographic facets from density functional calculations. Physiol Chem Phys 17(6):4193–4198

Ramanathan M, Luckarift HR, Sarsenova A, Wild JR, Ramanculov ER, Olsen EV, Simonian AL (2009) Lysozyme-mediated formation of protein-silica nano-composites for biosensing applications. Colloids Surf B Biointerfaces 73:58–64

Ren G, Hu D, Cheng EWC, Vargas-Reus MA, Reip P, Allaker RP (2009) Characterisation of copper oxide nanoparticles for antimicrobial applications. Int J Antimicrob Agents 33:587–590

Salamone P, Wodzinski R (1997) Production, purification and characterization of a 50-kDa extracellular metalloprotease from *Serratia marcescens*. Appl Microbiol Biotechnol 48:317–321

Shankar SS, Ahmad A, Pasricha R, Sastry M (2003) Bioreduction of chloroaurate ions by geranium leaves and its endophytic fungus yields gold nanoparticles of different shapes. J Mater Chem 13:1822–1826

Shao Y, Jin Y, Dong S (2004) Synthesis of gold nanoplates by aspartate reduction of gold chloride. Chem Commun 9:1104–1105

Sirelkhatim A, Mahmud S, Seeni A, Kaus NHM, Ann LC, Bakhori SKM, Hasan H, Mohamad D (2015) Review on zinc oxide nanoparticles: antibacterial activity and toxicity mechanism. Nanomicro Lett 7:219–242

Syed MA, Manzoor U, Shah I, Bukhari SHA (2010) Antibacterial effects of Tungsten nanoparticles on the *Escherichia coli* strains isolated from catheterized urinary tract infection (UTI) cases and *Staphylococcus aureus*. New Microbiol 33:329–335

Tao Y, Ju E, Ren J, Qu X (2015) Bifunctionalized mesoporous silica-supported gold nanoparticles: intrinsic oxidase and peroxidase catalytic activities for antibacterial applications. Adv Mater 27:1097–1104

Thakker JN, Dalwadi P, Dhandhukia PC (2013) Biosynthesis of gold nanoparticles using *Fusarium oxysporum* f. sp. cubense JT1, a plant pathogenic fungus. ISRN Biotechnology, 2013 (515091): 5 pages.

Tsuruta T (2004) Biosorption and recycling of gold using various microorganisms. J Gen Appl Microbiol 50:221–228

Vago A, Szakacs G, Safran G, Horvath R, Pecz B, Lagzi I (2015) One-step green synthesis of gold nanoparticles by mesophilic filamentous fungi. Chem Phys Lett 645:1–4

Venkatpurwar VP, Pokharkar VB (2010) Biosynthesis of gold nanoparticles using therapeutic enzyme: *in-vitro* and *in-vivo* efficacy study. J Biomed Nanotechnol 6(6):667–674

Verma ML (2017a) Fungus-mediated bioleaching of metallic nanoparticles from agro-industrial by-products. In: Prasad R (ed) Fungal nanotechnology. Springer, Cham, pp 89–102

Verma ML (2017b) Nanobiotechnology advances in enzymatic biosensors for the agri-food industry. Environ Chem Lett 15(4):555–560

Verma ML (2017c) Enzymatic nanobiosensors in the agricultural and food industry. In: Ranjan S, Dasgupta N, Lichfouse E (eds) Nanoscience in food and agriculture 4 (Sustainable agriculture reviews), vol 24. Springer, Cham, pp 229–245. ISBN 978-3-319-53111-3

Verma ML (2018) Critical evaluation of toxicity tests in context to engineered nanomaterials: an introductory overview. In: Kumar V, Dasgupta N, Ranjan S (eds) Nanotoxicology: toxicity evaluation, risk assessment and management. CRC Press, Boca Raton, pp 1–17

Verma ML, Barrow CJ (2015) Recent advances in feedstocks and enzyme-immobilised technology for effective transesterification of lipids into biodiesel. In: Kalia V (ed) Microbial factories. Springer, pp 87–103

Verma ML, Barrow CJ, Kennedy JF, Puri M (2012) Immobilization of β-galactosidase from *Kluyveromyces lactis* on functionalized silicon dioxide nanoparticles: characterization and lactose hydrolysis. Int J Biol Macromol 50:432–437

Verma ML, Rajkhowa R, Barrow CJ, Wang X, Puri M (2013a) Exploring novel ultrafine Eri silk bioscaffold for enzyme stabilisation in cellobiose hydrolysis. Bioresour Technol 145:302–306

Verma ML, Naebe M, Barrow CJ, Puri M (2013b) Enzyme immobilisation on amino-functionalised multi-walled carbon nanotubes: structural and biocatalytic characterisation. PLoS One 8(9):e73642

Verma ML, Chaudhary R, Tsuzuki T, Barrow CJ, Puri M (2013c) Immobilization of β-glucosidase on a magnetic nanoparticle improves thermostability: application in cellobiose hydrolysis. Bioresour Technol 135:2–6

Verma ML, Barrow CJ, Puri M (2013d) Nanobiotechnology as a novel paradigm for enzyme immobilisation and stabilisation with potential applications in biofuel production. Appl Microbiol Biotechnol 97:23–39

Verma ML, Puri M, Barrow CJ (2016) Recent trends in nanomaterials immobilised enzymes for biofuel production. Crit Rev Biotechnol 36(1):108–119

Virkutyte J, Varma RS (2011) Green synthesis of metal nanoparticles: biodegradable polymers and enzymes in stabilization and surface functionalization. Chem Sci 2(5):837–846

Wang L, Hu C, Shao L (2017) The antimicrobial activity of nanoparticles: present situation and prospects for the future. Int J Nanomedicine 12:1227–1249

Whaley SR, English DS, Hu EL, Barbara PF, Belcher AM (2000) Selection of peptides with semiconductor binding specificity for directed nanocrystal assembly. Nature 405(6787):665–668

Wright GD (2005) Bacterial resistance to antibiotics: enzymatic degradation and modification. Adv Drug Deliv Rev 57:1451–1470

Xie J, Lee JY, Wang DIC, Ting YP (2007) High-yield synthesis of complex gold nanostructures in a fungal system. J Phys Chem C 111:16858–16865

Xin Y, Yin M, Zhao L, Meng F, Luo L (2017) Recent progress on nanoparticle-based drug delivery systems for cancer therapy. Cancer Biol Med 14:228–241

Zeng J, Ma Y, Jeong U, Xia Y (2010) Au(I): an alternative and potentially better precursor than Au(III) for the synthesis of Au nanostructures. J Mater Chem 20:2290–2301

Zhang X, He X, Wang K, Yang X (2011) Different active biomolecules involved in biosynthesis of gold nanoparticles by three fungus species. J Biomed Nanotechnol 7:245–254

Zhang Y, Arugula MA, Wales M, Wild J, Simonian AL (2015) A novel layer-by-layer assembled multi-enzyme/CNT biosensor for discriminative detection between organophosphorus and non-organophosphorus pesticides. Biosens Bioelectron 67:287–295

Chapter 8
Microbial Nanobionic Engineering: Translational and Transgressive Science of an Antidisciplinary Approximation

Juan Bueno

"Life did not take over the world by combat, but by networking."

—Lynn Margulis (1938–2011), Microcosmos: Four Billion Years of Microbial Evolution

"It necessarily follows that chance alone is at the source of every innovation, and of all creation in the biosphere. Pure chance, absolutely free but blind, at the very root of the stupendous edifice of evolution: this central concept of modern biology is no longer one among many other possible or even conceivable hypotheses. It is today the sole conceivable hypothesis, the only one that squares with observed and tested fact. And nothing warrants the supposition – or the hope – that on this score our position is ever likely to be revised. There is no scientific concept, in any of the sciences, more destructive of anthropocentrism than this one."

—Jacques Monod (1910–1976), Chance and Necessity

Contents

J. Bueno (✉)
Research Center of Bioprospecting and Biotechnology for Biodiversity Foundation (BIOLABB), Bogotá, Colombia
e-mail: juanbueno@biolabb.org

R. Prasad (ed.), *Microbial Nanobionics*, Nanotechnology in the Life Sciences,
https://doi.org/10.1007/978-3-030-16383-9_8

8.1 Introduction

Biomimicry has been defined as the search for technological innovations based on the exploration of nature, and constitutes a new integral branch of knowledge that in an antidisciplinary approach is looking for the integrality that favors bioinspiration to obtain high-impact products (Kennedy and James 2015). In this way the field of biological design comprises several stages, including biomimicry, biofabrication, and biohybrid systems (Raman and Bashir 2017). Likewise, biomimicry can be a source of technological transfer, a topic of interest for mega-diverse countries (Barthlott et al. 2016). In the same way, biofabrication, which uses living organisms in the production of both organic and inorganic materials, constitutes the application of this type of reverse engineering of biomolecular systems (REBMS) (Groll et al. 2016; Dahoumane et al. 2017; Nagamune 2017). However, it is interesting to note that for an anti-discipline such as REBMS, the integration of the omic information is necessary in a multi-omics model that we have called the "integrome," to reproduce the entire functionality of the biological system under study (Quo et al. 2012; Bueno 2018a). At this time, when it is imperative to integrate, to innovate, natural systems, in addition to ecosystems and their sustainability, the development of biological designs with which to solve current human problems is of fundamental importance (Kennedy et al. 2015; Blok and Gremmen 2016). But these designs, which are far from imitating nature, must express the coherent interpretation of a functional model of solution; otherwise, it is the blind act of reproducing without knowing how to apply (Hunter 2017). For this reason, to fully understand a process that may be susceptible to biomimicry, this must be assumed from their symbiotic interactions and not as the simple use of a certain metabolic function (Lin et al. 2017), i.e., bio-inspiration has more links to coevolutionary interactions than bio-utilization; thus, the two terms should not be confused (Lin et al. 2015).

In this order of ideas, microorganisms have become a great example of biomimicry because of their great capacity for resilience in the face of environmental changes (Stone et al. 2016). This resilience is determined by the communication networks present, such as the case of quorum sensing (QS) in the life cycle of bacteria (Niu et al. 2013). Likewise, microbial metabolism has been considered a source of new products, such as microbial fuel cells (MFCs), which produce electricity by the reduction of metal ions, the transfer of electrons, and biofilms that function as conduction structures (Calignano et al. 2015; You et al. 2017; Shah et al. 2019; Siegert et al. 2019). Additionally, the morphogenesis of microbial biofilms has been a study model for the development of new structures (Rinaldi 2007). On the other hand, a fascinating approach is the use of the technology of genetically engineered microbial whole-cell biosensors (GEMWCBs) for use as chemosensors in environmental monitoring (He et al. 2016), in addition to the search for and identification of new antibiotics (Kobras et al. 2017).

The aim of this chapter is to provide tools for discovering integral solutions through biomimicry and microbial bionics, not to create a new discipline, but on the contrary, to transcend the simple content in the transformation of the supra-specialized task of research, which affects innovation and the change of paradigms.

8.2 Biodesign

The following must be absolutely clear: to duplicate nature, the functioning of a biological system must be understood entirely, and to this end it is important to abandon all reductionism and determinism and to assume the study of complex systems (Mazzocchi 2008), for which it is required not to fragment the system and analyze it in its entirety as functionality, because it is explained and applied when observing it in interaction and communication (Van Regenmortel 2004). An example of this fact is bacterial chemotaxis, regulated by signal transduction systems (Wuichet et al. 2007), and the largest sample of social cooperation, survival, and ecological success, as biofilms are (Flemming et al. 2016). This great capacity for persistence has been applied in models of microbial biocementation that exploit the phenomenon of microbiologically induced calcite precipitation to obtain new construction materials (Abo-El-Enein et al. 2013; Ng et al. 2012) that have a great impact on soil engineering and geotechnics (Mujah et al. 2017). In this way, Biobrick® has used this field to heal cracks in buildings with appreciable results (Deniz and Keskin-Gundogdu 2018). Another interesting application is the production of bioenergy based on methane from the anaerobic digestion of biomass by fungi, which reproduces the process that occurs in the intestine of herbivores (Haitjema et al. 2014).

One step beyond the understanding of a biological system is its synthesis or improvement, which is the main premise of the rise of synthetic biology for the development of new biotherapeutics, biomaterials, biosensors, and solutions in bioenergy, and in bioremediation (Densmore and Hassoun 2012). It is also necessary to implement computational models to predict the behavior of multicellular populations integrating transcription, intercellular signaling, and cellular biophysics to understand the functioning of the complex systems that have been the subject of the design (Rudge et al. 2012). With the advent of the possibility of using genetic circuits for the development of microorganisms with biotechnological applications, the use of algorithms for the design and projection of initiatives in nanobionics has become important (Misirli et al. 2011; Jang et al. 2012), these metadata from the integrome in interaction with the environmentome are crucial for the development of products from the biomimicry of natural resources and their systems (Bueno 2018b).

8.3 Molecular Biomimicry

The description of the genes encoding the metabolic pathways that produce bioactive molecules in microorganisms opens up possibilities for the design and use of microbial factories that allow a large amount of metabolites to be obtained from the degradation of biomass (Weber and Kim 2016). For this reason, the search for biosynthetic gene clusters (BGCs) has stimulated research into natural products as a source of chemically diverse compounds from ecological niches as varied as the human microbiome and extremophiles (Wilson et al. 2017). Owing to the identification of the BGCs from organisms with a wide chemical diversity in combination with the new techniques of synthetic biology, it is possible to develop new microbial factories for bioactive secondary metabolites; in this way, biofactories using yeasts as a source of new metabolites from biomass are an impactful biotechnological application, integrating the bioprospecting of biodiversity with genetic engineering (Smanski et al. 2016; Guzmán-Trampe et al. 2017). Likewise, metagenomics, which correlates genomes with metabolomes using bioinformatics tools, is an important approach to the study and description of promising BGCs (Harvey et al. 2015). The search for new BGCs provides an opportunity to develop extensive bioprospecting projects with which to implement a small molecule biotechnology industry, and is the reason why the protection, use, valuation, and due recognition of biodiversity is part of the contemporary scientific agenda (Naman et al. 2017).

Additionally, each BGC from microbial sources can be consigned and compared in databases such as antiSMASH, which is useful for exploring the biosynthetic potential of isolated microorganisms (Blin et al. 2016). An equally important approach is the stimulation of silent or cryptic gene clusters to look for genes that are expressed by environmental factors for use in synthetic expression systems (Reen et al. 2015). These clusters of genes are part of the communication that occurs in microbial ecosystems, which has a great impact on the study of the biomimicry of complex biological systems (Netzker et al. 2015).

8.4 Green Synthesis in Nanotechnology

Currently, the manipulation of matter at the nanometer scale is the basis of nanotechnology and consolidates it as an industry of high innovation and impact. In this way, the search for nanofactories for inorganic nanoparticles (NPs; ZnO, ZnS, CdS), metallic NPs (Au, Ag, Cu, Al), magnetic NPs (Co, Fe, Ni), and carbon NPs (fullerenes, quantum dots, carbon nanotubes) becomes a necessity for an industry increasingly keen for biological systems that replace toxic synthesis platforms (Rafique et al. 2017). For that reason, biogenic methods have been developed using bacteria, fungi, and plant extracts to obtain nanomedicines, nanocarriers, and nanomaterials (Netala et al. 2015; Ramos et al. 2018; Prasad 2014, Prasad et al. 2016, 2018a, b). In this order of ideas, several genera have shown the ability to produce

NPs, among them *Morganella*, *Pseudomonas*, *Bacillus*, *Lactobacillus*, *Geobacillus*, *Escherichia*, *Vibrio*, *Salmonella*, *Klebsiella*, *Streptomyces*, *Rhodococcus*, *Cladosporium*, *Fusarium*, *Aspergillus*, *Penicillium*, and *Candida* (Ebrahiminezhad et al. 2016). This nanobiotechnological approach is an exciting bionanoscience where fungi are highly tolerant in the production and accumulation of NPs; thus, they are considered to be an advantageous model (Ottoni et al. 2017; Prasad 2016, 2017; Abdel-Aziz et al. 2018; Prasad et al. 2018a). Microorganisms such as bacteria and fungi use biosorption and bioreduction mechanisms to perform NP biosynthesis (Johnson and Prabu 2015; Kitching et al. 2015; Prasad et al. 2016, 2018a; Cui et al. 2017). This bioreduction process is employed by live microorganisms to reduce environmental toxicity (Ayangbenro & Babalola, 2017); likewise, DNA can assemble nanostructures through electrostatic interactions and in situ reduction (Huang et al. 2015). In addition, several enzymes have been identified to be involved in microbial reduction, e.g.: c-Type cytochromes, nitrate reductase, NADH-dependent reductases, and sulfite reductase (Huang et al. 2015). On the other hand, organisms such as actinomycetes perform this bioreduction mediated by specific proteins with masses between 10 and 80 kDa; also, yeasts use other routes such as detoxification mediated by glutathione (Hulkoti and Taranath 2014). However, it is very important to analyze the capacity of sociomicrobiological associations such as biofilms for the synthesis of NPs because of their great capacity for bioreduction and the possibility of making a more industrially scalable biosynthesis (Tanzil et al. 2016), thus it is a biological system to take into consideration in biomimicry.

8.5 Biofabrication

Biofabrication is a multidisciplinary technology that employs living biotemplates to obtain complex constructions (Ullah et al. 2017). In this order of ideas, microorganisms are a great bioinspiration for the development of hybrid materials, devices, and nanorobots (Shi et al. 2017). In addition, flagellar nanorobots, known as "nano-micro-swimmers" have been configured as synthetic molecular machines consisting of 20 atoms or less; likewise, magnetotactic bacteria have been used as natural nanorobots that exploit the magnetotaxis process (Ali et al. 2017; Shi et al. 2017).

Exploring this interesting field, the use of microbial biotemplates for the synthesis of inorganic nano-/microstructures, with which it is possible to design nanomaterials with a defined geometry and thus obtain essential patterns for the development of devices, is therefore as microbial structures as the whole cell, the flagellum and the pili, in addition to the yeasts and the hyphae, have been used to build nanowires for biosensing (Selvakumar et al. 2014). The use of these scaffolds is very versatile and it is possible to obtain structures in three dimensions based on carbon, interconnected by pores, which improves the transfer of ions in electrodes (Ozden et al. 2017). Equally, this model has been useful for designing NP microtubules in which the hyphae structures from filamentous fungi are used as template with added metallic

NPs with the end to obtain microtubular structures, an important effort in joining biological structures with inorganic nanomaterials (Kubo et al. 2016).

On the other hand, an important approach in biofabrication is the use of bioprinting, which consists of positioning living cells on a matrix, for the development of tissues, organs, and biosensors. Thus, the microorganisms *Saccharomyces cerevisiae* var. *bayanus* and *Chlorella vulgaris* have been printed on hydrogel for use and study (Taidi et al. 2016). Thereby, the production of living materials with the capacity to perform controlled metabolic functions using 3D bioprinting makes it possible to obtain medical devices and perform bioremediation processes (Schaffner et al. 2017).

In conclusion, the model of biofabrication based on microorganisms is a promising research topic that has the advantages of offering a biotemplate that is uniform in size and shape, with functional groups useful for cell surface interaction and porosity, which allows ion exchange (Ullah et al. 2017).

8.6 Biomimicry of Antimicrobial Peptides

Host defense peptides (HDPs) are antimicrobial compounds with defined chemical characteristics, such as being cationic, linear, and amphipathic, that produce disruption of the membranes in the target cells. For this reason, they can be used for the development of new antimicrobial and anticancer medicines (Ergene et al. 2018). Approximately 2700 of these peptides from different natural sources have been identified, providing a large number of scaffolds for synthesis models (Bondaryk et al. 2017). In this way, the use of peptoids (Poly-N-substituted glycines) that mimic this type of peptide with a related isomeric structure has been evaluated as an important strategy in obtaining peptides with pharmacological activity (Mojsoska et al. 2015). These peptoids can be synthesized on an automated peptide synthesizer whose chemical functionality can be incorporated into the change of the primary amine (Huang et al. 2014). Also, it is possible to conjugate antimicrobial peptides with NPs to increase their microbicidal activity in biocidal products (Vitiello et al. 2018). With these new materials, novel nanofibers and surfaces with promising applications in burns and skin infections can be developed (de Castro and Franco 2015).

Additionally, peptoids as biomimetic polymers can be arranged in different structures as bilayers in the crystalline phase, which is useful in drug delivery (Ma et al. 2017). Thus, natural cationic polymers are an interesting model in biomimicry for the use of cationic antimicrobial peptides in drug discovery (Phillips et al. 2017).

8.7 Synthetic Microbial Communities

Microorganisms have evolved together under interaction and communication to compete for space. In this way, microbes have developed biological systems to produce and disseminate antimicrobial compounds (Little et al. 2008). In these natural ecosystems, intercellular communication is determined by the production, release, and exchange of several molecules, including peptides, small molecules, and natural products (Johns et al. 2016). Thereby, synthetic microbial communities can be generated to replicate microbial interaction patterns to develop biotechnological applications, because this microbial consortium has a greater capacity to perform metabolic conversions than other systems (Boyle and Silver 2012; Großkopf and Soyer 2014). These engineered microbial consortia can be applied for bioremediation, biofuel production, and MFC design (Hays et al. 2015). However, it is important to consider that to design these synthetic microbial communities it is necessary to obtain synthetic microbial ecosystems that replicate functionality first, to achieve resilience, diversity, and spatial organization (De Roy et al. 2014). Also, the assembly platform for microbial populations acquires a fundamental importance, in determining the optimal environmental conditions. The most frequently used techniques for creating controlled environmental parameters are cell immobilization and microfluidics (Goers et al. 2014).

This multi-omics model with the ability to predict the patterns of cooperation and competition among microbial populations, in addition to determining their symbiotic relationships, has grown in recent years as a fundamental application in biotechnology that can lead to the improvement of processes for obtaining and developing new products (Bosi et al. 2017).

8.8 Bio-nano-things

As important as the design and development of new bioactive compounds, and the improvement of bioprocesses, is the possibility of engineering devices from biochemical molecules and biological systems (Chude-Okonkwo et al. 2015). These bionanodevices, as part of the field of biomolecular engineering, and because of their small size, allow the development of hybrid nanomachines that combine nanomaterials with biological molecules useful for therapy and diagnosis (Nagamune 2017). This is how this chemical functionalization of materials can be made possible by combining the knowledge of the biological activity of molecules with electronic devices capable of transducing the signal in a message or required effect (Kulkarni et al. 2014). Also, these devices can be integrated with living organisms, as in the case of plants, to monitor the surrounding ecosystems (El-Din and Manjaiah 2017). In this way, the search for bioactive molecules (DNA, RNA, and proteins) from microbial sources capable of coupling with the nanointerface (e.g., yeast cytochrome c, glucose oxidase, and laccase) becomes necessary to design more and diverse nano-things (nanowires, nanotubes, and NPs) (Bonanni et al. 2004; Audette et al. 2011).

8.8.1 Biosensors

Microbial biosensors are non-expensive alternatives of high efficiency for the detection of different analytes that can be used as bioreporters of the ecosystems (Cerminati et al. 2015). In addition, microorganisms work as excellent indicators of serum and air quality, in addition to human diseases such as obesity and cancer (Turgeman 2015). These whole-cell biosensors can be constituted by bacteria, fungi, microalgae, and protozoa; among them, the fungus *Saccharomyces cerevisiae* has advantages for this type of device because it has physico-chemical tolerance and the ability to be genetically transformed to obtain GEMWCBs, although it has a cell wall that that is a barrier to entry and prevents proper signal transduction (Gutiérrez et al. 2015). Similarly, these biosensors can be linked to functional molecules for the detection of specific molecules to increase speed and sensitivity (Li et al. 2018; Faraz et al. 2018). On the other hand, MFCs used to generate energy can also be efficient biosensors, because the microorganisms associated with electrodes can detect physico-chemical changes such as an alteration of the electronic transfer (Bereza-Malcolm and Franks 2015).

8.8.2 Molecular Communications and Nanonetworks

An exciting model is the development of communication systems that exchange information through molecular transmission and reception. In this way, molecular communication (MC) is a promising paradigm that can study nanonetworks in bacteria containing DNA plasmids with information (Felicetti et al. 2016). This information is communicated using bacteria as a carrier and a node that passes the DNA molecule (Cobo and Akyildiz 2010). The behavior of these molecular interactions provide valuable data about intrabody biosensing and drug delivery (Unluturk et al. 2016). Also, intrabody nanonetworks are present performing MC in living beings on a nanometric scale, which opens up the possibilities of new diagnostic and treatment techniques that seek to maintain the stability of nanonetworks (Akan et al. 2017). In this order of ideas, nano-communications can be of two types: electromagnetic and molecular. Equally, this communication network can interact with nanomaterials such as graphene and carbon nanotubes to implement nano-scale antennas and transmitters (Abbasi et al. 2016). This is how these new biohybrid devices can be genetically designed to combine different functionalities with which to obtain a certain behavior (Cobo and Akyildiz 2010). In this transdisciplinary field of knowledge, biomimicry is a key tool for the development of innovations and biofilms and becomes a study model with extrapolatable results because of the great MC and exchange of genetic material using nanotubes and plasmids that exist within this biological system, mainly in polymicrobial biofilms (Hasan et al. 2015).

8.8.3 Microbial Organic Bioelectronics

Continuing with this order of ideas, the concept of organic bioelectronics makes an appearance. This field includes organic devices designed to perform the translation of signals from biological systems to electronic systems (Simon et al. 2016),

Using these semiconductors, which contain organic polymers such as poly(3,4-ethylenedioxythiophene), it is possible to perform van der Waals interactions that more easily allow chemical modifications (Rivnay et al. 2013). Interestingly, these carbon polymer devices are useful for the development of novel in vitro assays that biomimetize the environmental conditions of an infection, which is useful for the discovery of new drugs; also, it can be used for the release of antimicrobial drugs to prevent the appearance of biofilms in hospital environments (Löffler et al. 2015a, b). In addition, microorganisms such as *Shewanella oneidensis* and *Geobacter* sp. can be adhered to this screening platform to monitor complex chemicals (Zhou et al. 2017). The bioelectronic era opens up the way to a new series of clinical treatments, or "electroceuticals," that can monitor and alter the cellular interface by releasing medicines sensitive to electrical impulses (Löffler et al. 2017).

8.8.4 Microbial Fuel Cells

Microbial fuel cells consist of an anode and a cathode separated by compartments embedded in a substrate (organic/inorganic) with suspended microorganisms. When catabolic processes are performed, these MFCs generate electricity (Rahimnejad et al. 2015; Shah et al. 2019). MFCs have various applications, such as power generator, biosensor, bioremediation, robotics, and bioproduction (Mathuriya and Yakhmi 2016). Thus, the microorganisms that surround the electrodes in the MFCs accumulate as biofilms known as electrochemically active biofilms (EABs). The use of EABs increases the energy power and catalytic capabilities, which improves performance (Babauta et al. 2012). In addition, these devices can be organized in a uniform way in a cellular automaton, which allows the development of computational devices based on the information obtained by each MFC. Maybe in the future this will allow the development of hybrid computing systems (Tsompanas et al. 2017).

8.8.5 Microfluidics

Microfluidic systems consist of fluid devices whose flow is nonturbulent and are 10–100 μm in size; from this principle, useful techniques for chemical synthesis, high-efficiency biological tests, and the simulation of the functioning of organs known as microfluidics have been developed (Chiu et al. 2017). In addition, these techniques can synthesize NP libraries and evaluate in vitro biological activity that mimics in vivo conditions in a high-throughput manner, which is useful in drug screening (Ahn et al. 2018). In this way, microfluidic chips can offer in vitro models with greater predictive capacity in vivo than other assays developed to date, because they offer a platform for evaluation, reproducible, automatable, and to a greater extent, controllable (Esch et al. 2015). This technology of microscale devices allows microenvironments to be created in which mimicking the biological systems of tissues and organs can be coupled with obtaining bio-nano-things, thus developing models of microfluidic MC systems (Huh et al. 2012; Kuscu and Akan 2018). Similarly, the fact that this type of technique can replace the use of preclinical studies in experimental animals, because of its ability to reproduce organoids, multicellular spheroids, and bioprinted micro tissues, deserves special mention (Liu et al. 2017).

8.9 Conclusions and Perspectives

8.9.1 Megadiverse Countries and Biomimicry Production Systems

Because of what was previously outlined during the development of this chapter, it is necessary to debate how mega-diverse countries are going to make a fundamental source of this new technological shift in which nature enters as a model, and it is possible to glimpse once again a union between nature and technology (Dicks 2017). This is how a modern bioeconomy can claim the science of imitating life as a source of development in medicine, agriculture, and industry (Igor and Khaustova 2015). Also, it becomes important in a country with a large number of ecosystems and species to reconcile the bionics, which studies biological systems from their control, with biomimicry, which seeks an ecological design for sustainability (Wahl 2006). Biomimicry within its parameters seeks nature as a model, measure, and mentor, but from a transdisciplinary approach that tends toward antidiscipline, that is, it does not specialize and becomes integral (McGregor 2013), as it is a priority that the interface between the physical and biological sciences disappears so that the living systems can be studied in depth (Iouguina et al. 2014). On the other hand, the strategy to be used for the introduction of a production system based on bionics/biomimicry should contemplate and be directed toward aspects of the evolution of the biological system, starting from a problem to be solved in addition to the functionality of systems (Lotfi 2017). Consequently, to assume the functionality, any productive process that includes bionics/biomimicry should use the following levels – form, process, and interaction – to be able to implement the respective biological model (Jacobs 2014). We are facing an important interface where it is possible to interconnect communities and ecosystems for sustainable development, with a greater understanding of our biological world and our biodiversity. This is how to overcome supra-specialization becoming a priority and transcend the disciplines to break paradigms to get new knowledge of a relevant need.

Acknowledgement The author thanks I. Ritoré for her collaboration and invaluable support during the writing of this chapter.

References

Abbasi QH, Yang K, Chopra N, Jornet JM, Abuali NA, Qaraqe KA, Alomainy A (2016) Nano-communication for biomedical applications: a review on the state-of-the-art from physical layers to novel networking concepts. IEEE Access 4:3920–3935

Abdel-Aziz SM, Prasad R, Hamed AA, Abdelraof M (2018) Fungal nanoparticles: A novel tool for a green biotechnology? In: Fungal Nanobionics (eds. Prasad R, Kumar V, Kumar M, and Shanquan W), Springer Nature Singapore Pte Ltd. 61–87

Abo-El-Enein SA, Ali AH, Talkhan FN, Abdel-Gawwad HA (2013) Application of microbial biocementation to improve the physico-mechanical properties of cement mortar. HBRC J 9:36–40

Ahn J, Ko J, Lee S, Yu J, Kim Y, Jeon NL (2018) Microfluidics in nanoparticle drug delivery; from synthesis to pre-clinical screening. Adv Drug Deliv Rev 128:29–53

Akan OB, Ramezani H, Khan T, Abbasi NA, Kuscu M (2017) Fundamentals of molecular information and communication science. Proc IEEE 105:306–318

Ali J, Cheang UK, Martindale JD, Jabbarzadeh M, Fu HC, Kim MJ (2017) Bacteria-inspired nanorobots with flagellar polymorphic transformations and bundling. Sci Rep 7:14098

Audette GF, Lombardo S, Dudzik J, Arruda TM, Kolinski M, Filipek S, Mukerjee S, Kannan AM, Thavasi V, Ramakrishna S, Chin M, Somasundaran P, Viswanathan S, Keles RS, Renugopalakrishnan V (2011) Protein hot spots at bio-nano interfaces. Mater Today 14:360–365

Ayangbenro A, Babalola O (2017) A new strategy for heavy metal polluted environments: a review of microbial biosorbents. Int J Environ Res Public Health 14(1):94

Babauta J, Renslow R, Lewandowski Z, Beyenal H (2012) Electrochemically active biofilms: facts and fiction. A review. Biofouling 28:789–812

Barthlott W, Rafiqpoor MD, Erdelen WR (2016) Bionics and biodiversity-bio-inspired technical innovation for a sustainable future. In: Knippers J, Nickel K, Speck T (eds) Biomimetic research for architecture and building construction. Springer International Publishing, Cham, pp 11–55

Bereza-Malcolm LT, Franks AE (2015) Coupling anaerobic bacteria and microbial fuel cells as whole-cell environmental biosensors. Microbiol Aust 36:129–132

Blin K, Medema MH, Kottmann R, Lee SY, Weber T (2016) The antiSMASH database, a comprehensive database of microbial secondary metabolite biosynthetic gene clusters. Nucleic Acids Res 45:D555–D559

Blok V, Gremmen B (2016) Ecological innovation: biomimicry as a new way of thinking and acting ecologically. J Agric Environ Ethics 29:203–217

Bonanni B, Alliata D, Andolfi L, Bizzarri AR, Delfino I, Cannistraro S (2004) Yeast cytochrome on gold electrode: a robust hybrid system for bio-nanodevices. In: Nanotechnology 4th IEEE conference, 574–576

Bondaryk M, Staniszewska M, Zielińska P, Urbańczyk-Lipkowska Z (2017) Natural antimicrobial peptides as inspiration for design of a new generation antifungal compounds. J Fungi (Basel) 3:46

Bosi E, Bacci G, Mengoni A, Fondi M (2017) Perspectives and challenges in microbial communities metabolic modeling. Front Genet 8:88

Boyle PM, Silver PA (2012) Parts plus pipes: synthetic biology approaches to metabolic engineering. Metab Eng 14:223–232

Bueno J (2018a) Looking for the integrome, When metabolomics talk about the interactome. J Microb Biochem Technol 10:e134

Bueno J (2018b) Biofilm environmentome: A survival experiment. J Biochem Physiol 1:1

Calignano F, Tommasi T, Manfredi D, Chiolerio A (2015) Additive manufacturing of a microbial fuel cell-a detailed study. Sci Rep 5:17373

Cerminati S, Soncini FC, Checa SK (2015) A sensitive whole-cell biosensor for the simultaneous detection of a broad-spectrum of toxic heavy metal ions. Chem Commun (Camb) 51:5917–5920

Chiu DT, Di Carlo D, Doyle PS, Hansen C, Maceiczyk RM, Wootton RC (2017) Small but perfectly formed? Successes, challenges, and opportunities for microfluidics in the chemical and biological sciences. Chem 2:201–223

Chude-Okonkwo UA, Malekian R, Maharaj BT, Chude CC (2015) Bio-inspired approach for eliminating redundant nanodevices in internet of bio-nano things. In: IEEE Globecom Workshops (GC Wkshps), 1–6

Cobo LC, Akyildiz IF (2010) Bacteria-based communication in nanonetworks. Nano Commun Networks 1:244–256

Cui J, Zhu N, Kang N, Ha C, Shi C, Wu P (2017) Biorecovery mechanism of palladium as nanoparticles by *Enterococcus faecalis*: from biosorption to bioreduction. Chem Eng J 328:1051–1057

Dahoumane SA, Mechouet M, Wijesekera K, Filipe CD, Sicard C, Bazylinski DA, Jeffryes C (2017) Algae-mediated biosynthesis of inorganic nanomaterials as a promising route in nanobiotechnology – a review. Green Chem 19:552–587

De Castro AP, Franco OL (2015) Modifying natural antimicrobial peptides to generate bioinspired antibiotics and devices. Future Med Chem 7:413–415

De Roy K, Marzorati M, Van den Abbeele P, Van de Wiele T, Boon N (2014) Synthetic microbial ecosystems: an exciting tool to understand and apply microbial communities. Environ Microbiol 16:1472–1481

Deniz I, Keskin-Gundogdu T (2018) Biomimetic design for a bioengineered world. In: Kokturk G (ed) Interdisciplinary expansions in engineering and design with the power of biomimicry. IntechOpen Limited, London, pp 57–74

Densmore D, Hassoun S (2012) Design automation for synthetic biological systems. IEEE Des Test Comput 29:7–20

Dicks H (2017) The poetics of biomimicry: the contribution of poetic concepts to philosophical inquiry into the biomimetic principle of nature as model. Environ Philos 14:191–219

Ebrahiminezhad A, Bagheri M, Taghizadeh SM, Berenjian A, Ghasemi Y (2016) Biomimetic synthesis of silver nanoparticles using microalgal secretory carbohydrates as a novel anticancer and antimicrobial. Adv Nat Sci Nanosci Nanotechnol 7:015018

El-Din HE, Manjaiah DH (2017) Internet of nano things and industrial internet of things. In: Acharjya DP, Geetha MK (eds) Internet of things: novel advances and envisioned applications. Springer International Publishing, Cham, pp 109–123

Ergene C, Yasuhara K, Palermo EF (2018) Biomimetic antimicrobial polymers: recent advances in molecular design. Polym Chem 9:2407–2427

Esch EW, Bahinski A, Huh D (2015) Organs-on-chips at the frontiers of drug discovery. Nat Rev Drug Discov 14:248–260

Faraz M, Abbasi A, Naqvi FK, Khare N, Prasad R, Barman I, Pandey R (2018) Polyindole/CdS nanocomposite based turn-on, multi-ion fluorescence sensor for detection of Cr^{3+}, Fe^{3+} and Sn^{2+} ions. Sensors Actuators A 269:195–202. https://doi.org/10.1016/j.snb.2018.04.110

Felicetti L, Femminella M, Reali G, Liò P (2016) Applications of molecular communications to medicine: a survey. Nano Commun Networks 7:27–45

Flemming HC, Wingender J, Szewzyk U, Steinberg P, Rice SA, Kjelleberg S (2016) Biofilms: an emergent form of bacterial life. Nat Rev Microbiol 14:563–575

Goers L, Freemont P, Polizzi KM (2014) Co-culture systems and technologies: taking synthetic biology to the next level. J R Soc Interface 11:20140065

Groll J, Boland T, Blunk T, Burdick JA, Cho DW, Dalton PD, Derby B, Forgacs G, Li Q, Mironov VA, Moroni L, Nakamura M, Shu W, Takeuchi S, Vozzi G, Woodfield TB, Xu T, Yoo JJ, Malda J (2016) Biofabrication: reappraising the definition of an evolving field. Biofabrication 8:013001

Großkopf T, Soyer OS (2014) Synthetic microbial communities. Curr Opin Microbiol 18:72–77

Gutiérrez JC, Amaro F, Martín-González A (2015) Heavy metal whole-cell biosensors using eukaryotic microorganisms: an updated critical review. Front Microbiol 6:48

Guzmán-Trampe S, Ceapa CD, Manzo-Ruiz M, Sánchez S (2017) Synthetic biology era: improving antibiotic's world. Biochem Pharmacol 134:99–113

Haitjema CH, Solomon KV, Henske JK, Theodorou MK, O'Malley MA (2014) Anaerobic gut fungi: advances in isolation, culture, and cellulolytic enzyme discovery for biofuel production. Biotechnol Bioeng 111:1471–1482

Harvey AL, Edrada-Ebel R, Quinn RJ (2015) The re-emergence of natural products for drug discovery in the genomics era. Nat Rev Drug Discov 14:111–129

Hasan M, Hossain E, Balasubramaniam S, Koucheryavy Y (2015) Social behavior in bacterial nanonetworks: challenges and opportunities. IEEE Netw 29:26–34

Hays SG, Patrick WG, Ziesack M, Oxman N, Silver PA (2015) Better together: engineering and application of microbial symbioses. Curr Opin Biotechnol 36:40–49

He W, Yuan S, Zhong WH, Siddikee MA, Dai CC (2016) Application of genetically engineered microbial whole-cell biosensors for combined chemosensing. Appl Microbiol Biotechnol 100:1109–1119

Huang W, Seo J, Willingham SB, Czyzewski AM, Gonzalgo ML, Weissman IL, Barron AE (2014) Learning from host-defense peptides: cationic, amphipathic peptoids with potent anticancer activity. PLoS One 9:e90397

Huang J, Lin L, Sun D, Chen H, Yang D, Li Q (2015) Bio-inspired synthesis of metal nanomaterials and applications. Chem Soc Rev 44:6330–6374

Huh D, Torisawa YS, Hamilton GA, Kim HJ, Ingber DE (2012) Microengineered physiological biomimicry: organs-on-chips. Lab Chip 12:2156–2164

Hulkoti NI, Taranath TC (2014) Biosynthesis of nanoparticles using microbes – a review. Colloids Surf B Biointerfaces 121:474–483

Hunter P (2017) From imitation to inspiration: biomimicry experiences a revival driven by a more systematic approach to explore nature's inventions for human use. EMBO Rep 18:363–366

Igor M, Khaustova V (2015) Modern trends on bioeconomy development in the world: the introduction of NBIC-technologies in biomedicine. Int J Br 2:103–118

Iouguina A, Dawson J, Hallgrimsson B, Smart G (2014) Biologically informed disciplines: a comparative analysis of bionics, biomimetics, biomimicry, and bio-inspiration among others. Int J Des Nat Ecodyn 9:197–205

Jacobs S (2014) Biomimetics: a simple foundation will lead to new insight about process. Int J Des Nat Ecodyn 9:83–94

Jang SS, Oishi KT, Egbert RG, Klavins E (2012) Specification and simulation of synthetic multicelled behaviors. ACS Synth Biol 1:365–374

Johns NI, Blazejewski T, Gomes AL, Wang HH (2016) Principles for designing synthetic microbial communities. Curr Opin Microbiol 31:146–153

Johnson I, Prabu HJ (2015) Green synthesis and characterization of silver nanoparticles by leaf extracts of *Cycas circinalis*, *Ficus amplissima*, *Commelina benghalensis* and *Lippia nodiflora*. Int Nano Lett 5:43–51

Kennedy B, James JKN (2015) Integrating biology, design, and engineering for sustainable innovation. In: Integrated STEM Education Conference (ISEC), 88–93

Kennedy E, Fecheyr-Lippens D, Hsiung BK, Niewiarowski PH, Kolodziej M (2015) Biomimicry: a path to sustainable innovation. Des Issues 31:66–73

Kitching M, Ramani M, Marsili E (2015) Fungal biosynthesis of gold nanoparticles: mechanism and scale up. Microb Biotechnol 8:904–917

Kobras CM, Mascher T, Gebhard S (2017) Application of a *Bacillus subtilis* whole-cell biosensor (Plial-lux) for the identification of cell wall active antibacterial compounds. Methods Mol Biol 1520:121–131

Kubo AM, Gorup LF, Amaral LS, Filho ER, Camargo ER (2016) Kinetic control of microtubule morphology obtained by assembling gold nanoparticles on living fungal biotemplates. Bioconjug Chem 27:2337–2345

Kulkarni A, Seo HH, Lee TK, Kim HC, Kim T, Park SH, Lee MH, Moh SH (2014) Photoresponse of silicon nanowire array field effect transistors controlled by mycosporine-like amino acids. J Nanoeng Nanomanufact 4:247–251

Kuscu M, Akan OB (2018) Modeling convection-diffusion-reaction systems for microfluidic molecular communications with surface-based receivers in Internet of Bio-Nano Things. PLoS One 13:e0192202

Li N, Huang X, Zou J, Chen G, Liu G, Li M, Dong J, Du F, Cui X, Tang Z (2018) Evolution of microbial biosensor based on functional RNA through fluorescence-activated cell sorting. Sensors Actuators B Chem 258:550–557

Lin N, Jing S, Liang X, Yuan W, Chen H (2015) Biomimicry of symbiotic multi-species coevolution for global optimization. Metall Min Indust 7:187–194

Lin N, Chen H, Jing S, Liu F, Liang X (2017) Biomimicry of symbiotic multi-species coevolution for discrete and continuous optimization in RFID networks. Saudi J Biol Sci 24:610–621

Little AE, Robinson CJ, Peterson SB, Raffa KF, Handelsman J (2008) Rules of engagement: interspecies interactions that regulate microbial communities. Annu Rev Microbiol 62:375–401

Liu Y, Gill E, Shery Huang YY (2017) Microfluidic on-chip biomimicry for 3D cell culture: a fit-for-purpose investigation from the end user standpoint. Future Sci OA 3:FSO173

Löffler S, Libberton B, Richter-Dahlfors A (2015a) Organic bioelectronic tools for biomedical applications. Electronics 4:879–908

Löffler S, Libberton B, Richter-Dahlfors A (2015b) Organic bioelectronics in infection. J Mater Chem B 3:4979–4992

Löffler S, Melican K, Nilsson KPR, Richter-Dahlfors A (2017) Organic bioelectronics in medicine. J Intern Med 282:24–36

Lotfi NG (2017) Evolutionary design: the application of biological strategies in the product design process. Int J Des Nat Ecod 12:204–213

Ma X, Zhang S, Jiao F, Newcomb CJ, Zhang Y, Prakash A, Liao Z, Baer MD, Mundy CJ, Pfaendtner J, Noy A, Chen CL, De Yoreo JJ (2017) Tuning crystallization pathways through sequence engineering of biomimetic polymers. Nat Mater 16:767–775

Mathuriya AS, Yakhmi JV (2016) Microbial fuel cells – applications for generation of electrical power and beyond. Crit Rev Microbiol 42:127–143

Mazzocchi F (2008) Complexity in biology: exceeding the limits of reductionism and determinism using complexity theory. EMBO Rep 9:10–14

McGregor SL (2013) Transdisciplinarity and biomimicry. Transd J Eng Sci 4:57–65

Misirli G, Hallinan JS, Yu T, Lawson JR, Wimalaratne SM, Cooling MT, Wipat A (2011) Model annotation for synthetic biology: automating model to nucleotide sequence conversion. Bioinformatics 27:973–979

Mojsoska B, Zuckermann RN, Jenssen H (2015) Structure-activity relationship study of novel peptoids that mimic the structure of antimicrobial peptides. Antimicrob Agents Chemother 59:4112–4120

Mujah D, Shahin MA, Cheng L (2017) State-of-the-art review of biocementation by microbially induced calcite precipitation (MICP) for soil stabilization. Geomicrobiol J 34:524–537

Nagamune T (2017) Biomolecular engineering for nanobio/bionanotechnology. Nano Conv 4:9

Naman CB, Leber CA, Gerwick WH (2017) Modern natural products drug discovery and its relevance to biodiversity conservation. In: Kurtböke I (ed) Microbial resources. Academic Press, Cambridge, MA, pp 103–120

Netala VR, Kotakadi VS, Nagam V, Bobbu P, Ghosh SB, Tartte V (2015) First report of biomimetic synthesis of silver nanoparticles using aqueous callus extract of *Centella asiatica* and their antimicrobial activity. Appl Nanosci 5:801–807

Netzker T, Fischer J, Weber J, Mattern DJ, König CC, Valiante V, Schroeckh V, Brakhage AA (2015) Microbial communication leading to the activation of silent fungal secondary metabolite gene clusters. Front Microbiol 6:299

Ng WS, Lee ML, Hii SL (2012) An overview of the factors affecting microbial-induced calcite precipitation and its potential application in soil improvement. World Acad Sci Eng Technol 62:723–729

Niu B, Wang H, Duan Q, Li L (2013) Biomimicry of quorum sensing using bacterial lifecycle model. BMC Bioinf 14(Suppl 8):S8

Ottoni CA, Simões MF, Fernandes S, Dos Santos JG, da Silva ES, de Souza RFB, Maiorano AE (2017) Screening of filamentous fungi for antimicrobial silver nanoparticles synthesis. AMB Express 7:31

Ozden S, Macwan IG, Owuor PS, Kosolwattana S, Autreto PAS, Silwal S, Vajtai R, Tiwary CS, Mohite AD, Patra PK, Ajayan PM (2017) Bacteria as bio-template for 3D carbon nanotube architectures. Sci Rep 7(1):9855

Phillips DJ, Harrison J, Richards SJ, Mitchell DE, Tichauer E, Hubbard ATM, Guy C, Hands-Portman I, Fullam E, Gibson MI (2017) Evaluation of the antimicrobial activity of cationic polymers against mycobacteria: toward antitubercular macromolecules. Biomacromolecules 18:1592–1599

Prasad R (2014) Synthesis of silver nanoparticles in photosynthetic plants. J Nanopart: 963961. https://doi.org/10.1155/2014/963961

Prasad R (2016) Advances and applications through fungal nanobiotechnology. Springer, Cham

Prasad R (2017) Fungal nanotechnology: applications in agriculture, industry, and medicine. Springer International Publishing, Cham

Prasad R, Pandey R, Barman I (2016) Engineering tailored nanoparticles with microbes: quo vadis. WIREs Nanomed Nanobiotechnol 8:316–330. https://doi.org/10.1002/wnan.1363

Prasad R, Kumar V, Kumar M, Shanquan W (2018a) Fungal nanobionics: principles and applications. Springer, Singapore. https://www.springer.com/gb/book/9789811086656

Prasad R, Jha A, Prasad K (2018b) Exploring the Realms of Nature for Nanosynthesis. Springer International Publishing (ISBN 978-3-319-99570-0). https://www.springer.com/978-3-319-99570-0

Quo CF, Kaddi C, Phan JH, Zollanvari A, Xu M, Wang MD, Alterovitz G (2012) Reverse engineering biomolecular systems using -omic data: challenges, progress and opportunities. Brief Bioinform 13:430–445

Rafique M, Sadaf I, Rafique MS, Tahir MB (2017) A review on green synthesis of silver nanoparticles and their applications. Artif Cells Nanomed Biotechnol 45:1272–1291

Rahimnejad M, Adhami A, Darvari S, Zirepour A, Oh SE (2015) Microbial fuel cell as new technology for bioelectricity generation: a review. Alex Eng J 54:745–756

Raman R, Bashir R (2017) Biomimicry, biofabrication, and biohybrid systems: the emergence and evolution of biological design. Adv Healthc Mater 6:1700496

Ramos MA, Da Silva PB, Spósito L, De Toledo LG, Bonifácio BV, Rodero CF, Dos Santos KC, Chorilli M, Bauab TM (2018) Nanotechnology-based drug delivery systems for control of microbial biofilms: a review. Int J Nanomedicine 13:1179–1213

Reen FJ, Romano S, Dobson AD, O'Gara F (2015) The sound of silence: activating silent biosynthetic gene clusters in marine microorganisms. Mar Drugs 13:4754–4783

Rinaldi A (2007) Naturally better: science and technology are looking to nature's successful designs for inspiration. EMBO Rep 8:995–999

Rivnay J, Owens RM, Malliaras GG (2013) The rise of organic bioelectronics. Chem Mater 26:679–685

Rudge TJ, Steiner PJ, Phillips A, Haseloff J (2012) Computational modeling of synthetic microbial biofilms. ACS Synth Biol 1:345–352

Schaffner M, Rühs PA, Coulter F, Kilcher S, Studart AR (2017) 3D printing of bacteria into functional complex materials. Sci Adv 3:eaao6804

Selvakumar R, Seethalakshmi N, Thavamani P, Naidu R, Megharaj M (2014) Recent advances in the synthesis of inorganic nano/microstructures using microbial biotemplates and their applications. RSC Adv 4:52156–52169

Shah S, Venkatramanan V, Prasad R (2019) Microbial fuel cell: Sustainable green technology for bioelectricity generation and wastewater treatment. In: Shah S, Venkatramanan V, Prasad R (eds) Sustainable Green Technologies for Environmental Management. Springer Nature, Singapore, 199–218

Shi Z, Shi X, Ullah MW, Li S, Revin VV, Yang G (2017) Fabrication of nanocomposites and hybrid materials using microbial biotemplates. Adv Compos Hybrid Mater 1:79–93

Siegert M, Sonawane JM, Ezugwu CI, Prasad R (2019) Economic assessment of nanomaterials in bio-electrical water treatment. In: Prasad R, Karchiyappan T (eds) Advanced Research in Nanosciences for Water Technology. Nanotechnology in the Life Sciences. Springer, Cham, 1–23

Simon DT, Gabrielsson EO, Tybrandt K, Berggren M (2016) Organic bioelectronics: bridging the signaling gap between biology and technology. Chem Rev 116:13009–13041

Smanski MJ, Zhou H, Claesen J, Shen B, Fischbach MA, Voigt CA (2016) Synthetic biology to access and expand nature's chemical diversity. Nat Rev Microbiol 14:135–149

Stone W, Kroukamp O, Korber DR, McKelvie J, Wolfaardt GM (2016) Microbes at surface-air interfaces: the metabolic harnessing of relative humidity, surface hygroscopicity, and oligotrophy for resilience. Front Microbiol 7:1563

Taidi B, Lebernede G, Koch L, Perre P, Chichkov B (2016) Colony development of laser printed eukaryotic (yeast and microalga) microorganisms in co-culture. Int J Bioprint 2:146–152

Tanzil AH, Sultana ST, Saunders SR, Shi L, Marsili E, Beyenal H (2016) Biological synthesis of nanoparticles in biofilms. Enzym Microb Technol 95:4–12

Tsompanas MAI, Adamatzky A, Sirakoulis GC, Greenman J, Ieropoulos I (2017) Towards implementation of cellular automata in Microbial Fuel Cells. PLoS One 12:e0177528

Turgeman YJ (2015) Microbial mediations: cyber-biological extensions of human sensitivity to natural and made ecologies. Doctoral dissertation, Massachusetts Institute of Technology

Ullah MW, Shi Z, Shi X, Zeng D, Li S, Yang G (2017) Microbes as structural templates in biofabrication: study of surface chemistry and applications. ACS Sustain Chem Eng 5:11163–11175

Unluturk BD, Balasubramaniam S, Akyildiz IF (2016) The impact of social behavior on the attenuation and delay of bacterial nanonetworks. IEEE Trans Nanobioscience 15:959–969

Van Regenmortel MH (2004) Reductionism and complexity in molecular biology: Scientists now have the tools to unravel biological complexity and overcome the limitations of reductionism. EMBO Rep 5:1016–1020

Vitiello G, Silvestri B, Luciani G (2018) Learning from nature: bioinspired strategies towards antimicrobial nanostructured systems. Curr Top Med Chem 18:22–41

Wahl DC (2006) Bionics vs. biomimicry: from control of nature to sustainable participation in nature. Des Nat III 87:289–298

Weber T, Kim HU (2016) The secondary metabolite bioinformatics portal: computational tools to facilitate synthetic biology of secondary metabolite production. Synth Syst Biotechnol 1:69–79

Wilson MR, Zha L, Balskus EP (2017) Natural product discovery from the human microbiome. J Biol Chem 292:8546–8552

Wuichet K, Alexander RP, Zhulin IB (2007) Comparative genomic and protein sequence analyses of a complex system controlling bacterial chemotaxis. Methods Enzymol 422:1–31

You J, Preen RJ, Bull L, Greenman J, Ieropoulos I (2017) 3D printed components of microbial fuel cells: towards monolithic microbial fuel cell fabrication using additive layer manufacturing. Sustainable Energy Technol Assess 19:94–101

Zhou AY, Baruch M, Ajo-Franklin CM, Maharbiz MM (2017) A portable bioelectronic sensing system (BESSY) for environmental deployment incorporating differential microbial sensing in miniaturized reactors. PLoS One 12:e0184994

Chapter 9
Microbial Nanobionics: Application of Nanobiosensors in Microbial Growth and Diagnostics

Monica Butnariu and Alina Butu

Contents

M. Butnariu (✉)
Banat's University of Agricultural Sciences and Veterinary Medicine
"*King Michael I of Romania*" from Timisoara, Timis, Romania

A. Butu
National Institute of Research and Development for Biological Sciences, Bucharest, Romania

R. Prasad (ed.), *Microbial Nanobionics*, Nanotechnology in the Life Sciences,
https://doi.org/10.1007/978-3-030-16383-9_9

9.1 Background

In 1975, nanobiosensors (NBSs) take a new connotation when it is suggested that bacteria can be used as a biological element in a microbial electrode to measure alcohol concentration. The article marks the beginnings of research efforts in Japan and later in all countries of NBS's applications in environmental, biotechnology and microbial nanobionics. Microorganisms (the term "microbe") are microscopic plant or animals small organisms, invisible to the naked eye (but which can be seen by optical and electron microscopy), generally single-celled, with relatively simple internal structure.

They live in the soil, in the water, in the air and in the body of the plants or animals and may be saprophytic or pathogenic for the organism in which they develop. Microorganisms include a large and heterogeneous group of organisms with different morphology, biological activity and systematic position: bacteria, archaea, microscopic fungi (molds and yeasts), microalgae (microscopic algae) and protozoa. Viruses and infectious subviral agents (viroids, virins, virusoids and prions) are not considered as microorganisms by many microbiologists. Bacteria are the most numerous and include, in particular, as with other microorganisms, species with red, green, yellow, etc. pigment, which are protective against sunlight. Actinomycetes, though frequent in the soil, are rarely present on the leaves. Microfungi, especially yeasts, have an efficient dispersing mechanism from one leaf to the other, respectively, the ballistospore. Numerous examples demonstrate that saprophytic microorganisms on foliar surfaces can effectively control the development of plant pathogens. This effect can be achieved through competition in their early stages of action, by the synthesis of metabolites, which diminishes their virulence or stimulates the resistance of the host. Some microorganisms synthesize phytoalexins, gibberellins, auxinic substances, etc.

Some endophyte fungi, such as *Acremonium coenophialum*, produce repellents that protect the host plant (*Fescues arundinacea*) by rejecting or limiting aphids' feeding ability or causing digestive disturbances to herbivorous animals. Transmitted seeds of fungus flies find shelter and nutrients in the host plant, and these benefit by removing potential consumers (Safarpour et al. 2012).

The most important perspective for combating crop disease, vegetable plants, vines, etc., is the production of microbial preparations based on bacteria and fungi, a direction that requires isolation and selection studies of microorganisms with antagonistic effect, studies technology, toxicology and tests on bacteria and pathogenic fungi. Bacterioses are diseases caused by phytopathogenic bacteria such as *Xanthomonas campestris*, *Erwinia carotovora*, *P. lacrymans*, *A. tumefaciens*, etc. The microbial nanobionic research on phytosanitary protection of the vine, vegetable plants and other crops is aimed at finding the means of obtaining the maximum harvest in conditions of active and dynamic conservation and potentiation of the exploited natural system. Studies on the integrated and rational combating of diseases and pests of horticultural plants are developed by increasing the share of sanitary, biological and biotechnological phytoprotection for the purpose of eliminating toxic pesticides.

There are various microorganisms used as microbial antagonists for bacteria including *Bacillus subtilis*, *Bacillus cereus*, *Bacillus amyloliquefaciens*, *Bacillus licheniformis*, *Bacillus megaterium*, etc. containing lipoproteins with antibiotic effect. *Pseudomonas fluorescens* and *Pseudomonas syringae* have a broad spectrum of biocontrol activity against several pathogens attacking mono- and dicotyledonous plants. Bioproducts for agricultural use based on microorganisms are an important orientation in microbial nanobionics due to the advantages they present, namely, reducing pollution of the aquatic and terrestrial environment, are well tolerated by living organisms and have no side effects on humans and animals in relation to chemical products (Yang et al. 2015).

Microorganisms are able to produce lipids (microorganisms containing more than 20–25% lipids are named *oleaginous*). Only valuable oils are obtained through biotechnological processes, because in the case of ordinary oils, the cost price is superior to similar products obtained by classical methods. Bacteria accumulate especially waxes and polyesters. Lipid accumulation capacity was associated with lipase citrate activity, enzyme from the ATP cycle involved in citrate acetyl CoA formation, correlated with the equation:

$$\text{citrate} + \text{ATP} + \text{CoA} \rightarrow \text{acetylCoA} + \text{oxaloacetate} + \text{ADP} + \text{Pi}.$$

Citrate lyase activity is a very important indicator of the ability of a microorganism to accumulate lipids, especially since citrate is not present in non-oleaginous yeasts nor in non-oleaginous strains of some species to which normally belong to oleaginous strains such as *Lipomyces starkeyi* and *L. lipofer*. The biochemistry of lipid formation in microorganisms has been demonstrated by studies showing that oil yeasts do not differ from other yeasts in terms of glucose consumption, lipid turnover or acetyl–CoA carboxylase activity – the first enzyme in fatty acid biosynthesis. Differences in regulation of citric acid cycle and acetyl–CoA production have been observed. Unlike other yeasts that are converting excess carbon source into reserve polysaccharides, the oleaginous yeasts metabolize the carbon source by synthesizing lipids (Jin et al. 2016).

Structurally, acetyl–CoA carboxylase is a complex enzyme, and three functional components were identified, namely, biotin carboxyl carrier protein (BCCP), biotin carboxylase and carboxyl transferase.

$$\text{Biotin–BCCP} + \text{HCO}_3^- + \text{H}^+ + \text{ATP} \rightarrow \text{Carboxybiotin–BCCP} + \text{ADP} + \text{P}_\text{i}$$

$$\text{Carboxybiotin–BCCP} + \text{Acetyl–CoA} \rightarrow \text{Biotin–BCCP} + \text{Malonyl–CoA}$$

Thus, the enzyme catalyses the carboxylation of acetyl–CoA to synthesize malonyl–CoA, the two-step reaction being dependent on the presence of ATP. Several mechanisms have been proposed for regulating the activity of the enzyme: activation by intermediates of the tricarboxylic acid cycle, inhibition by acyl–CoA esters with long-chain fatty acids or reversible phosphorylation. Allosteric activation of acetyl–CoA carboxylase by citrate is possible. Another regulatory mechanism has been proposed based on the observation that long-chain acyl–CoA esters inhibit acetyl CoA carboxylase from various sources. However, the inhibitory action of these esters may be unspecific, due to their surfactant properties.

Malonyl–CoA, obtained from the acetyl–CoA carboxylase reaction, is used in the synthesis of long-chain saturated fatty acids through a series of reactions requiring NADPH and acetyl–CoA catalysed by the fatty acid synthase complex (Jin et al. 2016). The structure of the enzyme complex is different depending on the source of origin, but the sequence of reactions is similar for all organisms:

Acetyl transacylation (initiation reaction):

$$\text{MeCOSCoA} + \text{HS–enzyme} \rightarrow \text{MeCOS–enzyme} + \text{CoASH}$$

Malonyl transacylation:

$$\text{HO}_2\text{C–CH}_2\text{COSCoA} + \text{HS–enzyme} \rightarrow \text{HO}_2\text{C–CH}_2\text{COS–enzyme} + \text{CoASH}$$

Condensation with b-ketoacyl synthase:

$$\begin{aligned} &\text{HO}_2\text{C} - \text{CH}_2\text{COS–enzyme} + \text{MeCOS–enzyme} \\ &\rightarrow \text{HS–enzyme} + \text{CO}_2 + \text{MeCOCH}_2\text{S–enzyme} \end{aligned}$$

Reduction of b-ketoacyl:

$$\text{MeCOCH}_2\text{S–enzyme} + \text{NADPH} \rightarrow \text{MeCHOHCH}_2\text{COS–enzyme} + \text{NADP}^+$$

Dehydration of b-hydroxyacyl:

$$\text{MeCHOHCH}_2\text{COS–enzyme} \rightarrow \text{H}_2\text{O} + \text{MeCH} = \text{CH–COS–enzyme}$$

Reduction of 2,3-trans-enoyl acyl

$$\text{MeCH} = \text{CH–COS–enzyme} + \text{NAD}(\text{P})\text{H} \rightarrow \text{MeCH}_2\text{CH}_2\text{COS–enzyme} + \text{NAD}(\text{P})$$

Since the final product becomes a substrate where it reacts with another malonyl–S-enzyme molecule, the concrete result of each reaction cycle is the introduction of two additional carbon atoms into the initial acetyl group. These additional carbon atoms are themselves obtained from acetyl–CoA substrate of acetyl CoA carboxylase (Jin et al. 2016). After seven iterations of the cycle, palmitoyl–CoA is synthesized with the following stoichiometry:

$$\begin{aligned} &\text{MeCOSCoA} + 7\left(\text{HO}_2\text{CCH}_2\text{COSCoA}\right) + 14\text{NADPH} + 14\text{H}^+ \\ &\rightarrow \text{Me}\left(\text{CH}_2\right)_{14}\text{COSCoA} + 7\text{CO}_2 / \text{CoASH} + 14\text{NADP}^+ 6\text{H}_2 \end{aligned}$$

The nature of the final products of the complex is largely determined by the β-ketoacyl synthase substrate specificity.

For example, propionyl–CoA may replace acetyl CoA in the initiation reaction, resulting in the synthesis of uncharged fatty acids. Fatty acid synthases of eukaryotic microorganisms are structurally and functionally organized into indistinguishable complexes.

These type I synthases consist of multiple copies of only two multifunctional polypeptides. The presence of acyl transporter protein has been confirmed in type I synthases by the detection of 4–phosphopantetheine which is the acyl–acyl carrier protein. Citrate cleavage appears to be the rate-determining step in lipid biosynthesis, being subject to feedback inhibition by fatty esters of acetyl CoA.

The transport of citrate outside the mitochondria is also inhibited by these compounds, the two stages (the transport and cleavage of the citrate) being closely linked (Jin et al. 2016).

9.2 Microbial NBSs

Classes of microbial NBSs use the same principle of measuring metabolism activity in the presence of the analyte: using immobilized microorganisms from which the metabolism resulting products are measured has become commonly defined as microbial NBSs; measure the electrical activity of the metabolism of microorganisms when consuming a "biofuel", for example, glucose, having the generic name of bioelectrochemical cells. Being developed as a separate area of electrochemical cells which generate electricity, based on the consumption of biofuels, their ability to be true NBSs has been neglected (Jin et al. 2016).

We will treat these two classes separately by highlighting their performances and capabilities in terms of the requirements of being classified as NBSs. Microbial NBSs contain immobilized microorganisms and a much more diversified transduction chain. Generally, they are used for a single biochemical process.

Those microbial NBSs have the following advantages: lower sensitivity to inhibition or impurity of the analyte; higher tolerance for pH and temperature; longer lifetime than enzymatic electrodes; cheaper than enzymatic electrodes; high variability because they adapt easily to specific environmental conditions; cofactor independence; physiological response to toxic products; and easy preparation because the cultivation of microorganisms is simple. Among the disadvantages of these devices are as follows: they have a longer response time than enzymatic electrodes; and reusing them in a new measurement takes longer time (Jin et al. 2016).

The pattern of microbial NBSs is identical to that of an enzyme NBSs. The scheme consists in immobilizing intact cells in intimate contact with a specific transducer that converts the products or the effects of metabolism into a biochemical signal. Microbial NBSs are practically applied in biochemical and microbiological industrial processes in areas such as drug production, food industry, waste water treatment and energy production, where fermentation and electron transfer reactions play an important role. In the field of clinical laboratories – microbiology and parasitology – where culture media are used intensively, there is an important potential for implementation of microbial NBSs (Schenkmayerová et al. 2015).

Most materials in culture media cannot be determined by spectrophotometric methods being optically inactive or nontransparent. A combination of cultivation media–immobilization substrates with a suitable coupling to a transducer is feasible with advantages over reducing investigation times and responding to a specific diagnosis. The continuous, rapid, sensitive monitoring and control of the variables already mentioned allow the evaluation of substrates and metabolites, the number of viable cells present in cultures and the types of microorganisms (Park et al. 2013).

9.2.1 Functioning Principle of Electrochemical Microbial NBSs

Microbial NBSs contain microorganisms immobilized in a membrane and an electrochemical device. Their classification is based on the type of respiratory activity measured or on the type of electrochemical, metabolic, optical activity. When the respiratory activity of microorganisms immobilized in a substrate is modified, it is detected by an oxygen electrode. By measuring respiratory activities or oxygen consumption, substrate concentration is estimated. Other electrochemical methods are used for measurements of metabolic products. Electrochemical microbial NBSs show significant differences from enzymatic NBSs, being composed of living components that produce a physiological response, while enzymatic NBSs include alteration of the analyte. The microorganisms used in this case are aerobic. NBSs are introduced into an oxygen-saturated buffer. After addition to the substrate, the respiratory activity of the microorganisms increases, producing a decrease in oxygen concentration near the membrane. By using the oxygen electrode, the substrate concentration can be measured by detecting the decrease in oxygen concentration (Jin et al. 2016).

Aerobic and anaerobic microorganisms may be used. Most microbial NBSs are based on respiratory activity measurement and use the activity of aerobic microorganisms. The transducer of this type of NBSs can be electrochemical (potentiometric, amperometric), photodetector, thermistor or ISFET. It is based on the optical phenomena of microorganisms in metabolic processes: photoluminescence, electroluminescence, chemiluminescence and specific optical activity (chirality-dependent polarization, refractive index, absorbance, etc.). The intensity of luminescence of bacteria (photobacteria) is dependent on metabolic activity. It is possible to build highly sensitive microbial NBSs by combining these photobacteria with a photodetector. Luminescence is strongly affected by changes in external conditions for bacteria, which can lead to accidental accumulation of intracellular concentrations of NAD(P)H, FMN, H_2, ATP and aldehydes (enzyme synthesis of aldehydes requires ATP cofactor). Using this principle, the food of these microorganisms (glucose and amino acids) and their inhibitors (toxins and heavy metals) can be detected. The intensity of luminescence is a more sensitive parameter for metabolic activity than for heat generation. Moreover, the metabolic response of cells can cause luminescence in the bacteria (Zhang et al. 2015).

9.2.1.1 Microbial Thermistor

Placement of immobilized microorganisms is immediately near a thermistor that measures the heat absorbed or released by microorganisms in metabolic processes. The microbial thermistor has the following characteristics: it is based on the general principle of enthalpy exchange; many metabolic reactions are accompanied by a considerable evolution of heat–calorimetry (enthalpy of metabolic reactions, specific heat and latent). Although many cases of calorimetry applications have been reported in biochemical analyses, the principles of calorimetry cannot be applied due to the high cost of sensitive and complex instruments. An example of a microbial thermistor is the beer yeast gel which is inserted into the glass separator column constituting the thermistor outlet.

The solution to be analysed is pumped through a heat exchanger located in the thermostatically controlled water bath and then through the glass column. This method can measure glucose, fructose and casein. A microbial microthermistor consists of assembling all components directly on the semiconductor surface of the oxide. Microbial NBSs can be constructed by placing immobilized microorganisms in the immediate vicinity of a thermistor that measures the absorbed or ceased metabolic heat produced by them. This type of NBSs is based on the general principle of measuring enthalpy exchange. Since many metabolic reactions are accompanied by a considerable evolution of heat, calorimetry can be applied to measure a wide variety of analytes. The lack of specificity due to the principle of detection in general is adequately compensated by the use of specific, immobilized biocatalysts. The greater part of the heat released in the metabolic reactions can be lost in solution without being detected by the thermistor and limiting the sensitivity of this technique.

The thermistors and the transducer element are the most used, and they are made of semiconductor materials (most of them are oxide materials and have a negative temperature coefficient). Although many cases of calorimetry applications have been reported in biochemical analyses, it cannot be used as a routine analysis due to the high cost of the sensitive and complex instruments to be used for this type of measurement (Jin et al. 2016).

As an example, the beer yeast gel is introduced into the glass separator column constituting the thermistor outlet. The solution to be analysed is pumped through a heat exchanger located in the thermostatically controlled water bath and then through the glass column. By this procedure, glucose, fructose and casein can be measured. Recently microminiaturization and introduction of microfluidics has led to a new generation of microbial thermistors that can sense thermal effects from a relatively small number of microorganisms. A microbial microterminal consists of the assembly of all components directly on the semiconductor surface of the respective oxide (Zhang and Keasling 2011).

9.2.2 The Physiological Response of Microbial NBSs

It includes a number of factors: the substrate is absorbed through the cell membrane, breathing, luminescence and electrochemistry of bacterial metabolism, secretion or separation of metabolic and non-metabolic products and intracellular alterations or degradations of the substrate by metabolic sequences or enzymes not involved in the process. The behaviour of microorganisms is due to their physiological state, and this is due to the extreme limitation of nutrients. Metabolism is a condition that ensures the survival of cells.

The first critical condition for the formation of a microbial NBSs signal is the use of the substrate. The solute can enter cells through the specific translocation system, first through active transport or forced diffusion. Passive transport achieved only through diffusion due to oxygen and nutrient concentration gradients is of little importance. The active transport ensures the accumulation of nutrients in the substrate directly correlated with the concentration gradient. This makes the substrate accumulate proteins of high specificity that are consumed in energy metabolism.

The coupling of cells to the energy transduction system, especially in the respiratory chain, is an important aspect of active transport, being crucial for signal formation in microbial NBSs, the use of glucose, sucrose and other oligopeptides. After its use, the substrate is specifically degraded by the enzymatic sequences which are immobilizing the cells. According to aerobic conditions, oxygen is consumed.

Organic acids such as lactic acid, carbon dioxide, ammonium ions and hydrogen sulphide are secreted and accumulate in cellular metabolism products, highlighting the importance of Fe in bacteria biology, and especially for Gram-positive bacteria. Effective multiplication of pathogenic bacteria in the tissues of the host plant is conditioned by the presence of fertile Fe in a form accessible to their metabolism. Fe is necessary for certain enzymatic reactions essential to growth but is also a key component of the electron transport chain in the cell membrane–cellular respiration (D'Souza 2001).

Microorganisms cultivated for the preparation of microbial NBSs must come from cultures in metabolically balanced environments with respect to ferric Fe to ensure good cellular respiration of the microorganisms and good functioning of the Clark electrode.

Other types of electrodes can also be used in microbial NBSs, combined with electrodes for H_2, CO_2 and NH_3. At constant temperature and constant pressure, the proportion of oxygen in the air in volume is 20.9%. In aqueous solution and in equilibrium with atmospheric air, the proportion of oxygen is 20.9% of the total dissolved gas. In both cases, in air and in solution, the nitrogen balance is higher than the carbon dioxide balance.

The concentration of dissolved gases in the solution varies with temperature and may be affected by the solute or other solvents present in the solution. For example, a warm solution retains less oxygen than a cold solution. The presence of ethanol significantly increases the ability of an aqueous solution to retain oxygen.

Specialist studies show that small temperature differences can affect the balance of the environment with room temperature, as well as some reagents that we add to the working cell (e.g. ethanol). The selection of the transduction device is given by the physiology of the cells that we use in the NBSs structure (breathing, photoluminescence and gravimetry). The amperometric/potentiometric type, Clark electrode and selective ion electrodes of the ISFET type are also used. It is also used the piezoelectric crystal microbalance, optoelectronic detector, thermistors or ISFET, but the oxygen electrode predominates in microbial NBSs (Schenkmayerová et al. 2013).

Metabolic products such as lactate or pyruvate, carbon dioxide, ammonium ions and hydrogen sulphide are potentiometric determined with selective ion electrodes (ISE). The direct combination of microorganisms with a field effect transistor (FET) has been developed to estimate glucose of *Acetobacter* alcohol and xylose based on *Gluconobacter oxydans* cells.

It has been found that a series of soluble redox mediators, such as phenazine methyl sulphate, ferricyanide or ferricyanide in combination with benzoquinone, have the possibility of direct measurement of electrons following the metabolic activity of cells. Insoluble mediators such as ferrocenes, tetrathiafulvalene and tetracyanochinodimetan have been used which have been incorporated into carbon pastes in contact with *Paracoccus denitrificans*.

Another microbial NBSs technique was created by the luminescence of the bacteria connected with an optical detector. Thus, microbial NBSs was developed and described for the determination of metal ions and aromatic ions. Genetically engineered, the genes responsive to the emission of light from the *Vibrio* bacteria have been transferred to the genetic structure of *Escherichia coli* or *Serratia marcescens* (Jin et al. 2016).

The future of transducers in microbial NBSs belongs to the piezoelectric crystals, quartz microbalances with one of the functionalized electrodes. By combining enzyme microorganisms, it is possible to increase the selectivity of microbial NBSs (polymers such as starch, proteins and lipids). For this, microorganisms were combined with hydrolases (Cheng et al. 2011).

It resulted in an amperometric NBSs for the determination of NAD^+, based on *E. coli* cells and NAD-ase. Urea and creatinine were determined using a combination of nitrifying bacteria and urease or creatinine.

From experiments with NBS BOD, it was concluded that several species of microorganisms can be combined and multiple substrates can be used in the same experiment. Another typical example of NBSs containing mixed populations of different species of microorganisms having specific metabolic capacities is NBSs with nitrifying bacteria. These NBSs which were developed especially for wastewater investigation contained mixed crops of *Nitrosomonas* sp. and *Nitrobacter* sp. (Jin et al. 2016).

An amperometric measuring system for the determination of ammonia was used. Oxygen control is required to measure the ammonia concentration in the sample to be measured. Because NBSs react with other nitrites or urea, the amounts of nitrifying substances are added together. The impediment is the surplus of substrate (ammonia or urea).

This process, the combination of different species of microorganisms, is also used in toxicology. An inhibitor can be used when the substrate concentration is too high. One possibility is the use of dialysis membranes when the substrate is in excess (Su et al. 2011).

9.3 Electrochemical–Microbial NBSs: Gas Detector

This type of NBSs is easy to manufacture and use, because obtaining an electrochemical signal is the most common way of obtaining the response signal. Potentiometric devices measure the charge density accumulated at the surface of an electrode. An example of NBSs using cyclic voltammetry is gas NBSs. NBS CO_2 uses autotrophic bacteria. It is the most widespread from commercial point of view but has the disadvantage that various ionic acids and volatile organic or inorganic acids affect the potential within the cell, the pH and the permeable gas membrane covering the electrode (Jin et al. 2016).

The autotrophic bacterial strain, *Pseudomonas* S–17, which can only grow in the presence of carbonates, is the source of carbonate production. It is incubated under aerobic conditions at 30 °C for 1 month and immobilized on the tip of the oxygen electrode.

To increase the selectivity range, the cell is covered with a semipermeable PTFE membrane. Table 9.1 lists several practical applications of the microbial NBSs.

The sensitivity of the measurements is obtained using an oxygen-saturated buffer solution of pH 6.5 containing metal ions and 20 µmol of glucose. NBS's lifetime is more than a month. Being microbial NBSs, it is used in food testing (Cho et al. 2014).

Table 9.1 Microbial NBSs applications

NBSs	Microorganism immobilized	Device	Minimum response time	Measuring range $mg \cdot dm^{-3}$
Sugar assimilated	*Brevibacterium lactofermentum*	Electrode O_2	10	10–200
Glucose	*Pseudomonas fluorescens*	Electrode O_2	10	2–2510
Acetic acid	*Trichosporon brassicae*	Electrode O_2	10	3–60
Ethanol	*Trichosporon brassicae*	Electrode O_2	10	3–25
Methanol	*Unidentified bacteria*	Electrode O_2	10	5–2510
Formic acid	*Citrobacter freunca*	Combustion cell	30	$10–10^3$
Methane	*Methylomonas flagellata*	Electrode O_2	2	0–6.6[a]
Glutamic acid	*Escherichia coli*	Electrode O_2	5	8–800
Cephalosporin	*Citrobacter freundii*	Electrode pH	10	$100–55 \cdot 10^2$
BOD	*Trichosporon cutaneum*	Electrode O_2	15	3–60
Lysine	*Escherichia coli*	Electrode O_2	5	$10–10^7$
Ammonium	*Nitrifying bacteria*	Electrode O_2	10	0.05–1
Nitrogen dioxide	*Nitrifying bacteria*	Electrode O_2	3	0.51–255[b]
Nystatin	*Saccharomyces cerevisiae*	Electrode O_2	1 h	0.5–0.54[c]
Acid nicotinic	*Lactobacillus arabiensis*	Electrode pH	1 h	$10^{-5}–5$
Vitamin B1	*Lactobacillus fermentum*	Combustion cell	6 h	$10^3–10^7$
Community microorganisms	–	Combustion cell	15	$10^8–10^9$[d]
Mutagens	*Bacillus subtilis*	Electrode O_2	1 h	$1.6–2.85 \cdot 10^3$

[a]mmol; [b]ppm; [c]unit/cm^3; [d]no/cm^3

9.4 Biofuel Cells

A major challenge for device development is finding the right power sources. They must be capable of generating electricity for long periods of time. Biofuel cells (BFCs) promise much in this regard by their functioning based on the pair reactions of glucose oxidation–reduction of molecular oxygen to water.

Under ideal conditions, the by-products of BFC would be carbon dioxide and water. Glucose and oxygen are present in the cells and tissues of all eukaryotic organisms, including humans. It is possible to introduce them between plant resources, even the metabolic properties of cells to generate enough energy to feed clinical devices such as biocide transport systems (Virolainen and Karp 2014).

9.4.1 Microbial Biofuel Cells (MFC)

The use of microorganisms as microreactors in combustion cells eliminates the need for isolation of individual enzymes and allows active biomaterials to work in conditions close to their natural environment, resulting in high efficiency. Microorganisms are difficult to handle or require special conditions to be active, and their electrochemical contact with electrodes is virtually impossible. In MFC without mediators, it is possible to functionalise the anode with biocompatible materials with microorganisms in the anode compartment. In this case, microorganisms will form biofilms on the surface of the anode, and electron transfer is carried out through the cell membrane interface – the surface of the anode.

The MFC cell consists of two compartments an anode and a cathode separated by a proton exchange membrane (PEM) of perfluorosulphonic polymer (Nafion, DuPont). The electrodes are made of carbon paper that can be functionalized with different biocompatible materials. Cathodic compartment contains a neutral pH buffer. An anodic compartment was prepared with a glucose solution of 200 g/L in which various microorganisms were introduced. A 1 K ohm load was maintained throughout the experiments. Microorganisms (*E. coli*, *Klebsiella* and *S. aureus*) respond differently to what makes MFC a biosensor to identify and analyse their behaviour. The bacteria present in the household water were used in the MFC. As it was demonstrated, bacteria from household water prove to be suitable biocatalysts for electricity generation. Domestic water had a pH between 7.3 and 7.6, and a chemical oxygen content (COD) of 200–300 mg/L also has been used as glucose medium (170–1200 mg/L) with the following content (per litter): NH_4Cl, 310 mg; KCl, 130 mg; NaH_2PO_4, H_2O, 4.97 g; Na_2HPO_4, H_2O, 2.75 g; minerals, 12.5 mL; and vitamins, 12.5 mL (Cho et al. 2014).

Microorganisms have the ability to produce electrochemical active substances, substances that can be metabolic intermediates or end products of anaerobic respiration. For the purpose of generating energy, these fuel substances can be produced in one place and transported to biofuel cells to be used as fuel. In this case, the biocatalytic microbial reactor produces biofuel, and the biological part of the apparatus does not have direct contact with the electrochemical part. This scheme allows the electrochemical part to operate under conditions that are not compatible with the biological part of the apparatus. These two parts can even be separated in time, functioning completely individually. The most used fuel in this scheme is hydrogen gas, which allows the good development and efficiency of the H_2/O_2 combustion cells in operation alongside the bioreactors. With a microbial bioreactor, fuel is delivered directly into the fuel cell anode compartment. According to another approach, the microbiological fermentation process proceeds directly into the anode compartment of a combustion cell, feeding the anode with generated fermentation products in situ. In this case, the operating conditions in the anode compartment are dictated by the biological system, so there is a significant difference between them and conventional combustion cells.

There is a difference between a true biofuel cell and a simple combination of a bioreactor with a conventional combustion cell. This latter configuration is also frequently based on the biological production of hydrogen, but the electrochemical oxidation of H_2 occurs in the presence of mild bio-compounds. Other metabolic products (alcohols, H_2S) are also used in this type of system (Goswami et al. 2013).

9.4.2 Bioelectrochemical Cells Based on Enzymatic Electrodes

An additional methodology for the development of bioelectrochemical cells involves the application of redox enzymes for the oxidation and reduction of specific substrates of fuels and oxidants to electrodes and the generation of electrical energy. Consequently, the design of integrated enzyme electrodes that increase electrical contact is essential.

The detailed characterization of the electronic transfer rates at the interface, biocatalytic speeds and internal electrical resistances is essential for the construction of biofilm cells. The chemical modification of redox enzymes with synthetic units that improve electrical contacts with electrodes provides general means for increasing the electrical energy of biocombustion cells. The modification of the specific site of redox enzymes and the functionalization of electrodes are new and attractive means. Effective electrical connection of proteins with electrodes suggests that future efforts may be directed to the development of structural mutants of redox proteins. Nanoengineering of electrodes surfaces with biocatalytic unit cofactors of organic synthesis transfer allows the control of electronic transfer in cascades. Adjusting potential redox leads to increased power generation of biocombustion cells. The biocombustion cell configurations discussed above can theoretically be extended to other redox enzymes and fuel substrates, allowing numerous technological applications. An important potential of biocombustion cells is their use in assemblies and locations of human body fluids, e.g. raw and elaborated sap. The electrical power obtained can be used to power the implanted devices such as pacemakers, pumps, NBSs and prostheses (Choi et al. 2013).

9.4.2.1 MFC Electrochemistry

Similar to any galvanic cell in which chemical energy generation reactions occur, i.e. heat, MFC is subject to the same laws of thermodynamics, but the mechanisms are a bit different. For simplicity, we will limit ourselves to a thermodynamic analysis of the cell taking into account reversible reactions. The electrical power of the cell has two components corresponding to the two operating modes.

The first regime refers to the generation of electricity by the load accumulation until a maximum voltage is reached. It is assumed that this is constant for the second working regime by connecting it to a resistive consumer (Cho et al. 2014).

Theoretically in MFC, the yield can be evaluated as the mechanical work consumed to carry loads in the outer circuit to the potential difference measured relative to the maximum mechanical work performed by the cell due to the reactions from the cathode and the anode.

The measurement of electrode potential for the determination of potentials for anode, EA, and cathode, EC, is related to a reference electrode located in the vicinity. Consequently, the experimental values do not coincide. In practice, for an ideal cell with a couple of reversible reactions, it is considered a measured electric voltage and a standard equilibrium electric tension. Excessive free energy in the system induced by the addition of oxidants and fuel relative to equilibrium values will result in a higher overpotential of Ee. To maximize cell performance and achieve optimal voltage, it is necessary to maximize and reduce internal resistance values. The combustion cell density is related to the area of the electrodes or to the equivalent area of the catalysts (Alonso-Lomillo et al. 2010).

9.5 NBSs in Microbial Nanobionics

The field is relatively new and has evolved with the increasing need to find new materials with the greatest biocompatibility to reduce the intolerance of living organisms.

Primary metabolites are compounds linked by the synthesis of cellular components that occur during the growth phase (trophophase) of microorganisms. This group of metabolites includes amino acids, nucleotides and certain end products of metabolism such as ethanol and organic acids. During the trophophase, various enzymes are synthesized, especially exoenzymes that are of practical importance. Secondary metabolites accumulate during the phase following the active growth stage, called idiophase (Cho et al. 2014).

The compounds synthesized in this phase are not directly related to essential cellular materials and normal growth. Most antibiotics and mycotoxins are produced during this phase. Cultivated under ideal conditions without environmental limitations, microorganisms tend to form large amounts of biomass and accumulate less certain compounds.

By inducing mutations, specialists have developed methods of "trickling" microorganisms of interest so that they can produce excessively useful compound. Cultivation of microorganisms and their identification by traditional methods remain "golden standard" in the diagnosis of infectious diseases; however, in many cases, serological diagnosis allows rapid orientation. Serological diagnosis involves detecting antigens directly from the biological product, and qualitative or quantitative detection of antibodies present in response to a microbial infection. If it is discovered early, the diseases are easy to control. Unlike insect pests, pathogens are difficult to see, as well as cultivation issues, nutritional imbalances or air pollution, as the conditions that cause them are not visible (Chen and Park 2016).

9.5.1 Biodegradation of Pollutants Under the Action of Microorganisms in Microbial Nanobionics

Large-scale production, chemical processing and use have resulted in severe soil and subsoil contamination with a wide range of hazardous and toxic hydrocarbons. Such hydrocarbons, synthesized in large quantities, are polychlorinated biphenyls (PCBs), trichloroethylene (TCE) and others, which differ greatly from organic compounds by chemical structure, and are xenobiotic substances because they cannot readily biodegrade; polycyclic aromatic hydrocarbons (PAHs), which are toxic and because of the high molecular mass (with four or more aromatic rings in the structure) are non-biodegradable or hardly biodegradable. Polycyclic aromatic hydrocarbons (PAHs), produced from incomplete combustion of natural organic materials and hydrocarbons, appear in the soil as a result of natural forest fires. The intensification of energy-producing industrial processes and the inevitable production of residues and by-products such as PAH have led to soil contamination. Research studies identified numerous species of microorganisms that can degrade TCE, PCB, PAH and trichloroethylenes (TCE) (Sun et al. 2015).

Bacterial strains: *Trichloroethylene* (TCE): *Desulfitobacterium hafniense, Burkholderia cepacia, Burkholderia kururiensis, Desulfuromonas chloroethenica, Janibacter terrae. Pseudomonas putida, Rhodococcus ruber, Wautersia numazuensis.* Biphenyl: *Burkholderia xenovorans, Pseudomonas pseudoalcaligenes, Cupriavidus necator, Rhodococcus opacus, Rhodococcus ruber, Novosphingobium aromaticivorans, Sphingomonas* sp., *Novosphingobium stygium, Novosphingobium subterraneum, Sphingobium yanoikuyae, Thamnostylum piriforme.* Biphenyl: *Yarrowia lipolytica.* PAH: *Mycobacterium vanbaalenii*

Fungal strains: PAH: *Coriolopsis floccosa, Dichomitus squalens, Fomitopsis spraguei, Ganoderma lucidum, Irpex lacteus, Lentinus crinitus, Lentinus* sp., *Oligoporus* sp., *Phanerochaete chrysosporium, Phellinus gilvus, Pleurotus eryngii, Stropharia rugosoannulata, Trametes villosa, Trichaptum byssogenus.* The inability of microorganisms to mineralize specific contaminants and their ability to partially transform them are the evidence that these organisms require other substrates for growth. In such situations, contaminants are transformed, for example, into "cometabolism".

The wide range of structures for PAHs molecules requires that the degrading microorganisms possess an enzyme capable of accepting PAHs as a substrate or possess a small number of enzymes specific to substrates with PAH. Over time, in a particular bioremediation system, some of the PAHs may not convert at all or can only be partially catabolized to final products.

Cometabole aerobic and anaerobic processes participate in the degradation of halogenated organic compounds. Although the mechanisms of aerobic degradation of synthetic nitrogen-based chemical contaminants have not been fully elucidated, many researches are being carried out to determine the degradation mechanisms involved. Types of participant catalytic reactions include deamination, nitroreducers, N-desalkylating, deesterification, decolourization and hydrolysis. In the first stages of catabolism, monooxigenases and dioxigenases, nitroreductases and esterases are highly involved. Microorganisms have developed a variety of biochemical pathways to degrade or detoxify hydrocarbons. Hydrolase and oxygenase are the most important classes of enzymes, which are responsible for catalysing biotransformation reactions (Cho et al. 2014).

Hydrolase (hydrolase, esterase and amidase) requires no factors and is stable at a high pH and temperature variations. Microbial communities have a role in biogeochemical cycles; it is essential to analyse the structure of microbial communities and the changes that occur during the bioremediation process. Research is needed to characterize the role of hydrocarbon-metabolizing organisms in the degradation of petroleum substrates present in oil-contaminated soils. In the characterization of microbial communities, we use both cultural and independent methods of cultivation. Time and space changes in microbial communities during bioremediation can be determined using complex molecular methods.

Recent discoveries in molecular techniques, combined with genomic information, help microbiologists to learn the mysteries of the various roles that microorganisms play in communities. Genes for catabolism have the ability to spread with great frequency within microbial communities (Jin et al. 2016).

Microorganisms capable of degrading xenobiotic substances are present in the polluted environments, but natural biodegradation occurs at low rates. Therefore, various bioremediation technologies have been developed involving: knowing the ways to optimize the biodegradation conditions; knowledge of the behaviour and effects of chemicals introduced into the soil against the ecosystem; selection of microorganisms with superior degrading abilities; research on the identification and characterization of bacteria and fungi that proliferate in soils polluted with oil and salt water; were identified major bacteria that contribute to the degradation of oil (*Pseudomonas, Flavobacterium*, *Corynebacterium*, etc.); were elucidated the conditions under which these bacteria proliferate or disappear and how they can be stimulated; studies were carried out on pioneer plants to be installed on polluted lands that can be used in the remediation process; a study was carried out to characterize the quantitative and qualitative distribution of heterotrophic bacteria and filamentous fungi in soils affected by pollution from the oil extraction fields; in the soil, although they represent a small fraction of total weight (0.35%), microorganisms are important, being the active element, soil fertility effector; the activity of microorganisms in the soil is influenced by the presence of microhabitats, soil type, structure and texture, organic matter and nutrients, environmental factors, etc. (Cho et al. 2014).

Depending on the factors that act, the activities of the microorganisms such as metabolic and nutritional activity are biologically variable (the alternation of the vegetative phases with the latency phases). Soil microorganisms have the ability to use gaseous, solid and liquid hydrocarbons in the aliphatic and aromatic series as the sole source of carbon and energy by decomposing them to lower molecular weight compounds or carbon dioxide and water (Kumar and D'Souza 2010).

Widespread in natural and numerically significant in the environment, active microorganisms attack various compounds such as petroleum, kerosene, mineral oils, paraffin, light gas, natural and synthetic rubber, cooling oils, asphalted surfaces, underground pipes and electrical cables protected by corrosion by means of paraffin-impregnated materials, elastomers or different hydrocarbon derivatives. The first observations on this process of hydrocarbons degrading date back to 1895, when it was observed that thin layers of paraffin (considered biologically inert) were penetrated by the hyphae of *Botrytis cinerea*. Subsequently, several soil microfungi, including *Penicillium glaucum*, have been shown to attack paraffin, decomposing it and using it as the sole source of carbon and energy.

The importance of the phenomenon in nature has been signalled on the basis of the high frequency of soil active microorganisms and the inability to accumulate hydrocarbons synthesized by plants or waxes produced by insects. The ability to degrade hydrocarbons is widespread in the world of microorganisms, being encountered in bacteria (including actinomycetes), yeasts, filamentous fungi and algae (Cho et al. 2014).

Present in soil, freshwater and seawater and some sediments, in a range of conditions, these microorganisms have the ability to synthesize an enzyme spectrum that ensures the degradation of individual hydrocarbons and the potential for removing or converting oil from the environment. While in non-polluted ecosystems the number of microorganisms using hydrocarbons can account for only 0.1% of the total, the polluted can reach up to 100% of the number of viable microorganisms. This is due to the fact that in the highly polluted environment the composition of microbiota is modified being removing those that are sensitive to the pollutant (Jin et al. 2016).

More than 200 species of microorganisms capable of metabolizing hydrocarbons were identified, and these microorganisms were named hydrocarbonoclastic species. As a result of the microbiological analysis of samples of soils polluted with oil in some excess concentrations (>30%), bacterial strains with a survival rate were isolated under these severe pollution conditions. Bacterial strains were multiplied, isolated and purified by successive decimal dilutions of soil on solidified nutrient media (Schneider et al. 2016).

9.5.2 Estimating the Degree of Pollution and the Efficiency of Remediation Technologies with NBSs

The first stage of any polluting soil bioremediation process is to assess the level of pollution as accurate as possible: to determine the nature of the pollutants and their quantity in order to develop the most appropriate remediation biotechnologies.

If determinations of the nature and concentrations of pollutants are the responsibility of the relevant laboratories, an assessment of the effect pollutants have on the biological potential of affected soils, as defined by the intensity of microbial and enzymatic activity, is made within the microbiology laboratory. This potential is appreciated by means of quality bacterial and enzymatic indicators (microbial NBSs). Microbiology and enzymology research is confronted with the difficulty of comparing the microbial and enzymatic potential of habitats, starting only from individual parameters (a group of bacteria, an enzymatic activity).

To overcome these difficulties, two original formulas have been developed for the calculation of quality synthetic NBSs, which have proved to be of practical importance: a bacterial soil quality indicator and an enzymatic indicator of soil quality. Finding synthetic NBSs of the biological quality of some habitats is a constant of the concerns of researchers in various fields. Knowing the microbial potential of habitats is of great interest, since the physiological and implicitly enzymatic activity of microorganisms depends on the speed at which the organic matter is reintroduced in the biogeochemical circuit, as well as the rate of removal of any pollutant from the soil. However, the relevance of individual parameters (a group of bacteria or an enzyme) is limited, because different ecophysiological groups coexist in the same habitat, heterotrophic bacteria, nitrifying chemotrophs, denitrifying agents, sulphating agents, desulphorizing agents, ferrooxidants, Fe-reducing, etc., with the number of which varies by order of magnitude (Nigam and Shukla 2015).

Since the number of bacteria belonging to different physiological groups differs with several orders of magnitude from one group to another, comparing the results based on a single synthetic value in the form of an indicator is very difficult. A formula has been developed, based on the premise that the logarithmic function offers this possibility of bringing comparable very different values between them (Jin et al. 2016).

Enzymatic activity is diverse as the intensity of each activity. In order to assess the enzymatic potential of the soil, a formula has been formulated which takes into account the individual actual value of each studied activity, finally obtaining a single number, which can take values between 0 (when the actual existence of any of the studied activities is not recorded) and 1 (when all activities have the values equal to the theoretical maximum values). The formula provides an equal weight of each activity, thereby achieving a dynamic balance between the activities taken into account.

The advantage of the formula is that it makes it possible to compare the potential enzymes of different habitats established by different researchers under identical reaction conditions. The results obtained through the use of these two quality NBSs in various research demonstrate the utility of NBSs, including the determination of the effect of various pollutants on natural habitats. Synthetic NBSs are useful for assessing the effectiveness of some biotechnologies for the remediation of habitats affected by human activity as well as for estimating the overall biological potential of other habitats (Cho et al. 2014).

The overall biological potential of the studied soils and plants will be assessed at each stage based on bacterial and enzymatic quality NBSs.

The presence of a pollutant in the soil is evidenced by the variation in the number of bacteria belonging to different ecophysiological groups also by the variation in activity intensity of different enzymes. The literature shows that the increase in the concentration of a pollutant is followed by a decrease in the microbial and enzymatic potential, just as the decrease in the concentration of the pollutant is followed by the increase in the number of bacteria and the intensification of the enzymatic activity. All these are synthesized in the calculated values of enzyme and microbial NBSs soil quality (Tepper and Shlomi 2011).

9.5.3 Ecotoxicity Tests for NBSs

Estimation of soil pollution and efficiency of remediation biotechnologies can be done by performing ecotoxicity tests: growth inhibition test of *Pseudomonas putida* (SR EN ISO: 2001) and determination of the water inhibitory effect on luminescence at *Vibrio fischeri* (SR EN ISO 11348 (1–4): 2003). Direct reporting is made to European standards aimed at maintaining the environment at controllable levels. Bacteria can decompose, directly or indirectly (cometabolically), all-natural organic substances and many synthetic products, including toxic substances, for the other components of ecosystems.

The degradative capacity of heterotrophic bacteria, called "microbial infallibility", is questionable due to xenobiotic chemicals resistant to microbial degradation. Degradation of organic substances can be done directly, metabolically or cometabolically. In metabolic degradation, decomposed substances serve as the source of nutrients and electron donors for the very bacteria that make it (Espinosa-Urgel et al. 2015).

In cometabolic degradation, some substances resulting from the degrading action of particular bacteria are not used for nutrition by the bacteria but can be degraded by other bacteria or under the influence of metabolic products of any other bacteria (organic acids).

In the context of NBSs, it is of interest to find, isolate and select bacterial strains from the polluted habitats, thus increasing the chance of identifying strains already adapted to the pollution conditions, so finding more efficient microorganisms in the soil pollutant removal action. The selected strains will be subjected to laboratory treatments to increase their efficiency in the depollution process and then be reintroduced into the affected habitats or experimental models imagined for soil contamination (Cho et al. 2014).

Bacteria belonging to the genus *Pseudomonas* (*Gammaproteobacteria*, order IX *Pseudomonadales*, family I *Pseudomonadaceae*) have chemoorganoheterotrophic nutrition; are polar flagellates of bacilli, with strict aerobic respiration, having molecular oxygen as the ultimate acceptor of electrons in the respiratory chain; and do not grow to pH below 4.5.

They produce siderophores, chelating compounds that bind and immobilize Fe, but also other metal ions: Al, Cr, Zn, Cu, Mn, Pb, Cd, etc. The main siderophore produced by *Pseudomonas putida* as well as other species of the same genus (*P. aeruginosa*, *P. fluorescens*, *P. chlororaphis*), useful in the processes of decontamination of polluted soils, is *pyoverdine*.

The action of siderophores is of ecological importance given the poor accessibility of Fe^{3+} ion for living organisms due to the insolubility of its compounds. Siderophores bind Fe^{3+} compounds to form complexes that are transported inside cells where they can be used for respiration by anaerobic microorganisms such as Fe-reducing bacteria (*G. bremensis*, *G. pelophilus*, *G. sulfurreducens*). Some *Poaceae* (wheat, barley) produce a class of compounds, the so-called phytosiderophores, with action similar to siderophores, produced by bacteria.

Another way of enhancing the solubility of Fe^{3+} compounds in soil or other habitats is acidification of the environment, a process involving also the microbiota of the habitat through organic acids resulting from autolysis, respectively, the excretions of plant roots.

Pseudomonas genus members are recognized as having a large spreading and highest aerobic degradation capacity of a large number of hydrocarbons, aromatic compounds and derivatives, of which many natural compounds and end products or intermediates of industrial activity. From the variety of compounds that various strains of *Pseudomonas* can degrade metabolically, there are benzoate, phthalate, salicylate, polycyclic compounds, toluene, camphor, xylene, p-cresol, phenylacetate, naphthalene, 1,2,4-trimethylbenzene, nicotine and 3-chlorobenzoate. Genes responsible for enzyme synthesis involved in the degradation of these compounds with environmental impact are placed on plasmids, of which the most extensively studied is the plasmid TOL (also called pWWO) and its derivatives, originating in a *P. putida* strain (a plasmid having the genes required for the decomposition of toluene, m- and p-xylene) (Espinosa-Urgel et al. 2015).

Other plasmids carry genes responsible for the degradation of other high-impact compounds: CAM (camphor), OCT (octane) and NAH (naphthalene). Other important properties of the *Pseudomonas* species determined by plasmid genes: resistance to antibiotics (chloramphenicol, tetracycline, streptomycin, tobramycin, gentamicin, carbenicillin, etc.), to bacteriophages and to bacteriocins and especially resistance to various physicochemical agents that are harmful to other microorganisms, including UV radiation, borates and chromates, as well as many metal ions (Mulchandani and Rajesh 2011).

These properties are useful to the researcher in attempting to isolate the respective strains from their natural environment: the presence in the culture medium of the antibiotic at which they exhibit resistance ensures the elimination of other bacterial strains susceptible to the antibiotic. *Pseudomonas* species also exhibit some resistance to the action of metals, which for other bacterial species are toxic. In *P. aeruginosa*, as in *P. fluorescence*, Cu^{2+} resistance is encoded by chromosomal genes. At a strain of the *P. putida*, the Cu^{2+} ion may be accumulated in concentration by 6.5% of the dry weight of the bacteria. The storage capacity was higher in precultivated crops under conditions where the SO_4^{2-} ion was limiting.

In *P. stutzeri* the resistance to high silver concentrations may be due to the formation of silver sulphide complexes because no formation of any complex with polyphosphates or chelating proteins has been reported. In a strain of this species, isolated from a silver mine, three plasmids were found, the largest (~ 50 Md) possessing genes responsible for metal resistance. *P. fluorescence* detoxifies aluminium by producing a metabolite that complexes the metal. When Fe is also present in the medium, the two trivalent ions are immobilized in a lipid-rich complex, which also contains P (phosphatidylethanolamine). As in the case of Al and Fe, resistance to Zn, Ca and Ga is due to association with phosphatidylethanolamine found in lipid-rich complexes. It is believed that half of the plasmids known in *P. aeruginosa* confers resistance to Hg^{2+} ion. A plasmid of the *P. stutzeri* species (pPB) confers resistance to Hg^{2+} and organomercury compounds (Kaur et al. 2013).

Resistance to boron, chromium and tellurium is also determined by plasmid genes. After growth in the presence of increased concentrations of tellurite, *P. aeruginosa* and *P. putida* strains, carriers of plasmids responsible for metal resistance, accumulate structures containing tellurium in the periplasmic space. Also, on plasmids are genes that determine resistance to arsenic. Arsenatians are reduced to arsenites, which are eliminated outside the cell through an export system. A chromosomal operon involved in the regulation of this metabolic pathway has also been identified. The operon has hybridized with chromosomal segments of some enterobacteriaceae, respectively, *P. aeruginosa*. This operon appears to be the evolutionary precursor of plasmid operons, which, moreover, have the advantage of being present in several copies in the same cell, which gives them a greater resistance to metals.

The placement of genes responsible for the degradation of high-impact pollutants, as well as of some that confer resistance to the toxic action of heavy metals on transferable plasmas even interspecifically, is a great advantage. Other genes with a role in hydrocarbon degradation are located on transposable elements. Given advances in the molecular genetics field, plasmids or other genetic vectors possessing interest genes can be constructed to have interspecific mobility, and, implicitly, possessive bacteria exhibit great biotechnological efficiency in topical applications (Kaur et al. 2013).

Of interest for environmental remediation biotechnologies are members of the *Burkholderia* genus (*Betaproteobacteria*, ord. I *Burkholderiales*, family I *Burkholderiaceae*). Interest has increased due to its link with the chemical pollution of the environment and concerns about the exploitation of bacterial degradative abilities in remediation biotechnologies. The attention is retained by two species of the genus: *Burkholderia cepacia* (or *Pseudomonas cepacia*) and *Burkholderia multivorans*.

They are aerobic chemoorganoheterotrophic bacteria, but they can use nitrate as the ultimate acceptor of electrons in anaerobic respiration. All genus species have the ability to accumulate poly-β-hydroxybutyrate (PHB) as a C deposit, which can be degraded and used as a nutritional source when growth conditions become inappropriate. The ability to produce and store PHB is a great advantage in interspecific competition in the natural habitats of bacteria, usually poor in nutrients. The species of the genus produce countless siderophores. On Fe-deficiency environments, for example, *P. aeruginosa* and *P. fluorescens* synthesize salicylic acid, a high-capacity siderophore compound, for binding Fe and other metals.

Like the *Pseudomonas* species, *Burkholderia cepacia* is effective in degradation of toluene. *B. cepacia* grows lush on environments containing halogenated herbicides: 2,4,5-trichlorophenoxyacetic acid (2,4,5 T) and 2,4-dichlorophenoxyacetic acid (2,4D). Natural strains of *B. cepacia* are active in the degradation of polychlorinated biphenyls. Dehalogenation occurs frequently after cleavage of the ring, as with the metabolism of other haloaromatic compounds. In vitro hybrids have been constructed which can effect total degradation of 2-Cl, 3-Cl, 2,4-dichloro- and 2,4,5-trichlorobiphenyl.

The degradation of pollutant compounds in the environment is controlled by plasmids. A 50 kb plasmid carries the genes responsible for the degradation of p-nitrophenol by *B. cepacia*, following an oxidative pathway with formation of hydroquinone and nitrite. The plasmid is conjugative and can be transferred to other strains, an advantage in the context of the approach to finding the most suitable strains for decontamination of the environment as efficiently as possible. A new pathway for degradation of toluene is provided by enzymes whose synthesis is regulated by genes located on plasmids: an inducible degradation pathway, another constitutive, the latter being regulated by genes located on the TOM plasmid (108 kb). MOP plasmid carries genes involved in the catabolism of phthalate derivatives, and, on a plasmid of only 2 kb, there are genes responsible for the degradation of herbicides having phenylcarbamate in formula (Gao et al. 2016).

The mobility of various organic or inorganic metal compounds in soil, as well as in other habitats, is closely related to their oxidation state. The energy used by living systems to carry out vital activities is provided by the oxidoreduction reactions catalysed by enzymes from the first class, oxidoreductase, in biological systems generic named dehydrogenases, because in the electron transport chain the transfer of electrons is accompanied by a transfer of protons (H^+). The total oxidation reactions carried out at the various cellular components, whereby the cell acquires the energy needed to carry out its vital activity, constitute respiration. The tendency of a chemical compound to accept or yield electrons in biological oxidation reactions is quantitatively expressed by the oxidoreduction potential (redox) = Eh.

To visualize the meaning of the transfer of electrons (H^+) in the biological systems, an axis was imagined where the various redox couples are placed in the ascending order of their redox potential. This axis still bears the name of the electron tower (Wang et al. 2013).

The reduced form of the redox pair at the top of the tower (negative values; $2H^+/H_2$: Eh = −0.421 V) has the highest tendency to yield electrons, so to oxidize; an oxidized form of the torque at the base has the greatest tendency to accept electrons, so to reduce ($½O_2/H_2O$: Eh = +0.816 V). The redox couples located at the middle of the electron tower can act in two ways: as electron acceptors, which they take from couples with a more negative redox potential, located towards the top of the tower, and as electron donors, which gives couples with a more positive redox potential towards the base of the tower. In their living environments, including soils, different ecophysiological groups of bacteria carry out the oxidoreduction reactions according to the potential redox present in the electron tower. Example:

Desulfovibrio: oxidizes H_2 by way of

$$SO_4^{2-}; 2H^+/H^2 (Eh = -0.42\,V);$$
$$SO_4^{2-}/H_2S(Eh = -0.22\,V); \Delta Eh = 0.20V;$$
$$4H_2 + H_2SO_4 \rightarrow H_2S + 4H_2O + E(38\,KJ/mol);$$

Beggiatoa: oxidizes H_2S by way of

$$O_2: SO_4^{2-}/H_2S(Eh = -0.22\,V);$$
$$\tfrac{1}{2}O_2/H_2O(Eh = +0.816);\ \Delta Eh = 1.04\,V;$$
$$H_2S + 2O_2 \rightarrow H_2SO_4 + E(200\,KJ/mol)$$

The *Thiobacillus* genus (*Betaproteobacteria*, order II *Hydrogenophilales*, family I *Hydrogenophilaceae*) comprises Gram-negative bacteria of bacillary form, some species with flagellar mobility. They are chemolithoautotrophic bacteria, so-called unpigmented sulphurous bacteria, which obtain the energy needed to fix CO_2 from the oxidation reactions of the reduced sulphur compounds or even elemental sulphur. Their nutritional and respiratory capacities are complex. It preferably oxidizes S or thiosulphate, rather than hydrogen sulphide. The reaction of *T. thiooxidans*: $S + 1\tfrac{1}{2}O_2 + H_2O \rightarrow H_2SO_4 + E$.

As a result of sulphuric acid production, acidity in their natural environment may reach 1.5. They are known to be the most acid-tolerant bacteria. They develop optimally at pH 2–4 and are aerobic. Like *Beggiatoa*, they nourish themselves also heterotroph. The natural habitat of unpigmented sulphurous bacteria is waters and soils rich in reduced sulphur compounds, mine waters, sulphurous springs. Also aerobic is *T. ferrooxidans*, which oxidizes the Fe compounds in which it is found in the Fe^{2+} state, to compounds where the Fe is Fe^{3+}, as well as other ferrobacteria or ferruginous bacteria (*Gallionella*, *Sphaerotilus*):

$$2FeSO_4 + H_2SO_4 + \tfrac{1}{2}O_2 \rightarrow Fe_2(SO_4)_3 + H_2O + E$$
$$\text{and}\ 2Fe(OH)_2 + \tfrac{1}{2}O_2 + H_2O \rightarrow 2Fe(OH)_3 + E.$$

Ferruginous bacteria are met frequently in waters and rarely in soils. It is of practical importance for water supply systems. Metal pipes can be corroded under the action of ferruginous bacteria. The formation of oxidized compounds, in particular of ferric hydroxide $Fe(OH)_3$ insoluble compound, can lead to clogging of the metallic pipes by depositing them as red-brown layers, after which is recognized the presence of ferrobacteria in their natural life environments (Wang et al. 2013).

Denitrifying bacteria (reducing nitrate) have anaerobic respiration, using nitrate as the ultimate acceptor of electrons in the electron transport system. In the presence of O_2, they do aerobic breathing, even if nitrates are found in the environment due to the superior energy efficiency of aerobic respiration, as well as the O_2 repress of nitrate reductase synthesis. There are species that, through the reactions produced, fall into two distinct ecophysiological groups of bacteria at the same time. *T. denitrificans*, aerobes, in the presence of O_2 can carry out the oxidation reactions of the reduced sulphur compounds, but in the absence of O_2, for oxidation they use the oxygen from the nitrates present in the medium: $S+2HNO_3 \rightarrow H_2SO_4+N_2+O_2+E$. *T. denitrificans* sulphonating bacteria, which oxidizes sulphate-reduced sulphur compounds it is also a denitrifying bacterium because, under the conditions specified, it produces nitrate N_2. It is a typical example of anaerobic respiration, where the final electron acceptor (H) is not O_2 but an oxidized inorganic compound (NO^{3-}) (Hsieh and Chung 2014).

Some denitrifying bacteria can oxidize also reduced Fe compounds, others can ferment, so denitrifying bacteria have a wide range of options with regard to alternative mechanisms of energy metabolism. Desulphurization–sulphate-reducing bacteria are anaerobic bacteria, in which the final acceptor of electrons is sulphate (SO_4^{2-}). The final product of sulphate reduction is hydrogen sulphide (H_2S). In the assimilation reduction of sulphides, H_2S is converted to organic sulphur, consisting of amino acids (essential, cysteine and methionine; nonessential, glutathione, taurine and homocysteine). In the nonassimilable reduction is excreted in the environment. The typical desulphurizers species is *D. desulfuricans* (*Deltaproteobacteria*). Reaction produced is $4H_2 + H_2SO_4 \rightarrow H_2S + 4H_2O + E$.

Disproportionation or dismutation of sulphur is a chemical reaction whereby an element is simultaneously reduced and oxidized by forming two different products. Some reducing sulphate bacteria have the ability to cleave a sulphur compound in an intermediate oxidation state, into two compounds, one lower and other more oxidized than the original substrate. *D. sulfodismutans* can decompose thiosulphate (intermediate oxidation state) into (more oxidized) sulphate and (lower) hydrogen sulphide: $Na_2S_2O_3 + H_2O \rightarrow Na_2SO_4 + H_2S + E$.

The process has ecological significance because it provides a way for desulphurization bacteria to recover the energy of intermediate sulphur compounds resulting from incomplete oxidation of H_2S by sulphur-oxidizing bacteria (*Beggiatoa*, *Thiobacillus*) (Wang et al. 2013).

Transformation of metals under the action of microorganisms plays a role in metal cycles in the biosphere. Metal transformations are dominated by oxidoreduction reactions, complexation of organic and inorganic compounds and change between water soluble and insoluble forms. When a microorganism oxidizes or reduces a metal, it precipitates or becomes soluble. Thus Cr^{6+} is reduced to Cr^{3+}, which precipitates as oxides, sulphides or phosphates of chromium. Bacteria living in environments with high concentrations of metals have specific physiological mechanisms, i.e. they survive under these unfavourable conditions: extracellular precipitation, metal ion binding and cell surface elimination, intracellular seizure and intracytoplasmic inclusions (Saikia et al. 2014).

Cation binding to the cell surface has become one of the most attractive biotransformation models. Those metals that have an electron configuration containing 10–12 layers of electrons are toxic to organisms at relatively low concentrations. In this group enter Hg^{2+}, Ag^{+}, Pb^{2+}, Cd^{2+}, Zn^{2+}. Metal–microorganism interactions play a role in numerous biotechnologies, such as bioremediation, biomineralization, bioleaching and microbial corrosion, and therefore it is of interest. It is intended to use bacterial strains in pure cultures or in consortia capable of mobilizing/immobilizing metal ions. Microorganisms can mobilize metals by autotrophic or heterotrophic leaching, chelation by metabolites and siderophores, methylation and redox transformations. The heterotrophic leaching is when the microorganism acidifies the environment by proton efflux (proton motor force), resulting in the release of metallic cations. Autotrophic leaching is when the acidophilic bacterium obtains the energy needed to fix CO_2 by oxidation of reduced Fe inorganic compounds (Fe^{2+}) or reduced sulphur compounds. Siderophores are Fe^{3+}-specific ligands but can also bind other metals, such as Mn, Mg, Cr, etc. Methylation involves the methyl group that is enzymatically transferred to a metal, forming a number of different metalloids (Wang et al. 2013).

Redox transformations allow microorganisms to mobilize metals, metalloids and organometallic compounds. There are metal mobilization techniques, the technology being chosen and depending on the physical and chemical characteristics of the metal. As can occur in inorganic or organic forms. Arsenic trioxide, sodium arsenite and arsenic trichloride are the most common inorganic compounds.

Neurotoxicity for the central or peripheral nervous system may be due to inorganic compounds and begins with changes in sensitivity followed by weakening of muscle activity. Organic arsenic compounds may be in trivalent or pentavalent form and may occur in methylated form as a result of biomethylation in soil, water, sediment or living organisms.

Pentavalent forms of arsenic are pentoxide, arsenic acid and arsenates, for example, $PbHAsO_4$, used as a pesticide to combat Colorado beetle. Pentavalent forms of arsenic affect enzyme activity in higher animal and human organisms. There have been discovered bacteria that can use arsenic oxide compounds in anaerobic respiration, during which they are reduced to trivalent arsenic compounds ($As^{5+} \rightarrow As^{3+}$) (nonassimilation reduction): *S. arsenophilum* and *S. barnesii*. *Shewanella* bacteria (*Gammaproteobacteria*, *X. Alteromonadales*, fam. I. *Alteromonadaceae*) anaerobic chemoorganoheterotroph does not reduce As^{5+} compounds but can release the ion in the medium as well as reduce Fe compounds (Fe^{2+}). It has been found that *Pseudomonas arsenitoxidans* grows chemolithotroph, obtaining the energy needed to fix CO_2 from the oxidation reactions of arsenic reduced compounds. The bacterium is able to grow in the presence of organic substance, so it is just autotroph. Growth is more intense in the presence of arsenite. By this ability to obtain energy as a result of oxidation reactions of arsenite, *P. arsenitoxidans* strain is a unique organism in the world of prokaryotes. The discovery is important because other bacteria that oxidize arsenic (*B. arsenoxydans*, *A. faecalis*) cannot grow chemolithoautotroph. The discovery of this strain of *P. arsenitoxidans* represents an advance in understanding the interactions between microorganisms and arsenic compounds, known for their strong antibiotic action (Liu et al. 2011).

Chromium is present both in living organisms and in rock, water and soils. They are in nature only in the form of compounds, not in elemental form. The most common forms in nature are the compounds of bivalent (Cr^{2+}), trivalent (Cr^{3+}) and hexavalent (Cr^{6+}) chromium.

For the manufacture of steel, Cr^0 is used. Metallurgical industry's activities result in Cr^{2+} and Cr^{6+}, while Cr^{3+} occurs naturally in the environment. The toxicity of Cr compounds is due to the valence state of the metal. Absorption of Cr compounds by cells is more intense in the case of Cr^{6+} compounds, because chromium (CrO_4^{2-}) anion penetrates into cells by facilitated diffusion, while passive diffusion and phagocytosis are responsible for less efficient processes of penetration of Cr^{3+} compounds. From the point of view of the negative impact of Cr compounds on the health of the environment, Cr^{6+} compounds, most frequently encountered in contaminated sites, are of the greatest interest. Cr^{6+} can be reduced to Cr^{3+} by organic matter and S^{2-} and Fe^{2+} ions, under anaerobic conditions, frequently encountered in groundwater and flooded soils. In the presence of chromates (CrO_4^{2-}) and of dichromates ($Cr_2O_7^{2-}$), metallic cations such as lead ones precipitate. In the presence of chromates and dichromates, the Fe and aluminium oxides are absorbed by the soil particles. Toxicity and mobility of Cr depend on soil characteristics and the amount of organic matter incorporated by it. Hexavalent Cr is more toxic and more mobile than all other forms (Hou et al. 2013).

Trivalent Cr is mobile, but its mobility decreases with adsorption by clay minerals and when pH drops below 5. pH increase stimulates leaching and solubilisation of hexavalent Cr compounds. When Cr is discharged into natural waters, it accumulates in sediments, which can be subjected to bioremediation procedures. It assesses the effects of minerals present in subsoil layers on the efficiency of lead removal from groundwater using biofilms composed of sulphate-reducing microorganisms and examines the stability of metal deposits after biofilms have been temporarily exposed to air (Hou et al. 2013).

To quantify the effects, the lead was immobilized in *D. desulfuricans* biofilms, grown under anaerobic conditions in two bioreactors filled with one with hematite (redox-active) and the other with quartz (redox-inert). Biofilms grown on hematite were denser, thicker and more porous than those grown on quartz. Average H_2S concentrations were higher in quartz biofilm than in hematite biofilm. Lead has been more effectively immobilized in quartz biofilm than in hematite biofilm. Under the action of desulphurization bacteria, H_2S was produced which reacted with present Pb to form precipitating PbS. It has been shown that lead precipitates more in the presence of biofilm located near redox-inerts (quartz). Lead deposits have been partially reoxidized, especially in biofilms grown on hematite. In bioreactors, biofilms responded to the presence of O_2 by lowering their density and by increasing the rate of H_2S production (Wang et al. 2014).

Although the reduction of Fe^{3+} to Fe^{2+} was not quantified, it was found that Fe was continuously released from hematite throughout the experiment. Various acidothermophilic microorganisms from soil, sludge and water from polluted environments due to activities in the metallurgical industry have been isolated for use in the bioremediation of toxic metals. The tolerance of these microorganisms to high concentrations of Ag, As, Bi, Cd, Cr, Co, Cu, Hg, Li, Mo, Pb, Sn and Zn was improved (Wadhwani et al. 2016).

Isolation and adaptation of microorganisms to high concentrations of metals were done by growing them on autotrophic bacteria media under different incubation conditions (pH, 2–4.5; temperature, 40–65 °C) and concentrations of metals in culture media (10^{-3}–10^{-7} M). If at the concentration of 10^{-7} M metals 72 isolates were obtained, at the concentration of 10^{-3} M of the metals remained only 16 different isolates. Isolates have been used to test their biosolubilization capacity for metal sulphide ores. In the case of chalcopyrite ($CuFeS_2$), a copper solubilization of 85.82% was obtained in the case of covellite (CuS) 97.5% after 5 days of incubation in the presence of a metal concentration of 10^{-3} M, pH 2.5, at 55 °C. Bioadsorption was better in chalcopyrite: Ag 73%, Pb 35%, Zn 34%, As 19%, Ni 15% and Cr 9% (Williams et al. 2016).

9.5.4 Bioremediation Models with NBSs

After identifying the affected sites and selecting from nature, eventually acquiring bacterial strains adapted to the conditions of the studied soils from the world-recognized bacterial cultures (ATCC, American Type Culture Collection; DSMZ, Deutsche Sammlung von Mikroorganismen und Zellkulturen), lab experimental models are created to obtain the optimal conditions for bioremediation of soils contaminated with heavy metals, hydrocarbons, products resulting from emissions of sulphur dioxide, etc.

According to the literature, an experiment can be carried out for a contaminated soil plot with high concentrations of Cu, Zn, Cd, Pb and As by washing it with an acidified aqueous solution. For example, the initial soil pH was 5.4–6. The soil contained a rich indigenous microbiota, which had the most important contribution to the solubilisation of contaminants. Microbial activity has been improved by controlled changes in the wash water flow rate, oxygen content and soil nutrients (Hou et al. 2013).

In less than 18 months, the residual soil concentrations of metals (with the exception of lead) fell below the permitted level for the soil type analysed and some research were made:

- Research on the effect of different adsorbent surfaces on the microbial and enzymatic potential of habitats. Research has been carried out in numerous laboratory models, aiming at the production of analogues of therapeutic sludge from saline lakes. Zeolites (volcanic structures rich in clinoptilolite), clay minerals (kaolin), brown coal in the form of powder and peat were used as adsorbing surfaces. Considering the purpose of the research to obtain analogues of therapeutic sludge, the various experimental variants were inoculated with halophilic microorganisms in the form of sediments from different saline lakes suspended in waters from the same lakes and incubated for different periods (up to 8 years) under aerobic or anaerobic conditions, in light or in the dark. Periodically, the evolution of the microbial and enzymatic potential of the therapeutic sludge analogues was monitored. Without exception, there was a substantial improvement in the enzymatic and microbial potential of the systems, with differences between the experimental variants based on the nutrient additions (urea, KH_2PO_4). The conclusion is that clay minerals and humic substances exert a stabilizing and protective effect on enzymes from sediment and, unquestionably, can be said from soils. These substrates contribute to the survival of free enzymes accumulated in microsites formed by organomineral complexes, where they are protected from proteolysis and inactivation. The best conditions for enriching the enzymatic potential were provided by anaerobiosis incubation, whether the adsorbent was zeolite or kaolin. Enzymatic potential developed rapidly but subsequently declined rapidly in variants containing brown coal than in peat-containing varieties.

- A biotechnology to improve the quality of crude refractory clays has been developed with Fe impurities. It is known that refractory clays are of inferior quality due to their low content of Al_2O_3 and their high Fe_2O_3 content. The developed biotechnology consists in various experimental variants by incubating crude refractory clays under anaerobiosis conditions favouring the development of Fe-reducing bacteria. Fe-reducing bacteria reduce trivalent iron (Fe^{3+}) to ferrous iron (Fe^{2+}); by anaerobic respiration, trivalent Fe compounds are insoluble, while the divalent Fe are soluble in water. Microbial deferrating biotechnology proved feasible. The intensity of the reduction process of trivalent Fe was influenced by nutrient concentration, temperature and duration of incubation (30–90 days). Of the two types of clay treated, in one case the microbial deferase was 67% and in the other case 60%.
- The effectiveness of zeolites in immobilizing heavy metals in soil is tested. The soils examined were contaminated with large amounts of metals: Pb (1054–40375 ppm) in association with Zn (490–1175 ppm), Cd (13.2–24.2 ppm) and Cu (37.6–409.5 ppm) and Cu (360–7000 ppm) in association with Zn (1900–3100 ppm), Cd (40–80 ppm) and Pb (50–2000 ppm). Soils polluted with heavy metals and mixed with zeolites, primarily clinoptilolite (soil 83%–17% zeolite) and organic residues, were planted with *L. perenne*. Plant growth has shown the stimulating effect of the addition of zeolite. The better growth of plants was due to the fact that the mixture of organic matter + zeolite + contaminated soil provides essential nutrient plants (ammonium, humus, potassium, calcium) on the one hand, and on the other hand, heavy metals that inhibit plant growth are blocked by cation exchange, as a result of the fact that the metals penetrate the structure of the zeolite and no longer have direct access to the roots of the plants.
- In in situ bioremediation of soils contaminated with heavy metals, in the field of bioremediation of technogenic soils, tailings dumps result from mining operations of Pb, Zn and Fe. At the tailings heap, 14 experimental plots were installed in the first year, the other 2 in the following year, subjected to a different treatment and seeded with *L. perene* and *T. pratense*. Later on, on the heap were planted seedlings of *H. rhamnoides*. Applied biotechnologies have led to the formation of favourable conditions for the development of microorganisms, for plant growth and for intense and sustainable enzyme activity. The best technology

for tailings bioremediation containing Pb and Zn impurities has been covering with a layer of 10 cm natural soil from the vicinity of the heap, NPK mineral fertilization and seeding with a mixture of herbaceous plants or plants from the spontaneous flora from that region. The ascending evolution of the microbial and enzymatic potential of the soils from the experimental plots has been remarkable from year to year. Crops and juvenile plantations have expanded, so that after 20 years, the tailings heap, initially with a selenar appearance, with no trace of vegetation, especially on the upper terraces, is covered with the vegetation. The technogenic soil is being transformed; it already records an enzymatic and microbial potential comparable to natural soils. More than 2000 tree and shrub seedlings were planted on the terraces and slopes of the tailings pond at the Iara Fe mine, mostly by *H. rhamnoides*. There were 26 experimental plots, cultivated with the following herbaceous species: *F. rubra*, *F. arundinacea*, *D. glomerata*, *L. perenne* (Poaceae family), *O. viciifolia*, *T. repens*, *T. pratense*, *L. corniculatus* and *M. sativa* (Fabaceae family). Both parcels and planting seedlings have undergone differentiated treatments. The evolution of the vegetation and of the microbial and enzymatic potential of the soils from the experimental plots was monitored. Microbial potential assessment was based on bacterial NBSs of soil quality, calculated by taking into account the number of aerobic mesophilic heterotrophic, ammonifier, denitrifying, Fe-reducing and desulphurization bacteria. Enzymatic potential was assessed on the basis of the enzymatic NBSs of soil quality, calculated on the basis of the following enzymatic activities: catalase, sucrase, phosphatase and actual and potential dehydrogenase. The results obtained prove the efficiency of the applied technologies. After only 1 year of vegetation, a remarkable biological potential has developed in the soils of the experimental plots. The vegetation had a good evolution, and *H. rhamnoides* seedlings had a massive expanded, already providing a good cover for the tailings (Kumar and D'Souza 2011).

9.5.5 Technology of Bioabsorption of Heavy Metals from NBS Diluted Solutions

Concerning heavy metal pollution, the use of microbial biomass has emerged as a solution for the development of "environmental friendly" economic and technological processes used in the treatment of contaminated sites. Biomass of dead bacteria and living microbial biomass can capture metals from dilute solutions through the bioabsorption process from contaminated soil mixed with water. Bioabsorption technology has the advantage of low operating costs, is effective for dilute solutions and generates minimal effluent; in this process, dead microbial biomass functions as an ion exchanger by the qualities of various reactive groups available on the cell surface such as carboxyl, amines, phosphates, sulphates and hydroxyls. Bioabsorption is possible for both forms of biomass, but bioaccumulation is only facilitated by living biomass. Bioaccumulation is a growth-dependent process, and it is difficult to define the variety of effluents as opposed to bioabsorption that is an independent growth process.

Thus, microbial biomass can be used and exploited more effectively as a bioabsorbent than as a bioaccumulator (Schallmey et al. 2014).

Research in this field has identified a range of microorganisms and their metal binding capabilities. Biomass of fungi offers the advantage of having a cell wall material with excellent metal-binding properties. Many fungi and yeasts have demonstrated excellent bioabsorption potential of metals, especially *Rhizopus*, *Aspergillus*, *Streptoverticillum* and *Saccharomyces*. Among bacteria, the *Bacillus* species was identified as having the greatest retention potential of metals and was used in the preparation of commercial bioabsorbents.

There have been reports of metal bioabsorption through the use of *Pseudomonas*, *Zoogloea ramigera* and *Streptomyces*. Binding to cell walls is involved in the bioabsorption process, and their modification may affect the binding of metal ions. However, many things cannot be done to offset the conditions of the area that do not accept the bioremediation process or which do not provide ideal conditions for implementing a biorestoration system.

There are many soil properties that influence the bioremediation process, the most important being: soil type and permeability, distribution of granular structure, soil moisture content, pH and temperature.

The soil type is an important variable in designing the bioremediation process. Noncohesive soils, such as gravel and sand, are better for bioremediation than compact and dense soil. Soil permeability is a key factor in the success of the bioremediation process, by facilitating the transport and distribution of nutrients and acceptors. The more permeable the soil, the better the conditions for the successful application of the bioremediation process, this being true for both the unsaturated and the saturated area. Likewise, the air and water circulation in the soil is influenced by its permeability. In soils with high permeability, the introduction and movement of air through bioventilation are facilitated, this being possible for both in situ and ex situ bioremediation (Hou et al. 2013). The soil must be sufficiently permeable to prevent the microbial mass from causing pore clogging. In this respect, it is necessary for the sites to have a hydraulic conductivity higher than 10^{-4} cm/s, this value allowing the bioremediation to be carried out in situ with good results. An equally important role as permeability in the prevention of biopollution is the distribution of the granular soil structure.

Studies in this area show that highly porous materials with far granular structures are much more susceptible to biopollution than materials with increased porosity; a soil without drainage is more predisposed to biopollution than a well-drained material. Soil moisture is an important property in the unsaturated treatment systems because microorganisms need water as a support for the metabolic process. In the bioremediation process, the ideal soil moisture is 50% (Mahr et al. 2015).

9.5.6 Microbial Bioinsecticides with NBSs

A number of microbial species exhibit inhibitory properties towards different insect or nematode species. Insecticidal bioproducts obtained from bacteria, fungi or viruses can successfully replace treatments with various toxic chemicals.

The most well-known bioinsecticides are based on the use of *B. thuringiensis* bacteria, which is capable of producing a protein with toxic effects for insects during sporulation process. Studies on this species have shown that there are several varieties with specific action against certain types of insects: *B. thuringiensis* active var. *kurstaki* on coleoptera and lepidoptera, *B. thuringiensis* var. active *israelensis* on diptera, *B. thuringiensis* var. *tenebrionis* with specific effect against Colorado beetle, etc. (Espinosa-Urgel et al. 2015).

To obtain the biopreparation, the bacteria are cultivated in the fermenters until the spores are released by cell lysis, at which time the crystal protein (δ-endotoxin) is released into the environment (approximately 30 hour). After centrifuging the spores and the crystals, the sediment is dried and included in an inert material so as to obtain a wettable powder that can be applied to the plants. A series of bioproducts is based on pathogenic viruses for insects (*baculoviruses*), with a high specificity against specific insect. The most modern ones use wild-type baculoviruses or genetically engineered variants that act efficiently especially on insect larvae (Lu et al. 2013).

The morphogenetic changes caused by the infection of plants with *A. tumefaciens* or *A. rhizogenes* strains are due to the transfer of genes of bacterial origin into the plant cells. *Agrobacterium* genes exhibit an organized genetic system that allows the introduction of new genes into the plant cell, due to the presence in their cell of a plasmid Ti ("inducing tumour") or Ri ("root inducing"), which is the determining factor of tumour production, respectively, abnormal roots, in plants.

The transfer of a portion of these plasmids into the plant cell genome causes functional alterations that result from the occurrence of the tumour phenomenon. These plasmids are therefore an organized and complex system of transfer of natural genetic information into the plant cell without affecting the ecological balance. Most studies on the structure of *Agrobacterium* plasmids were performed with Ti plasmids isolated from *A. tumefaciens* strains (Espinosa-Urgel et al. 2015).

They are large (over 200 kb), with a small number of copies/bacterial cell, transferable from one cell to another, but only under special conditions. The Ti plasmids contain several regions with distinct functions: the DNA–T region (containing the tumorigenesis genes), the virulence region (viral genes), the incompatibility region (the genes inc), the region controlling the conjugative transfer of the plasmid (tra genes), etc., the first two being more important for the process of plant cell transformation (Gredell et al. 2012).

The virulence region comprises genes essential for the transfer of bacterial DNA into the plant cells without being transferred themselves. The viral region comprises a segment of about 30–40 kb (depending on the bacterial strain), consisting of more than 20 genes organized in at least 6 transcriptional units (operons): virA (one gene), virB (11 genes) virC (2 genes), virD (4 genes), virE (2 genes) and virG (one gene). Of these transcriptional units, virA, virB, virD and virG are essential for the DNA–T transfer, and virC and virE ensure an increase in transfer efficiency (Weising and Kahl 1996).

All these operons are induced in the presence of injured plant cells (which produce a series of phenolic compounds with inductive role), their functions being as follows: virA and virG encode proteins responsible for recognizing injured plant cells and inducing other viral functions; virC and virD encode proteins that provide the formation of a single-stranded DNA–T, transferable in the plant cell; virD and virE encode proteins that form a complex together with DNA–T guiding it towards the plant cell nucleus; they also appear to be involved in the process of integrating the DNA–T into the new host genome; and virB encodes proteins responsive to the transport of the DNA–T protein complex via the bacterial plasma membrane (Gredell et al. 2012).

The studies performed by the DNA heteroduplex analysis of viral regions derived from different types of Ti plasmids revealed an extensive homology of this region (although some small non-homologous regions it might be due to insertions or deletions occurring in the plasmid structure during evolution).

By comparing the viral region of a nopalinic Ti plasmid with that of one octopinic, it was observed that there are some differences: the first does not present the virF locus but instead contains the tsz gene, which allows the bacterium to produce the trans-zeatin phytohormone. This gene is located in the octopinic Ti plasmid at the left end of the region, along with the virA. The expression of genes located in the virulence region of Ti plasmids is inducible and is regulated by the action of external factors.

Experimentally, it has been shown that under the action of some signal molecules in plant exudates or wounds produced in plants, a strong activation of these genes occurs. These signal molecules are phenolic compounds, such as acetosyringone or α-hydroxyacetosiringone, which act on the virA and virG regulatory genes. The products of these genes are similar to other regulatory membrane proteins (Env Z, Omp R, etc.).

Along with the plasmid viral genes, some genes, located at the bacterial chromosome, have been identified and are involved in a lesser extent in the genetic transformation process (Espinosa-Urgel et al. 2015). Of these genes we mention the chvA and chvB genes involved in the synthesis and secretion of β-1,2 glucan, the chvE gene required for viral region induction and for bacterial chemotactis, the cel locus responsible for the synthesis of cellulosic fibrils involved in attaching bacteria to plant cells, the pscA gene (exoC) involved in the synthesis of cyclic glucan and succinoglycanic acid and the At locus involved in the synthesis of surface cell proteins. Functions of recognition and attachment of bacteria to the plant cell wall are not encoded by plasmid genes but by chromosomal genes.

These genes determine, inter alia, the structure of exopolysaccharides from the surface of bacteria involved in recognition processes (Weising and Kahl 1996).

The DNA–T region represents a portion of Ti (or Ri) plasmid that is transferred to the plant cell, integrating stably into the cell genome. This DNA–T region (from the transferred DNA) is about 20–25 kb in length (depending on the host bacterial strain) and contains all the genes (oncogenes) that transferred to the plant cell which will determine the characteristic phenotype of the transformed cells. Genes located on the DNA–T determines two characteristic properties for plant tumoural cells: growth on culture media in the absence of phytohormones and synthesis and secretion of opines, specific substances for these cell types. Eight genes at the DNA-T level were identified and mapped: aux1, aux2 and alpha, all of which are involved in encoding auxine synthesis, ipt coding the synthesis of cytokines, ops which determines the synthesis of the opines by transformed cells (ocs or nos genes, depending on the type of plasmid Ti), also the eighth gene interacting with the others, it seems that increase the susceptibility of auxin–transformed cells (Espinosa-Urgel et al. 2015).

The DNA–T content of all natural Ti plasmids is flanked by short, approximately 25 bp nucleotide sequences, directly repeated, essential for the plant cell transfer process. Analysis of these terminal sequences showed that they are conserved in the different *Agrobacterium* strains, being recognized by the enzymatic equipment involved in the transfer of DNA–T from bacterial cell to the plant cell. In the case of Ri plasmids, the analysis of the DNA–T region has been shown to contain a series of genes responsible for the "hairy" phenotype of transformed plant tissues.

The genes transferred to the plant cells are grouped at the edges of the DNA–T region. Thus, on the left side of the T-DNA are the role genes A, B, C and D that sensitize plant cells to auxin action. It has been shown that in certain plant species, the simple presence of role genes, even in the absence of other DNA-T genes, results in the appearance of the characteristic "hairy root".

On the right side of the DNA-T from plasmid Ri, there are two genes coding the synthesis of auxins (Indole-3-acetic acid) in the transformed cells; along with these genes are those responsible for the synthesis of opines (mannopine and agropina) (Shin 2010). Regenerated plants from plant tissue transformed with these bacteria have a modified appearance compared to regenerated plants in normal tissue.

9.6 Conclusions and Remarks

Microbial NASs contain immobilized microorganisms, and a transduction chain, in general, is used for a single biochemical process. NASs HIS (*High Integrated Systems*) with microorganisms show difference from other NASs: living cells regenerating, occurring undesirable changes during the operation of the biosensor, higher response time than enzymatic NASs, reusing them in another determination requires a duration longer time. Fields of use are food industry, drug production, wastewater treatment, energy production, fermentation reactions and electron transfer to signaling molecules (phytohormones).

Microbial NASs are used to determine the concentration of some *Bacillus subtilis* substances, ammonium ion determination, acceleration of the respiratory process by penetration of NH_4 into NH^{4+} permease system cells, active repression by nutrient limitation, decrease of NASs response time, determination of concentration of substances, cells of *Bacillus cadaveris* and *Proteus morganii*, determination of [L-aspartate] and [L-cysteine], regeneration of microbial sensors by placement of exhausted electrodes in a culture medium suitable for bacterial strains used, development of new cells directly on the surface of electrodes, regenerating initial activity and extending the life of the biosensor (autolysed cells remain at the surface of the electrode, limiting to several regeneration cycles), selection of mutagens. This paper is a bibliographic study which aims to enumerate of some NBS-related notions, symbiosis and micorizations as a manifestation of interactions between different groups of microorganisms or between microorganisms and higher plants. The study showed the close relationship between microorganisms and the physical, chemical and biological elements of the environment, the interrelations between different groups, but also their important role for plants. The beneficial role is noted in the growth and development processes of plants, in the intervention in soil nutrient absorption processes and in their resistance to diseases and drought. NBSs regulate microbial flora around plant roots, suppress pathogenic organisms and help the proliferation of beneficial fungi (*Cladosporium*, *Trichoderma* and *Gliocladium*) and nitrogen-fixing bacteria; improve germination; develop the root system; strengthen root formation; form additional secondary roots; facilitate the absorption of nutrients from soil; increase plant resistance to drought; improve plant yield; act on plants at the biochemical and cellular level; and change the biometric features of vegetative growth (germination, root system growth, twinning, time required for phenological phases, etc.) and crop structure characteristics. Despite the many problems that microorganisms can cause due to their degradation capacities, some species are of practical interest because they can lead to the removal of pollutants or xenobiotics from the environment, or restoring the quality of the environment.

References

Alonso-Lomillo MA, Domínguez-Renedo O, Arcos-Martínez MJ (2010) Screen–printed biosensors in microbiology: a review. Talanta 82(5):1629–1636

Chen J, Park B (2016) Recent advancements in nanobioassays and nanobiosensors for foodborne pathogenic bacteria detection. J Food Prot 79(6):1055–1069

Cheng MS, Lau SH, Chow VT, Toh CS (2011) Membrane–based electrochemical nanobiosensor for *Escherichia coli* detection and analysis of cells viability. Environ Sci Technol 45(15):6453–6459

Cho JH, Lee DY, Lim WK, Shin HJ (2014) A recombinant *Escherichia coli* biosensor for detecting polycyclic aromatic hydrocarbons in gas and aqueous phases. Prep Biochem Biotechnol 44(8):849–860

Choi O, Lee Y, Han I, Kim H, Goo E, Kim J, Hwang I (2013) A simple and sensitive biosensor strain for detecting toxoflavin using β–galactosidase activity. Biosens Bioelectron 50:256–261

D'Souza SF (2001) Microbial biosensors. Biosens Bioelectron 16(6):337–353

Espinosa-Urgel M, Serrano L, Ramos JL, Fernández–Escamilla AM (2015) Engineering Biological Approaches for Detection of Toxic Compounds: a New Microbial Biosensor Based on the *Pseudomonas putida* TtgR Repressor. Mol Biotechnol 57(6):558–564

Gao G, Qian J, Fang D, Yu Y, Zhi J (2016) Development of a mediated whole cell–based electrochemical biosensor for joint toxicity assessment of multi–pollutants using a mixed microbial consortium. Anal Chim Acta 924:21–28

Goswami P, Chinnadayyala SS, Chakraborty M, Kumar AK, Kakoti A (2013) An overview on alcohol oxidases and their potential applications. Appl Microbiol Biotechnol 97(10):4259–4275

Gredell JA, Frei CS, Cirino PC (2012) Protein and RNA engineering to customize microbial molecular reporting. Biotechnol J 7(4):477–499

Hou QH, Ma AZ, Zhuang XL, Zhuang GQ (2013) Construction and properties of a microbial whole–cell sensor CB10 for the bioavailability detection of Cr^{6+}. Environ Sci 34(3):1181–1189
Hsieh MC, Chung YC (2014) Measurement of biochemical oxygen demand from different wastewater samples using a mediator–less microbial fuel cell biosensor. Environ Technol 35(17–20):2204–2211
Jin X, Angelidaki I, Zhang Y (2016) Microbial Electrochemical Monitoring of Volatile Fatty Acids during Anaerobic Digestion. Environ Sci Technol 50(8):4422–4429
Kaur A, Kim JR, Michie I, Dinsdale RM, Guwy AJ, Premier GC (2013) Microbial fuel cell type biosensor for specific volatile fatty acids using acclimated bacterial communities. Biosens Bioelectron 47:50–55
Kumar J, D'Souza SF (2010) An optical microbial biosensor for detection of methyl parathion using *Sphingomonas* sp. immobilized on microplate as a reusable biocomponent. Biosens Bioelectron 26(4):1292–1296
Kumar J, D'Souza SF (2011) Immobilization of microbial cells on inner epidermis of onion bulb scale for biosensor application. Biosens Bioelectron 26(11):4399–4404
Liu Z, Liu J, Zhang S, Xing XH, Su Z (2011) Microbial fuel cell based biosensor for in situ monitoring of anaerobic digestion process. Bioresour Technol 102(22):10221–10229
Lu TK, Bowers J, Koeris MS (2013) Advancing bacteriophage–based microbial diagnostics with synthetic biology. Trends Biotechnol 31(6):325–327
Mahr R, Gätgens C, Gätgens J, Polen T, Kalinowski J, Frunzke J (2015) Biosensor–driven adaptive laboratory evolution of l–valine production in *Corynebacterium glutamicum*. Metab Eng 32:184–194
Mulchandani A, Rajesh R (2011) Microbial biosensors for organophosphate pesticides. Appl Biochem Biotechnol 165(2):687–699
Nigam VK, Shukla P (2015) Enzyme based biosensors for detection of environmental pollutants: a review. J Microbiol Biotechnol 25(11):1773–1781
Park M, Tsai SL, Chen W (2013) Microbial biosensors: engineered microorganisms as the sensing machinery. Sens (Basel) 13(5):5777–5795
Safarpour H, Safarnejad MR, Tabatabaie M, Mohsenifar A (2012) Development of high–throughput quantum dot biosensor against *Polymyxa species*. Commun Agric Appl Biol Sci 77(3):7–13
Saikia SK, Gupta R, Pant A, Pandey R (2014) Genetic revelation of hexavalent chromium toxicity using *Caenorhabditis elegans* as a biosensor. J Expo Sci Environ Epidemiol 24(2):180–184
Schallmey M, Frunzke J, Eggeling L, Marienhagen J (2014) Looking for the pick of the bunch: high–throughput screening of producing microorganisms with biosensors. Curr Opin Biotechnol 26:148–154
Schenkmayerová A, Bučko M, Gemeiner P, Katrlík J (2013) Microbial monooxygenase amperometric biosensor for monitoring of Baeyer–Villiger biotransformation. Biosens Bioelectron 50:235–238
Schenkmayerová A, Bertóková A, Sefčovičová J, Stefuca V, Bučko M, Vikartovská A, Gemeiner P, Tkáč J, Katrlík J (2015) Whole–cell *Gluconobacter oxydans* biosensor for 2–phenylethanol biooxidation monitoring. Anal Chim Acta 854:140–144
Schneider G, Kovács T, Rákhely G, Czeller M (2016) Biosensoric potential of microbial fuel cells. Appl Microbiol Biotechnol 100(16):7001–7019
Shin HJ (2010) Development of highly–sensitive microbial biosensors by mutation of the nahR regulatory gene. J Biotechnol 150(2):246–250
Su L, Jia W, Hou C, Lei Y (2011) Microbial biosensors: a review. Biosens Bioelectron 26(5):1788–1799
Sun JZ, Peter Kingori G, Si RW, Zhai DD, Liao ZH, Sun DZ, Zheng T, Yong YC (2015) Microbial fuel cell–based biosensors for environmental monitoring: a review. Water Sci Technol 71(6):801–809
Tepper N, Shlomi T (2011) Computational design of auxotrophy–dependent microbial biosensors for combinatorial metabolic engineering experiments. PLoS One 6(1):e16274
Virolainen N, Karp M (2014) Biosensors, antibiotics and food. Adv Biochem Eng Biotechnol 145:153–185

Wadhwani SA, Shedbalkar UU, Singh R, Chopade BA (2016) Biogenic selenium nanoparticles: current status and future prospects. Appl Microbiol Biotechnol 100(6):2555–2566

Wang X, Liu M, Wang X, Wu Z, Yang L, Xia S, Chen L, Zhao J (2013) p–Benzoquinone–mediated amperometric biosensor developed with *Psychrobacter* sp. for toxicity testing of heavy metals. Biosens Bioelectron 41:557–562

Wang J, Zheng Y, Jia H, Zhang H (2014) Bioelectricity generation in an integrated system combining microbial fuel cell and tubular membrane reactor: effects of operation parameters performing a microbial fuel cell–based biosensor for tubular membrane bioreactor. Bioresour Technol 170:483–490

Weising K, Kahl G (1996) Natural genetic engineering of plant cells: the molecular biology of crown gall and hairy root disease. World J Microbiol Biotechnol 12(4):327–351

Williams TC, Pretorius IS, Paulsen IT (2016) Synthetic evolution of metabolic productivity using biosensors. Trends Biotechnol 34(5):371–381

Yang H, Zhou M, Liu M, Yang W, Gu T (2015) Microbial fuel cells for biosensor applications. Biotechnol Lett 37(12):2357–2364

Zhang F, Keasling J (2011) Biosensors and their applications in microbial metabolic engineering. Trends Microbiol 19(7):323–329

Zhang J, Jensen MK, Keasling JD (2015) Development of biosensors and their application in metabolic engineering. Curr Opin Chem Biol 28:1–8

Chapter 10
Cancer Bionanotechnology: Biogenic Synthesis of Metallic Nanoparticles and Their Pharmaceutical Potency

Maluta Steven Mufamadi, Jiya George, Zamanzima Mazibuko, and Thilivhali Emmanuel Tshikalange

Contents

10.1 Introduction

Cancer is one of the leading causes of death worldwide; about 8.2 million people die as a result of cancer each year. Proliferation of abnormal cells and uncontrolled growth lead to the transformation of normal cells into cancerous cells that finally results in death (Chow 2010; Siegel et al. 2017). There has been a marginal increase

M. S. Mufamadi (✉) · J. George
Nanotechnology and Biotechnology, Nabio Consulting, Pretoria, South Africa
e-mail: steven@nabioconsulting.co.za

Z. Mazibuko
Knowledge Economy and Scientific Advancement, Mapungubwe Institute for Strategic Reflection (MISTRA), Johannesburg, South Africa

T. E. Tshikalange
Department of Plant and Soil Sciences, University of Pretoria, Pretoria, South Africa

R. Prasad (ed.), *Microbial Nanobionics*, Nanotechnology in the Life Sciences,
https://doi.org/10.1007/978-3-030-16383-9_10

in cancer cases in the last few years in sub-Saharan African countries (Pace and Shulman 2016). Current cancer treatments are facing many challenges, such as side effects and nonspecific systemic distribution that leads to damage not only to the tumor tissue but also healthy tissues and cells. Therefore, there is an urgent need to find effective and safe anticancer agents (Wicki et al. 2015). The development of nanoparticles has paved new pathways and provided new avenue in cancer therapy with a drug delivery system targeted at the tumor site. This has since advanced the bioavailability, stability, absorption, solubility, and therapeutic effectiveness of many anticancer agents (Sakamoto et al. 2010; Diaz and Vivas-Mejia 2013; Prasad et al. 2017a). Nanotechnology is an emerging field of science and engineering that is promising to generate new medical applications such early detection, rapid and real-time diagnostics, and advanced drug/gene delivery systems for diseases like cancer, diabetes, cardiovascular, and infectious diseases (Lara et al. 2010; Prasad et al. 2014, 2017a, b). The advantage of using this technology is that it deals with the manipulation and synthesis of particles at an atomic scale and molecular level, i.e., 1–100 nm (Rajput et al. 2017; Barkhade 2018). Although physical and chemical methods are more used during the production of nanoparticles, both methods use toxic chemical that are not suitable for medical or pharmaceutical applications as it may have adverse effect. Chemical methods have negative effects on humans and the environment, are very expensive, and require high temperature, energy, and pressure for nanoparticle formation (Bhattacharya and Mukherjee 2008).

The biological synthesis of silver and gold nanoparticles using microorganisms has attracted great attention for medical and pharmaceutical applications due to its capability to produce effective antibacterial, antifungal, antiviral, and anticancer agents (Park et al. 2016; Prabhu and Poulose 2012; Otari et al. 2015; Prasad 2014; Prasad et al. 2016, 2018a, b). Moreover, biological synthesis promises to offer an inexpensive method for the production of metal nanoparticles that are nontoxic and environmentally friendly (Sadowski et al. 2008; Narayanan and Sakthivel 2010; Hulkoti and Taranath 2014; Monika et al. 2015). Microbial production of silver nanoparticles involves mixing a silver nitrate ($AgNO_3$) solution with microbes, which can take place either intracellularly or extracellularly (Bhainsa and D'Souza 2006; Balaji et al. 2009; Natarajan et al. 2010; Kalishwaralal et al. 2010; Sunkar and Nachiyar 2012; Siddiqi et al. 2018; Prasad 2016, 2017; Prasad et al. 2016, 2018a, b). Microbes act as reducing and capping agents of metal compounds, resulting in nanoparticles with different shapes and particle sizes ranging from 1 to 100 nm. During intracellular synthesis, nanoparticles were shown to be synthesized inside the microbes' cell cytoplasm, while in the extracellular method, nanoparticles were shown to be synthesized outside microbes employing cell-free culture supernatants (Shivaji et al. 2011; Nanda et al. 2012; Das et al. 2014; Otari et al. 2015). However, intracellular synthesis of metallic nanoparticles is considered to be more expensive when compared with extracellular synthesis and requires additional steps post nanoparticle formation in order to release the synthesized nanoparticles inside the cell. As a result, this makes intracellular synthesis more difficult when it comes to nanoparticle isolation and identification in the presence of microbes, making this method highly biohazardous. Extracellular synthesis is cheap, faster, eco-friendly, and safer than the intracellular approach. During extracellular methods, metal ion synthesis occurs outside of the cells, employing culture supernatants with biomolecules such as enzyme nitrate reductase, proteins, and biosurfactants as reducing, capping, and stabilizing agents (Sanghi and Verma 2009; Saravanan et al. 2011; Sathiyanarayanan et al. 2013; Lee et al. 2016; Aziz et al. 2015, 2016, 2019).

This chapter will provide comprehensive detail on biogenic synthesis of silver and gold nanoparticles using bacteria, fungi, and yeast through intracellular and extracellular routes. Additionally, the influence of utilizing different microbes for production of metal nanoparticles on the particles size and shape will be discussed. This chapter will also demonstrate the potential benefits and the mechanism of toxicity of nanoparticles on cancer cells as the potential anticancer agent. Furthermore, the chapter will look at the setbacks and future prospect on biological production of silver and gold nanoparticles with up-scalable industrial production.

10.2 Microbial Synthesis of Silver and Gold Nanoparticles

Biological production of Ag NPs and Au NPs using bacteria, fungi, and yeast is promising to offer cheaper and eco-friendly methods. Microbial synthesis of metallic nanoparticles involving microbes consist of two routes: intracellular and extracellular (Mukherjee et al. 2001; Gade et al. 2008; Rajeshkumar et al. 2013; Sarangadharan and Nallusamy 2015; Sowani et al. 2016). Intracellular production of metallic nanoparticles occurs inside cells in the presence of biomolecules that are released by microbes as reduction agents (Siddiqi et al. 2018). The intracellular mechanism involves the transportation of ions into the cell wall which influence the enzymes to convert the toxic metal ions into nontoxic metal nanoparticles (Malarkodi et al. 2013). In the extracellular route, the reduction of metal ions to form nanoparticles occurs on the outer cell membrane in the culture supernatants media without microbial cells (Maliszewska and Puzio 2009; Singh et al. 2013; Abo-State and Partila 2015). As in the intracellular route, the extracellular mechanism also involves enzyme activity that facilitates the reduction and capping of metallic ions into metallic nanoparticles (Malarkodi et al. 2013). The presence of biomolecules such as enzyme nitrate reductase, proteins, and biosurfactants facilitates the reduction of metal ions which also act as capping and/or stabilizing agents (Saravanan et al. 2011; Sathiyanarayanan et al. 2013; Lee et al. 2016; Joshi et al. 2018). Nitrate reductase enzyme in microorganisms is mainly responsible for the reduction of metal ions, as well as stabilizing during the formation of nanoparticles (Kumar et al. 2007; Marambio-Jones and Hoek 2010). It has also been demonstrated that the reduction process using microbes is responsible for production of metal nanoparticles with unique shape and particle size distribution (Mohseniazar et al. 2011; Zomorodian et al. 2016). The formation of Ag NPs in the microbial culture medium is also shown to be responsible for color change, from clear/light yellow to a light/dark brown color (Eugenio et al. 2016; Dhoondia and Chakraborty 2012; Singh et al. 2017). Post color change observation, the formation of nanoparticles was validated with surface plasmon absorption band(s) at about 430 nm using UV-visible spectroscopy (Korbekandi et al. 2013; Kumari and Ajeet 2016; Abo-State and Partila 2018). The formation of metal nanoparticles through metal ion reduction in the presence of the biomolecules as capping agents, either through intracellular or extracellular route, was confirmed by Fourier transform infrared (FTIR) spectroscopy (Singh et al. 2017). Nanoparticle size and morphology were characterized by transmission electron microscope (TEM), scanning electron microscope (SEM), atomic force microscopy (AFM), and X-ray diffraction pattern (Singh et al. 2014, Kumar and Ghosh 2016, Ajah et al. 2018). Particle size distribution and polydispersity were measured employing dynamic light scattering (DLS)/Zetasizer Nano ZA and Nano tracking analyzer (Zomorodian et al. 2016; Ghiutta et al. 2018; Luo et al. 2018; Wypij et al. 2018).

10.3 Silver Nanoparticle Synthesis Using Microbes

Several microbial strains bacteria, fungi, and yeast have been used for Ag NPs synthesis, either via intracellular or extracellular route. Microbes are used for Ag NPs production because they are easy to handle, grow in low-cost medium, maintain safety levels, and have the potential of adsorbing the metal ions and reducing them into nanoparticles (Kathiresan et al. 2009; Menon et al. 2017; Prasad et al. 2016). Different microbes or extracts in culture medium facilitated the production of Ag NPs with different size and shape (Tables 10.1 and 10.2).

10.3.1 Silver Nanoparticle Synthesis Using Bacteria

Table 10.1, shows different bacteria strains used for biosynthesis of Ag NPs with different particle size and shape. Very recently, Abo-State and Partila (2018) have reported extracellular biosynthesis of Ag NPs from four bacteria isolates including *Pseudomonas aeruginosa*, *Bacillus cereus*, *Achromobacter xylosoxidans*, and *Ochrobactrum* sp. Nanoparticles were formed post incubation of Ag NO_3 in the cell-free extracts in the dark for about 4 hours. The bacterial isolates produced spherical Ag NPs with particle size ranging from 7.8 to 13.4 nm. Ghiutta et al. (2018) reported production of Ag NPs using two *Bacillus* species, namely, *Bacillus amyloliquefaciens* and *Bacillus subtilis*. Brown color and UV-vis spectrophotometry results with the absorbance peak at 418 nm and 414 nm validated the formation of Ag NPs. It was reported that Ag NPs formed with a spherical shape and with an average diameter smaller than 140 nm employing SEM and DLS. Rajeshkumar et al. (2013) investigated biological production of Ag NPs using *Vibrio alginolyticus* using both routes, intracellular and extracellular biosynthesis. The Ag NPs were formed with the diameters in range of 50–100 nm. In a recent study, Luo et al. (2018) investigated the influence of nutrient broth and Mueller-Hinton broth used as growth media of bacterial culture in the reduction of silver ions to form Ag NPs. The study demonstrated both culture media to be responsible for the reduction and

Table 10.1 Silver nanoparticles synthesis using bacteria

No.	Bacteria	Location	Size (nm)	Shape	References
1	*Pseudomonas aeruginosa, Bacillus cereus, Achromobacter xylosoxidans*, and *Ochrobactrum* sp.	Extracellular	7.8 to 13.4	Spherical	Abo-State and Partila (2018)
2	*Bacillus amyloliquefaciens* and *Bacillus subtilis*	Extracellular	140	Spherical	Ghiutta et al. (2018)
3	*Streptomyces xinghaiensis OF1 strain*	Extracellular	5 to 50	Spherical	Wypij et al. (2018)
4	*Haemophilus influenza*	Extracellular	80 to 101	Spherical	Ajah et al. (2018)
5	*Streptomyces* sp. AOA21	Extracellular	35 to 60	Spherical	Adiguzel et al. (2018)
6	*Escherichia coli*	Extracellular	50	Spherical	Shah et al. (2017)
7	*Pseudomonas aeruginosa*	Extracellular	33 to 300	Spherical	Peiris et al. (2017)
8	*Bacillus licheniformis Dahb1*	Extracellular	18.69 to 63.42	Spherical	Shanthi et al. (2016)
9	*Brevibacillus borstelensis*	Extracellular	5 to 15	Spherical	Kumar and Ghosh (2016)
10	*Bacillus* sp. HAI4	Extracellular	33 to 264	Spherical	Taran et al. (2016)
11	*Escherichia coli*	Extracellular	20 to 50	Spherical	Kushwaha et al. (2015)
12	*Bacillus licheniformis*	Extracellular	3 to 130	Spherical	Sarangadharan and Nallusamy (2015)
13	*Actinotalea* sp. MTCC 10637	Extracellular	5 to 80	Spherical	Suman et al. (2014)
14	*Deinococcus radiodurans*	Extracellular	16.82	Spherical	Kulkarni et al. (2015)
15	*Pseudomonas aeruginosa*	Extracellular	50 to 85	Spherical	Paul and Sinha (2014)
16	*Bacillus* sp. JAPSK2	Extracellular	21.9	Spherical	Singh et al. (2014)
17	*Ochrobactrum anthropi*	Intracellular	38 to 85	Spherical	Thomas et al. (2014)
18	*Vibrio alginolyticus*	Intracellular and extracellular	50 to100	Spherical	Rajeshkumar et al. (2013)
19	*Enterobacter aerogenes*	Extracellular	25 to 35	Spherical	Karthik and Radha (2012)
20	*Escherichia coli*	Intracellular	10 to 50	Spherical	Muthukkumarasamy et al. (2012)
21	*Xanthomonas oryzae*	Extracellular	14.86	Spherical	Narayanan and Sakthivel (2013)
22	*Streptococcus thermophiles*	Extracellular	28 to 122	Spherical	El-Shanshoury et al. (2011)

Table 10.2 Silver nanoparticles synthesis using fungi and yeast

No1.	Microbes	Strains	Location	Size (nm)	Shape	References
1	Fungi	*Fusarium oxysporum*	Extracellular	21.3 to 37.3	Spherical	Ahmed et al. (2018)
2	Fungi	*Penicillium* sp.	Extracellular	9 to 15	Spherical	Barkhade (2018)
3	Fungi	*Emericella nidulans EV4*	Extracellular	10 to 20	Spherical	Rajam et al. (2017)
4	Fungi	*Rhizopus stolonifer*	Extracellular	9.47	Spherical	Rahim et al. (2017)
5	Fungi	*Ganoderma enigmaticum*	Extracellular	15 to 25	Spherical	Gudikandula et al. (2017)
6	Fungi	*Guignardia mangiferae*	Extracellular	5 to 30	Spherical	Balakumaran et al. (2015)
7	Fungi	*Trichoderma viride*	Extracellular	1 to 50	Spherical	Elgorban et al. (2016)
8	Fungi	*Aspergillus fumigatus, A. clavatus, A. niger, and A. flavus*	Extracellular	5 to 185	Spherical	Zomorodian et al. (2016)
9	Fungi	*Aspergillus oryzae*	Intracellular and Extracellular	6 to 26	Spherical	Phanjom and Ahmed (2015)
10	Fungi	*Fusarium oxysporum, Aspergillus niger, Alternaria solani*	Extracellular	5 to 25	Spherical	Juraifani and Ghazwani (2015)
11	Fungi	*Aspergillus niger, Fusarium oxysporum*	Extracellular	50 to 100	Spherical	Khan et al. (2014)
12	Fungi	*Macrophomina phaseolina*	Cell-free filtrate	5 to 40	Spherical	Chowdhury et al. (2014)
13	Fungi	*Fusarium oxysporum*	Intracellularly	25 to 50	Spherical	Korbekandi et al. (2013)
14	Fungi	*Trichophyton rubrum, Trichophyton mentagrophytes,* and *Microsporum canis*	Extracellular	50 to 100	Spherical	Moazeni et al. (2012)
15	Fungi	*Trichoderma reesei*	Extracellular	5 to 50	Spherical	Vahabi et al. (2011)
16	Fungi	*Trichoderma viride*	Extracellular	5 to 40	Spherical	Fayaz et al. (2010)
17	Fungi	*Aspergillus niger*	Extracellular	3 to 30	Spherical	Jaidev and Narasimha (2010)
18	Fungi	*Mucor hiemalis*	Extracellular	5 to 15	Spherical	Aziz et al. (2016)
19	Yeast	*Saccharomyces cerevisiae*	Extracellular	10 to 60	Spherical	Sowbarnika et al. (2018)
20	Yeast	*Saccharomyces* sp. BDU-XR1	Intracellular	8 to 17	Spherical	Ganbarov et al. (2017)
21	Yeast	*Cornitermes cumulans*	Extracellular	2 to 10	Circular	Eugenio et al. (2016)
22	Yeast	*Saccharomyces cerevisiae*	Extracellular	5 and 20	Spherical	Niknejad et al. (2015)
23	Yeast	*Saccharomyces cerevisiae, Rhodotorula glutinis, and Geotrichum candidum*	Extracellular	2.5 to 20	Spherical	Zahran et al. (2013)
24	Yeast	MKY3	Extracellular	2 to 5	Spherical	Kowshik et al. (2013), Korbekandi et al. (2013)

capping of $AgNO_3$ ions to form Ag NPs. The size and shape of Ag NPs produced by reacting silver ions and bacterial culture was shown to be mainly dependent on the culture media. It was concluded that the growth media of bacterial culture have significant influence in the biological production of metallic nanoparticles. Wypij et al. (2018) have reported production of Ag NPs from cell-free supernatant of *Streptomyces xinghaiensis* OF1 strain at room temperature. The physicochemical characterization and surface charges of Ag NPs were validated employing TEM (5–50 nm), Zetasizer and NTA (64–69 nm), and DLS, and the particles were shown to have negative zeta potential (−15.7 mV). Ajah et al. (2018) have demonstrated a high production of Ag NPs from culture supernatant of *Haemophilus influenzae* isolate. The extracellular study reported silver nitrate to be responsible for bioreduction of silver ions into Ag NPs. The formation of Ag NPs was confirmed by a brown color in the bacteria supernatant and the appearance of an absorption peak at 425 nm. Kushwaha et al. (2015) also demonstrated extracellular synthesis of Ag NPs from *E. coli* bacteria isolated from urine with diameters in the range 20 to 50 nm. Shahverdi et al. (2007) have reported rapid and extracellular biosynthesis of Ag NPs from enterobacteria. The interaction of $AgNO_3$ with the bacteria was achieved within 5 minutes in culture supernatants of enterobacteria. Kalimuthu et al. (2008) confirm biosynthesis of Ag NPs with crystal structure employing bacteria stain *B. licheniformis.*

10.3.2 Silver Nanoparticle Synthesis Using Fungi

Extracellular and intracellular biosynthesis of Ag NPs using pathogenic and non-pathogenic fungi has been reported, namely, *Fusarium oxysporum* (Ahmed et al. 2018); *Emericella nidulans* EV-4 (Rajam et al. 2017); *Raphanus sativus* (Singh et al. 2017); *Rhizopus stolonifer* (Rahim et al. 2017); *Ganoderma enigmaticum*; *Trametes ljubarsky* (Gudikandula et al. 2017); *Trichoderma viride* (Elgorban et al. 2016); *Aspergillus* species such *A. fumigatus*, *A. clavatus*, *A. niger*, and *A. flavus* (Zomorodian et al. 2016); *Aspergillus oryzae* (Phanjon and Ahmed 2015); *Neurospora intermedia* (Hamedi et al. 2014); *Eucalyptus chapmaniana* (Vadlapudi and Kaladha 2014); and *Penicillium nalgiovense* (Maliszewska et al. 2014); see Table 10.2. Ahmed et al. (2018) recently reported biosynthesis of Ag NPs from the isolate of filamentous fungus *F. oxysporum*. This extracellular synthesis presented Ag NPs with a spherical to oval shape and particle size with a diameter ranged from 21.3 to 37.3 nm. The nanoparticles showed to be very stable and with a peak at 408–411 nm of surface plasmon resonance remained unchanged for period more than 4 months under 4 °C. In a study conducted by Zomorodian et al. (2016), the extracellular biosynthesis of Ag NPs using *Aspergillus* species, namely, *A. fumigatus*, *A. clavatus*, *A. niger*, and *A. flavus*, was reported. Nitrate reductase enzyme activity during biotransformation of silver ions to Ag NPs in the presence of fungal isolates was confirmed by nitrate reduction test kit (Fluka 73,426, Sigma-Aldrich). The tests exhibited high nitrate reductase enzyme activity with three stars in *A. fumigates*, intermediate enzymatic activity in *A. clavatus* and *A. niger*, and lowest enzymatic activity in *A. flavus*. The formation of Ag NPs was confirmed by UV-vis spectroscopy and AFM. Another study by Jaidev and Narasimha (2010) also reported extracellular production of Ag NPs using *A. niger* as reducing and capping agent. Rahim et al. (2017) reported biological production of Ag NPs using *R. stolonifer* (mold). The stable Ag NPs formed extracellularly with a spherical shape and were monodisperse with a mean size of 9.47 nm. Gudikandula et al. (2017) have investigated the production of Ag NPs from two white rot fungi, namely, *G. enigmaticum* and *T. ljubarsky*. The study reported rapid reduction time of aqueous Ag^{+} to form Ag NPs during the incubation of silver ions together with these two fungi isolates. The color of the supernatant solution changed from light yellow to a brown color within 12 hours post incubation period showing production of Ag NPs. TEM imaging was used to confirm spherical to round shape of the Ag NPs, with a particle size range of 15 to 25 nm. Another study by Chan and Don (2013) also reported biosynthesis of Ag NPs using white rot fungi. Hamedi et al. (2014) investigated using nonpathogenic fungus *N. intermedia* for Ag NPs synthesis extracellularly. Controlled biosynthesis of Ag NPs was conducted in culture supernatant and cell-free filtrate of the fungus in the presence and absence of light. The results revealed stable Ag NPs formation with monodispersity in cell-free filtrate. The presence of light during incubation of silver ions in the cell-free filtrate was reported to have an influence toward the rate of reduction of metal ions into Ag NPs.

10.3.3 Silver Nanoparticle Synthesis Using Yeast

The studies on biological synthesis of Ag NPs using yeast fungi strains have also gained attention in recent years (Saravanan et al. 2013; Moghaddam et al. 2015; Niknejad et al. 2015; Waghmare et al. 2015; Eugenio et al. 2016; Korbekandi et al. 2016; Ganbarov et al. 2017; Sowbarnika et al. 2018). The advantage of using yeast strains for Ag NPs production is that they are easy to control in laboratory conditions, show rapid growth, and are inexpensive to cultivate (Kumar et al. 2011; Skalickova et al. 2017). In addition, they have a high secretion of enzyme, protein, and other metabolites that can act as reducing and stabilizing agents during the production of Ag NPs. In another study, yeast strain *Saccharomyces cerevisiae* was reported to be responsible for protein secrete that facilitated silver ion reduction to produce Ag NPs stable in alkaline medium, pH 11–12 (Zahran et al. 2013). Saravanan et al. (2013) reported the production of Ag NPs using baker's yeast *S. cerevisiae*. The study showed baker yeast to have capability for production of Ag NPs with a particle size range of 60 to 80 nm. In a similar study by Jha et al. (2008), the synthesis of Ag NPs with a spherical shape and size ranging from 6 to 20 nm using baker's yeast *S. cerevisiae* was reported. Ganbarov et al. (2017) recently showed intracellular production of Ag NPs employing *Saccharomyces* sp. BDU-XR1 isolated from spontaneous yogurt. SEM study revealed Ag NPs with an average size range of 8–17 nm. Ortega et al. (2015) also exhibited extracellular production of Ag NPs using yeast

Cryptococcus laurentii in the culture supernatants. Kumar et al. (2011) showcased the use of novel marine yeast *Candida* sp. VITDKGB for biosynthesis of Ag NPs. UV-visible spectrum with a peak at 430 nm and the color of culture supernatant changing from light pale yellow to brown indicated the formation of Ag NPs. AFM analysis confirmed the formation of Ag NPs with size around 87 nm. Waghmare et al. (2015) reported biosynthesis of Ag NPs using yeast stain *Candida utilis*. The Ag NPs formed and were reported to have a spherical shape and a size in the range of 20–80 nm. Kowshik et al. (2013) reported extracellular synthesis of Ag NPs with a diameter in the range of 2 to 5 nm using yeast strain MKY3.

10.4 Gold Nanoparticle Biosynthesis Using Microbes

Similar to Ag NPs, biosynthesis of Au NPs using microbes as green synthesis route has gained a lot of attention (Mourato et al. 2011; Pantidos and Horsfall 2014; Monika et al. 2015; Rao et al. 2017). Biosynthesis offers biocompatibility and non-toxicity and is an easier and inexpensive procedure for Au NPs production. In addition, the biosynthesis of Au NPs is easy to scale up for larger-scale production of nanoparticles at a lower cost and is able to act as antimicrobial and anticancer agent (Ramesh and Armash 2015; Menon et al. 2017). The secretion of enzyme methionine sulfoxide reductases, proteins, amino acids, carbohydrates, and other metabolites by microbes facilitated the process of reduction of gold ions into Au NPs. The reduction process of Au NPs occurs during the incubation period of gold ions and microbial cell cultures, either via the intracellular or extracellular route employing cell-free culture supernatants (Narayanan and Sakthivel 2011; Glisic et al. 2012; Kar et al. 2014). Au NPs biosynthesis using bacteria, fungi, and yeast is summarized in Table 10.3. In a recent study by Montero-Silva (2017), it was reported that the mechanism of gold ion reduction during extracellular biosynthesis was influenced by oxidation of macromolecules that are located in the bacteria cellular membranes. The production of Au NPs in microbial culture medium was observed by color change from a light yellow to a dark red/purple color. Gold ions incubated with diverse microorganisms produced Au NPs with different shapes such as spherical, rod, quasihexagonal, pentagonal shape, and/or nanocrystal (Narayanan and Sakthivel 2011). Meysam et al. (2014) have reported biological production of Au NPs with a particle size range of 20–50 nm using bacteria strain *Streptomyces fulvissimus* isolate. Extracellular production of Au NPs was achieved by exposure of bacterial biomass with auric chloride (AuCL4) ions. The formation of Au NPs was characterized by color of the reaction solution that changed from a yellow color to a dark red color and UV-visible spectrum with a peak at 550 nm. A recent study by Ranjitha and Rai (2017) demonstrated good catalytic activity for the degradation of methylene blue using *Streptomyces griseoruber* for the production of Au NPs in a bacterial culture supernatant. UV-visible spectrophotometer results showed the presence of peak absorbance between 520 and 550 nm and confirmed the formation of Au NPs (5–50 nm) during bioreduction. Li et al. (2016) reported rapid

Table 10.3 Gold nanoparticles synthesis using bacteria, fungi, and yeast

No1.	Microbes	Species	Location	Size (nm)	Shape	Reference
1	Bacteria	*Streptomyces griseoruber*	Extracellular	5 to 50	Crystalline	Ranjitha and Rai (2017)
2	Bacteria	*Brevibacillus formosus*	Extracellular	5 to 12	Spherical	Srinath et al. (2017)
3	Bacteria	*Cupriavidus metallidurans* and *Escherichia coli*	Extracellular	20 to 60	Spherical and truncated triangular	Montero-Silva (2017)
4	Bacteria	*Salmonella enterica*	Extracellular	42	Crystalline	Mortazavi et al. (2017)
5	Bacteria	*Deinococcus radiodurans*	Extracellular	43.75	Spherical, triangular, and irregular	Li et al. (2016)
6	Bacteria	*Klebsiella pneumonia*	Intracellular	10 and 15	Spherical	Prema et al. (2016)
7	Bacteria	*Streptomyces fulvissimus*	Extracellular	20 to 50	Crystalline and spherical	Meysam et al. (2014)
8	Bacteria	*Rhodopseudomonas capsulata*	Microbial source	10 to 20	Spherical	Singh and Kundu (2014)
9	Bacteria	*Shewanella oneidensis*	Extracellular	2 to 50	Spherical	Suresh et al. (2011)
10	Fungi	*Thermophilic filamentous*	In/extracellular	6 to 40	Spherical and hexagonal	Molnar et al. (2018)
11	Fungi	*Fusarium oxysporum*	Extracellular	21.82	Spherical	Gitanjali and Ashok (2015)
12	Fungi	*Fusarium solani*	Extracellular	20 to 50	Spherical	Gopinath and Arumugam (2014)
13	Fungi	*Aspergillus terreus*	Extracellular	10 to 19	Spherical to rod	Priyadarshini et al. (2014)
14	Fungi	*Aspergillus sydowii*	Extracellular	8.7 to 15.6	Spherical	Vala (2014)
15	Fungi	*Aspergillus clavatus*	Intracellular	20 to 35	Triangular, spherical, and hexagonal	Verma et al. (2011)
16	Fungi	*Neurospora crassa*	Intracellular	3 to 100	Spherical	Castro-Longoria et al. (2011)
17	Fungi	*Phanerochaete Chrysosporium*	Intracellular	10 to 100	Spherical	Sheikhloo and Salouti (2011)
18	Fungi	*Fusarium semitectum*	Extracellular	10 to 35	Spherical	Barabadi et al. (2014)
19	Yeast	*Magnusiomyces ingens*	Cell-free extracts	28.3 to 20.3	Spherical and pseudo-spherical	Qu et al. (2018)
20	Yeast	*Phaffia rhodozyma*	Cell-free extracts	4 to 7	Spherical	Ronavari et al. (2018)
21	Yeast	*Candida cylindracea*	Extracellular	10 to 30	Spherical and triangular	Noorbatch et al. (2014)
22	Yeast	*Candida guilliermondii*	Extracellular	50 to 70	Spherical	Mishra et al. (2011)

biosynthesis of gold nanoparticles using *Deinococcus radiodurans* with the reaction time of 8 hours. Fungi strain, namely, *Thermophilic filamentous* was reported to produce HAu NPs with different sizes ranging from 6 to 40 nm (Molnar et al. 2018). Srinath et al. (2017) demonstrated on the used gold mine bacteria *Brevibacillus formosus* to produce Au NPs. The interaction of 1 mM hydrogen tetrachloroaurate ($HAuCl_4$) with the bacteria was done in the microwave oven to produce Au NPs with an average particle size of 5–12 nm and a spherical shape. Gitanjali and Ashok C (2015) used the fungus *F. oxysporum* for the biosynthesis of spherical and ellipsoid Au NPs with an average size of 21.82 nm. Another study by Gopinath and Arumugam (2014) reported biosynthesis of Au NPs using fungus *F. solani*. The interaction of the $HAuCl_4$ solution with *F. solani* was reported to occur through the extracellular route in the culture filtrate. The reduction and formation of Au NPs was observed by fungal culture filtrate medium color change from transparent liquid filtration to purple or pink post 24 hours mixture incubation at 30 °C. TEM analysis revealed the formation of stable Au NPs with a spherical shape and size ranging from 20 to 50 nm. Zhang et al. (2016) exhibited biosynthesized Au NPs using yeast strain *Magnusiomyces ingens* and the catalytic activity of nitrophenols, a reducing agent. SEM and TEM images were used for confirmation of nanoparticles with different shapes, including spherical, hexagon, and triangular shape. Many other studies were reported for biosynthesis of Au NPs using yeast strains: actinomycete *Gordonia amicalis* HS-11 (Sowani et al. 2016) and *Yarrowia lipolytica* (Nair et al. 2013).

10.5 The Mechanisms and Anticancer Activity of Biogenerated Metal Nanoparticles

The efficacy of biogenerated metal nanoparticles on various human cancer cells including cervical carcinoma, colon cancer, breast cancer, ovarian cancer, brain tumor, and prostate cancer cells has been studied and reported (Chithrani and Chan 2007; Zhang and Gurunathan 2016; Nagajyothi et al. 2017; Buttacavoli et al. 2018; Miri et al. 2018). It was also reported that nanoparticle size and shape influence distribution in the bloodstream and uptake in tumor tissues (Nativo et al. 2008; AshaRani et al. 2009; Ng et al. 2015; Truong et al. 2015; Bregoli et al. 2016). Although biogenerated metal nanoparticles possess higher anticancer properties, intensive investigation of the mechanism of uptake of nanoparticles to induce toxicity is needed in order to confirm efficacy and clinical suitability. According to AshaRani et al. (2009), the cell uptake involves molecules entering the cells through plasma membrane via active transport or passive diffusion. Active transport is energy dependent, while passive transport requires no energy.

The uptake of NPs into the cells mainly occurs through an active transport called endocytosis. During endocytosis, the NPs are engulfed in plasma membrane invaginations that are pinched off to form membrane-bound vesicles called endosomes (phagosomes, in case of phagocytosis). The endosomes then deliver the

NPs to various intracellular compartments. Endocytosis can be broadly divided into two categories: phagocytosis (the uptake of large particles) and pinocytosis (the uptake of fluids and solutes). Phagocytosis takes place in specialized cells such as macrophages, monocytes, neutrophils, and dendritic cells. Pinocytosis, on the other hand, is present in all cell types and has different forms depending on the cell origin and function. There are several different classifications of pinocytosis, and these include the (a) clathrin-mediated endocytosis, (b) caveolae-mediated endocytosis, (c) clathrin- and caveolae-independent endocytosis, and (d) macropinocytosis (Sahay et al. 2010). Clathrin- and caveolae-mediated endocytosis indicates receptor-mediated endocytosis. In receptor-mediated endocytosis (RME), specific molecules combine with receptor proteins embedded in the plasma membrane. A molecule that binds specifically to a receptor is called a ligand. Ligands such as folate, albumin, and cholesterol facilitate uptake through caveolae-mediated endocytosis, whereas ligands for glycoreceptors facilitate clathrin-mediated endocytosis. Macropinocytosis is enabled by the incorporation of cell-penetrating peptides, such as TaT (*trans*-activating transcriptional activator) peptide into the design of the engineered nanoparticle (Stern et al. 2012; Petros and DeSimone 2010; Sahay et al. 2010).

Clathrin- and caveolae-mediated endocytotic pathways are the preferred mode of uptake pathways for NPs because the plasma proteins are adsorbed onto the surface of NPs when exposed to physiological conditions. Generally, large particles (>500 nm) are taken up by phagocytosis, whereas smaller particles (<100 nm) enter via the RME pathways (Oh and Park 2014; Alkilany and Murphy 2010). NPs can be internalized into many types of cells through the plasma membrane. The internalization is highly dependent on the NPs size, shape, surface properties (charge and coating), as well as the cell type (Oh and Park 2014; Stern et al. 2012). It is important to know the exact mode of endocytosis because it determines the site of drug release, which can be the acidic lysosomal compartment, cytosol, or another organelle, which, in turn, determines the efficacy of the drug and its toxicity on cells. Co-localization of NPs with specific endocytosis markers or exclusion of specific endocytotic mechanism by inhibitors of endocytosis can be used to study the mode of uptake pathway of NPs. For example, actin filaments play a crucial role in the phagocytotic process of cellular uptake and inhibition of actin polymerization by cytochalasin D disrupts this phagocytotic mode of uptake (Sahay et al. 2010).

Ag NPs and Au NPs have shown promising results in cancer treatment. Easy synthesis, chemical stability, and ease of surface modifications on Au NPs make them highly suitable for targeted drug delivery, especially in cancer cells (Ng et al. 2015; Alkilany and Murphy 2010). Ag NPs have shown to be more toxic to normal cells as compared to Au NPs. Nonetheless, the antimicrobial, antiproliferative, and apoptosis-inducing properties of (Ag NPs) make them ideal candidates for anticancer therapy. For both these metallic NPs, RME has been proposed to be the major uptake pathway (Ng et al. 2015; Alkilany and Murphy 2010; AshaRani et al. 2009). Studies have shown that transferrin-coated Au NPs of different sizes (14, 50, and 74 nm) were taken up by the clathrin-mediated endocytosis in STO fibroblasts and HeLa cervical cancer and SNB19 brain tumor cell lines (Chithrani and Chan 2007). Nativo et al. (2008) showed that both clathrin- and caveolae-mediated endocytosis are involved in the uptake of PEG-modified 16 nm Au NPs in HeLa cells. Macropinocytosis was found to be the main uptake mechanism in a study conducted using gold nanorods and nanospheres (15–50 nm) by Bartneck et al. (2010). In this study they showed that gold nanorods were taken up more efficiently by macrophages as compared to the nanospheres. AshaRani et al. (2009) observed clathrin-mediated endocytosis and macropinocytosis as the main pathways in Ag NPs (6–20 nm) cellular uptake in human lung fibroblasts and human glioblastoma cells. In some instances, more than one entry point is involved in the uptake of NPs. Gliga et al. (2014) showed that 10 nm and 75 nm citrate-coated Ag NPs were taken up by a combination of endocytotic mechanisms (clathrin-mediated, caveolae-mediated, macropinocytosis, and phagocytosis) in human (BEAS-2B) lung cells.

Phagocytosis is a process where the cell ingests large solid particles such as bacteria and food. The ingested material is engulfed by the folds of plasma membrane to form a vacuole, which then fuses with lysosome to form phagolysosome. The ingested materials in the phagolysosomes are then degraded by enzyme-catalyzed hydrolysis. Clathrin-mediated endocytosis involves the formation of clathrin-coated pits on plasma membrane, which are pinched off to form endocytic vesicles (early and late endosomes) that fuse with lysosomes, an acidic environment prone to NPs degradation. Caveolae-mediated endocytosis is flask-shaped invaginations of plasma membrane forming caveolar vesicle. The advantage of this pathway is that they can bypass the lysosomal degradation pathway. Macropinocytosis occurs in many cells, including macrophages, and involves formation of actin-dependent membrane protrusions (similar to phagocytosis) and form large endocytic vesicle called macropinosomes. They eventually fuse with the lysosomal compartment or recycle their content to the surface. This pathway serves as a nonspecific entry point and internalizes large particles in cells, which lack phagocytosis (Sahay et al. 2010; Singh et al. 2010; Hillaireau and Couvreur 2009). Once the NPs are internalized, the subcellular distribution of NPs to different organelles inside the cell depends on the surface charge and protein-protein interactions (Singh et al. 2010). By surface modification of NPs (size, surface charge, coating, and ligands), it is possible to achieve targeted drug delivery to the desired organelles via the preferred mechanism of uptake. For example, caveolae were found to be enriched with aminopeptidase P (APP). Au NPs modified with antibody to APP were shown to be transported across endothelium to tumor lung tissue via this pathway (Sahay et al. 2010). Albumin-coated Ag NPs (uptake via albumin receptor-mediated endocytosis and localized in mitochondria) showed high potential in targeted drug delivery against an invasive human breast cancer cell line (Azizi et al. 2017). The cell type also plays a critical role in determining the nanoparticle entry and intracellular location in the cells. Because of the diversity of nanomaterials and cells used in the intracellular trafficking studies, it is challenging to find common factors that define the intracellular transport of nanomaterials (Sahay et al. 2010).

10.6 Future Prospect of Biological Synthesis of Metallic Nanoparticles

The possibility of future success in biosynthesis of metallic nanomaterials (Ag NPs and Au NPs) employing microbes promises to offer a new generation of antimicrobials and anticancer agents that are effective and inexpensive. However, the following setbacks on biosynthesis of Ag NPs and Au NPs still need intensive investigation:

1. The mechanism of biosynthesis of metallic nanoparticles is not clearly understood, both extracellular and intracellular synthesis. The ability to understand this mechanism fully could facilitate the manipulation of the biosynthesis process in order to produce nanoparticles that are stable with the desired shape, particle size, and monodispersity. Many studies showed the antibacterial and anticancer or uptake of Ag NPs and Au NPs by cancer cell to be influenced by the size (1–100 nm) and shape of these particles.
2. Although green synthesis is able to scale up for larger-scale production of metallic nanoparticles, industrial production with high yield will require optimization of the cell growth and bioreduction conditions in the reaction mixture.
3. The influence of microbial culture media nutrients and potential microbial contaminations need to be investigated as it could lead to false results.
4. Intracellular biosynthesis is associated with the following difficulties: nanoparticle isolation and identification requires extra processing phases like ultra-sonication in order to achieve maximum results. Due to these challenges, this method is considered by others to be a biohazardous.
5. In-depth study on Ag NPs and Au NPs properties in mammalian immune system and in vivo toxicity study are recommended in order to confirm safety and clinical significance.
6. Due to the ability of the biosynthesized Ag NPs and Au NPs employing microbes on medical and pharmaceutical applications, an increasing awareness and research investment in green route synthesis is crucial in order for its potential to be realized.

10.7 Conclusion

Biosynthesis of metallic nanoparticles using microbes has gained a lot of attention in the recent years. Regardless of the microbial strain used, namely, bacteria, fungi, and yeast, or the method used for fabrication of metallic nanoparticles, intracellular or extracellular route, all promised to offer inexpensive and environmentally friendly techniques. However, a large amount of research has been shown to focus on the extracellular route employing cell-free culture supernatants. The advantage of using this biological route over the intracellular route is that it is capable of rapid metal ion reductions during Ag NPs and Au NPs formation, and it does not require an extra processing step such as ultra-sonication for particles isolation. The formations of metallic nanoparticles were shown to be influenced by the release of microbial enzymes and/or oxidation of macromolecules that are located in their cellular membrane. As a potential anticancer agent, metallic nanoparticles showed to induce cytotoxicity that causes stress to cancer cells. The mechanism for metallic nanoparticle cytotoxicity is shown to be influenced by cancer cell uptake, such as clathrin-mediated endocytosis, caveolae-mediated endocytosis, macropinocytosis, and phagocytosis resulting in cancer cell damage. In order to understand the long-term effect of Ag NPs and Au NPs on human health, intensive in vivo studies are needed to confirm safety and clinical significance.

References

Abo-State MAM, Partila AM (2015) Microbial production of silver nanoparticles by *Pseudomonas aeruginosa* cell free extract. J Ecol Health Environ 3:91–98

Abo-State MAM, Partila AM (2018) Production of silver nanoparticles (AgNPs) by certain bacterial strains and their characterization. Novel Res Microbiol J 1(2):19–32

Adiguzel AO, Adiguzel SK, Mazmanci B, Tunçer M, Mazmanci MA (2018) Silver nanoparticle biosynthesis from newly isolated streptomyces genus from soil. Mater Res Express 5(4)

Ahmed A, Hamzah H, Maaroof M (2018) Analyzing formation of silver nanoparticles from the filamentous fungus *Fusarium oxysporum* and their antimicrobial activity. Turk J Biol 42:54–62

Ajah HA, Khalaf KJ, Hassan AS, Hassan Ali Aja HA (2018) Extracellular biosynthesis of silver nanoparticles by *Haemophilus influenzae* and their antimicrobial activity. J Pharm Sci Res 10(1):175–179

Alkilany AM, Murphy CJ (2010) Toxicity and cellular uptake of gold nanoparticles: what we have learned so far? J Nanopart Res 12(7):2313–2333

AshaRani PV, Hande MP, Valiyaveettil S (2009) Anti-proliferative activity of silver nanoparticles. BMC Cell Biol 10(1):65

Aziz N, Faraz M, Pandey R, Sakir M, Fatma T, Varma A, Barman I, Prasad R (2015) Facile algae-derived route to biogenic silver nanoparticles: synthesis, antibacterial and photocatalytic properties. Langmuir 31:11605–11612. https://doi.org/10.1021/acs.langmuir.5b03081

Aziz N, Pandey R, Barman I, Prasad R (2016) Leveraging the attributes of *Mucor hiemalis*-derived silver nanoparticles for a synergistic broad-spectrum antimicrobial platform. Front Microbiol 7:1984. https://doi.org/10.3389/fmicb.2016.01984

Aziz N, Faraz M, Sherwani MA, Fatma T, Prasad R (2019) Illuminating the anticancerous efficacy of a new fungal chassis for silver nanoparticle synthesis. Front Chem 7:65. https://doi.org/10.3389/fchem.2019.00065

Azizi M, Ghourchian H, Yazdian F, Bagherifam S, Bekhradnia S, Nyström B (2017) Anti-cancerous effect of albumin coated silver nanoparticles on MDA-MB 231 human breast cancer cell line. Sci Rep 7(1):5178

Balaji D, Basavaraja S, Deshpande R, Bedre M, Prabhakara B, Venkataraman A (2009) Extracellular biosynthesis of functionalized silver nanoparticles by strains of *Cladosporium cladosporioides* fungus. Colloids Surf B: Biointerfaces 68:88–92

Balakumaran MD, Ramachandran R, Kalaichelvan PT (2015) Exploitation of endophytic fungus, *Guignardia mangiferae* for extracellular synthesis of silver nanoparticles and their in vitro biological activities. Microbiol Res 178:9–17

Barabadi H, Honary S, Ebrahimi P, Ali MM, Alizadeh A, Naghibi F (2014) Microbial mediated preparation, characterization, and optimization of gold nanoparticles. *Braz J Microbiol* 451:493–150

Barkhade T (2018) Extracellular biosynthesis of silver nanoparticles using fungus *Penicillium* species. Int J Res – Granthaalayah 6(1):277–283

Bartneck M, Keul HA, Singh S, Czaja K, Bornemann J, Bockstaller M, Moeller M, Zwadlo-Klarwasser G, Groll J (2010) Rapid uptake of gold nanorods by primary human blood phagocytes and immunomodulatory effects of surface chemistry. ACS Nano 4(6):3073–3086

Bhainsa CK, D'Souza FS (2006) Extracellular biosynthesis of silver nanoparticles using the fungus *Aspergillus fumigatus*. Colloids Surf B: Biointerfaces 47:160–164

Bhattacharya R, Mukherjee P (2008) Biological properties of "naked" metal nanoparticles. Adv Drug Deliv Rev 60:1289–1306

Bregoli L, Movia D, Gavigan-Imedio JD, Lysaght J, Reynolds J, Prina-Mello A (2016) Nanomedicine applied to translational oncology: a future perspective on cancer treatment. Nanomedicine 12(1):81–103

Buttacavoli M, Albanese NN, Cara GD, Alduina R, Faleri C, Gallo M, Pizzolanti G, Gallo G, Feo S, Baldi F, Cancemi P (2018) Anticancer activity of biogenerated silver nanoparticles: an integrated proteomic investigation. Oncotarget 9(11):9685–9705

Castro-Longoria E, Vilchis-Nestor AR, Avalos-Borja M (2011) Biosynthesis of silver, gold and bimetallic nanoparticles using the filamentous fungus *Neurospora crassa*. Colloids Surf B Biointerfaces 83:42–48

Chan Y, Don MM (2013) Biosynthesis and structural characterization of ag nanoparticles from white rot fungi. Mater Sci Eng C 33:282–288

Chithrani BD, Chan WC (2007) Elucidating the mechanism of cellular uptake and removal of protein-coated gold nanoparticles of different sizes and shapes. Nano Lett 7(6):1542–1550

Chow AY (2010) Cell cycle control by oncogenes and tumor suppressors: driving the transformation of normal cells into cancerous cells. Nat Educ 3(9):7

Chowdhury S, Basu A, Kundu S (2014) Green synthesis of protein capped silver nanoparticles from phytopathogenic fungus *Macrophomina phaseolina* Goid with antimicrobial properties against multidrug-resistant bacteria. Nanoscale Res Lett 9:365

Das VL, Thomas R, Varghese RT, Soniya EV, Mathew J, Radhakrishnan EK (2014) Extracellular synthesis of silver nanoparticles by the *Bacillus* strain CS 11 isolated from industrialized area. Biotechnology 4:121–126

Dhoondia ZH, Chakraborty H (2012) *Lactobacillus* mediated synthesis of silver oxide nanoparticles. Nanomat Nanotechnol 2(15):1–7

Diaz MR, Vivas-Mejia PE (2013) Nanoparticles as drug delivery systems in cancer medicine: emphasis on RNAi-containing Nanoliposomes. Pharmaceuticals 6:1361–1380

Elgorban AM, Al-Rahmah AN, Sayed SR, Hirad A, Mostafa AA, Bahkali AH (2016) Antimicrobial activity and green synthesis of silver nanoparticles using *Trichoderma viride*. Biotechnol Biotechnol Equip 30(2):299–304

El-Shanshoury AER, ElSilk SE, Ebeid ME (2011) Extracellular biosynthesis of silver nanoparticles using *Escherichia coli* ATCC 8739, Bacillus subtilis ATCC 6633, and *Streptococcus thermophilus* ESh1 and their antimicrobial activities. ISRN Nanotechnol. https://doi.org/10.5402/2011/385480

Eugenio M, Muller N, Frases S, Almeida-Paes R, Lima LMR, Lemgruber L, Farina M, de Souza W, Sant'Anna C (2016) Yeast-derived biosynthesis of silver/silver chloride nanoparticles and their antiproliferative activity against bacteria. R Soc Chem 6:9893–9904

Fayaz M, Tiwary CS, Kalaichelvan PT, Venkatesan R (2010) Blue orange light emission from biogenic synthesized silver nanoparticles using *Trichoderma viride*. Colloids Surf B: Biointerfaces 75:175–178

Gade AK, Bonde P, Ingle AP, Marcato PD, Durän N et al (2008) Exploitation of *Aspergillus niger* for synthesis of silver nanoparticles. J Biobaased Mater Bioenergy 2:243–247

Ganbarov KG, Jafarov MM, Ramazanov MA, Agamaliyev ZA, Eyvazova GM (2017) Biosynthesis of silver nanoparticles using *Saccharomyces* sp. strain BDU-XR1species. Deutscher Wissenschaftsherold – German Science Herald, N 1/7–9. https://doi.org/10.19221/201712

Ghiutta I, Cristea D, Croitoru C, Kost J, Wenkrt R, Vyrides I, Anayiotos A, Munteanu D (2018) Characterization and antimicrobial activity of silver nanoparticles, biosynthesized using *Bacillus* species. Appl Surf Sci 438:66–73

Gitanjali H, Ashok C (2015) Synthesis, characterization and stability of gold nanoparticles using the fungus *Fusarium oxysporum* and its impact on seed germination. Int J Rec Sci Res 6(3):3181–3185

Gliga AR, Skoglund S, Wallinder IO, Fadeel B, Karlsson HL (2014) Size dependent cytotoxicity of silver nanoparticles in human lung cells: the role of cellular uptake, agglomeration and ag release. Part Fibre Toxicol 11(1):11

Glisic BD, Rychlewska U, Djuran MI (2012) Reactions and structural characterization of gold(III) complexes with amino acids, peptides and proteins. Dalton Trans 41:6887–6901

Gopinath K, Arumugam A (2014) Extracellular mycosynthesis of gold nanoparticles using *Fusarium solani*. Appl Nanosci 4:657–662

Gudikandula K, Vadapally P, Charya S (2017) Biogenic synthesis of silver nanoparticles from white rot fungi: their characterization and antibacterial studies. Open Nano:64–78

Hamedi S, Shojaosadati SA, Shokrollahzadeh S, Hashemi-Najafabadi S (2014) Extracellular biosynthesis of silver nanoparticles using a novel and non-pathogenic fungus, *Neurospora intermedia*: controlled synthesis and antibacterial activity. W J Microbiol Biotechnol 30(2):693–704

Hillaireau H, Couvreur P (2009) Nanocarriers' entry into the cell: relevance to drug delivery. Cell Mol Life Sci 66(17):2873–2896

Hulkoti NI, Taranath TC (2014) Biosynthesis of nanoparticles using microbes- a review. Colloids Surf B Biointerfaces 121:474–483

Jaidev LR, Narasimha G (2010) Fungal mediated biosynthesis of silver nanoparticles, characterization and antimicrobial activity. Colloids Surf B Biointerfaces 81(2):430–433

Jha AK, Prasad K, Kulkarni AR (2008) Yeast mediated synthesis of silver nanoparticles. IJNN 4(1):17–22

Joshi N, Jain N, Pathak A, Singh J, Prasad R, Upadhyaya CP (2018) Biosynthesis of silver nanoparticles using *Carissa carandas* berries and its potential antibacterial activities. J Sol-Gel Sci Techn 86(3):682–689. https://doi.org/10.1007/s10971-018-4666-2

Juraifani AAAA, Ghazwani AA (2015) Biosynthesis of silver nanoparticles by *Aspergillus niger, Fusarium oxysporum* and *Alternaria solani*. Afr J Biotechnol 14(26):2170–2174

Kalimuthu K, Babu RS, Venkataraman D, Bilal M, Gurunathan S (2008) Biosynthesis of silver nanocrystals by *Bacillus licheniformis*. Colloids Surf B: Biointerfaces 65(1):150–153

Kalishwaralal K, Deepak V, Pandian SRK, Kottaisamy M, BarathManiKanth S, Kartikeyan B, Gurunathan S (2010) Biosynthesis of silver and gold nanoparticles using *Brevibacterium casei*. Colloids Surf B: Biointerfaces 77:257–262

Kar PK, Murmu S, Saha S, Tandon V, Acharya K (2014) Anthelmintic efficacy of gold nanoparticles derived from a Phytopathogenic Fungus, *Nigrospora oryzae*. PLoS One 9(1):e84693. https://doi.org/10.1371/journal.pone.008469

Karthik C, Radha KV (2012) Biosynthesis and characterization of silver nanoparticles using *Enterobacter aerogenes*: a kinetic approach. Dig J Nanomater Biostruct 7:1007–1014

Kathiresan K, Manivannan S, Nabeel M, Dhivya B (2009) Studies on silver nanoparticles *synthesized by a marine fungus, Penicillium fellutanum* isolated from coastal mangrove sediment. Colloids Surf B Biointerfaces 71:133–137

Khan RH, Yasmeen K, Kishor K (2014) Biological synthesis and characterization of silver nanoparticles from *Fusarium oxysporum*. Der Pharmacia Sinica 5(5):112–117

Korbekandi H, Ashari Z, Iravani S, Abbasi S (2013) Optimization of biological synthesis of silver nanoparticles using Fusarium oxysporum. Iran J Pharm Res 12(3):289–298

Korbekandi H, Mohseni S, Mardani Jouneghani R, Pourhossein M, Iravani S (2016) Biosynthesis of silver nanoparticles using *Saccharomyces cerevisiae*. Artif Cells, Nanomed Biotechno 44(1):235–239

Kowshik M, Ashtaputre S, Kharrazi S et al (2013) Extracellular synthesis of silver nanoparticles by a silver-tolerant yeast strain MKY3. Nanotechnology 14(1):95–100

Kulkarni RR, Shaiwale NS, Deobagkar DN, Deobagkar DD (2015) Synthesis and extracellular accumulation of silver nanoparticles by employing radiation-resistant *Deinococcus radiodurans*, their characterization, and determination of bioactivity. Int J Nanomedicine 10:963–974

Kumar A, Ghosh A (2016) Biosynthesis and characterization of silver nanoparticles with bacterial isolate from Gangetic-alluvial soil. Int J Biotechnol Biochem 12(2):95–102

Kumar AS, Abyaneh MK, Gosavi SW, Kulkarni SK, Pasricha R, Ahmad A, Khan MI (2007) Nitrate reductase-mediated synthesis of silver nanoparticles from AgNO3. Biotechnol Lett 29:439–445

Kumar D, Karthik L, Kumar G, Roa KB (2011) Biosynthesis of silver nanoparticles from marine yeast and their antimicrobial activity against multidrug resistant pathogens. Pharmacol Online 3:1100–1111

Kumari J, Ajeet S (2016) Green synthesis of nanostructured silver particles and their catalytic application in dye degradation. J Genet Eng Biotechnol 14(2):311

Kushwaha A, Singh VK, Bhartariya J, Singh P, Yasmeen K (2015) Isolation and identification of *E. coli* bacteria for the synthesis of silver nanoparticles: characterization of the particles and study of antibacterial activity. European J Exp Biol 5(1):65–70

Lara HH, Ayala-nunez NV, Ixtepan-Turrent L, Rodriguez-Padilla C (2010) Mode of antiviral action of silver nanoparticles against HIV-1. J Nanobiotechnol 8:1–10

Lee JH, Lim JM, Velmurugan P, Park YJ, Park YJ, Bang KS, Oh BT (2016) Photobiologic-mediated fabrication of silver nanoparticles with antibacterial activity. J Photochem Photobiol B Biol 62:93–99

Li J, Li Q, Ma X, Tian B, Li T, Yu J, Dai S, Weng Y, Hua Y (2016) Biosynthesis of gold nanoparticles by the extreme bacterium *Deinococcus radiodurans* and an evaluation of their antibacterial properties. International J Nanomed 11:5931–5944

Luo K, Jung S, Park K, Kim Y (2018) Microbial biosynthesis of silver nanoparticles in different culture media. J Agric Food Chem 66(4):957–962

Maliszewska I, Puzio M (2009) Extracellular biosynthesis and antimicrobial activity of silver nanoparticles. Acta Phys Pol A 116:S60–S62

Maliszewska I, Juraszek A, Bielska K (2014) Green synthesis and characterization of silver nanoparticles using ascomycota fungi *Penicillium nalgiovense*. J Clust Sci 25:989–1004

Malarkodi C, Rajeshkumar S, Vanaja M, Paulkumar K, Gnanajobitha G, Annadurai G (2013) Eco-friendly synthesis and characterization of gold nanoparticles using *Klebsiella pneumoniae*. J Nanostruct Chem 3(1):30

Marambio-Jones C, Hoek EM (2010) A review of the antibacterial effects of silver nanomaterials and potential implications for human health and the environment. J Nanopart Res 12(5):1531–1551

Menon S, Rajeshkumar S, Kumar VS (2017) A review on biogenic synthesis of gold nanoparticles, characterization, and its applications. Res-Effic Technol 3:516–527

Meysam SN, Hosein SBG, Naimeh K (2014) Biosynthesis of gold nanoparticles using *Streptomyces fulvissimus* isolate. Acta Phys Pol A 2(2):153–159

Miri A, Darroudi M, Entezari R, Sarani M (2018) Biosynthesis of gold nanoparticles using *Prosopis farcta* extract and its in vitro toxicity on colon cancer cells. Res Chem Intermed 44:1–9

Mishra A, Tripathy SK, Yun SI (2011) Bio-synthesis of gold and silver nanoparticles from *Candida guilliermondii* and their antimicrobial effect against pathogenic bacteria. J Nanosci Nanotechnol 11(1):243–248

Moazeni M, Rashidi N, Shahverdi AR, Noorbakhsh F, Rezaie S (2012) Extracellular production of silver nanoparticles by using three common species of dermatophytes: *Trichophyton rubrum, Trichophyton mentagrophytes* and *Microsporum canis*. Iran Biomed J 16(1):52–58

Moghaddam AB, Namvar F, Moniri M, Tahir PM, Azizi S, Mohamed R (2015) Nanoparticles biosynthesized by fungi and yeast: a review of their preparation, properties, and medical applications. Molecules 20(9):16540–16565

Mohseniazar BM, Zarredar H, Alizadeh S, Shanehbandi M (2011) Potential of microalgae and *Lactobacilli* in biosynthesis of silver nanoparticles. Bio Impacts 1(3):149–152

Molnar Z, Bodai V, Szakacs G, Erdelyi B, Fogarassy Z, Safran G, Varga T, Konya Z, Toth-Szeles E, Szucs R, Lagz I (2018) Green synthesis of gold nanoparticles by *Thermophilic filamentous* fungi. Sci Rep 8:3943. https://doi.org/10.1038/s41598-018-22112-3

Monika B, Anupam B, Madhu S, Priyanka K (2015) Green synthesis of gold and silver nanoparticles. Res J Pharm, Biol Chem Sci 6(3):1710–1716
Montero-Silva F (2017) Synthesis of extracellular stable gold nanoparticles by *Cupriavidus* 1 *metallidurans* CH34 cells. BioRxiv. https://doi.org/10.1101/139949
Mortazavi SM, Jkatami M, Sharif I, Heli H, Kaykavousi K (2017) Bacterial biosynthesis of gold nanoparticles using *Salmonella enterica* subsp. *enterica* serovar Typhi isolated from blood and stool specimens of patients. J Clust Sci 28(5):2997–3007
Mourato A, Gadanho M, Lino AR, Tenreiro R (2011) Biosynthesis of crystalline silver and gold nanoparticles by extremophilic yeasts. *Bioinorg Chem Appl* 546074
Mukherjee P et al (2001) Extracellular biosynthesis of bimetallic au-ag alloy nanoparticles. Nano Lett 1:515–519
Muthukkumarasamy S, Sharadha A, Vignesh S, Dhanabalan K, Gurunathan K (2012) Extracellular synthesis of polygonal silver nanoparticles using extract of *Escherichia coli* ATCC25922 and its antibacterial activities. Dig J Nanomater Biostruct 7(4):1419–1426
Nair V, Sambre D, Joshi S, Bankar A, Ravi Kumar A, Zinjarde S (2013) Yeast-derived melanin mediated synthesis of gold nanoparticles. J Bionanosci 7(2):159–168
Nagajyothi PC, Muthuraman P, Sreekanth TVM, Kim DH, Shim J (2017) Green synthesis: in-vitro anticancer activity of copper oxide nanoparticles against human cervical carcinoma cells. Arab J Chem 10:215–225
Nanda A, Zarina A, Nayak B (2012) Extra / intracellular biosynthesis of silver nanoparticles from potential bacterial species. IEEE Xplore. https://doi.org/10.1109/ICONSET.2011.6168000
Natarajan K, Selvaraj S, Ramachandra MV (2010) Microbial production of silver nanoparticles. Dig J Nanomater Biostruct 5(1):135–140
Narayanan KB, Sakthivel N (2010) Biological synthesis of metal nanoparticles by microbes. Adv Colloid Interf Sci 156(1–2):1–13
Narayanan KB, Sakthivel N (2011) Facile green synthesis of gold nanostructures by NADPH-dependent enzyme from the extract of *Sclerotium rolfsii*. Colloids Surf A Physicochem Eng Asp 380:156–161
Narayanan KB, Sakthivel N (2013) Biosynthesis of silver nanoparticles by phytopathogen *Xanthomonas oryzae* pv. *oryzae* strain BXO8. J Microbiol Biotechnol 23:1287–1292
Nativo P, Prior IA, Brust M (2008) Uptake and intracellular fate of surface-modified gold nanoparticles. ACS Nano 2(8):1639–1644
Ng CT, Tang FMA, Li JJ, Ong C, Yung LLY, Bay BH (2015) Clathrin-mediated endocytosis of gold nanoparticles i*n Vitro*. Anat Rec 298(2):418–427
Niknejad F, Nabili M, Daie Ghazvini R, Moazeni M (2015) Green synthesis of silver nanoparticles: advantages of the yeast *Saccharomyces cerevisiae* model. Curr Med Mycol 1(3):17–24
Noorbatch IA, Zulkifli S, Salleh MH (2014) Green synthesis of gold nanoparticles using *Candida cylindracea*. J Pure Appl Microbiol 8:881–884
Oh N, Park J (2014) Endocytosis and exocytosis of nanoparticles in mammalian cells. Int J Nanomedicine 9(Suppl 1):51–63
Ortega FG, Fernández-Baldo MA, Fernández JG, Serrano MJ, Sanz MI, Diaz-Mochón JJ, Lorente JA, Raba J (2015) Study of antitumor activity in breast cell lines using silver nanoparticles produced by yeast. Int J Nanomedicine 10:2021–2031
Otari SV, Patil RM, Ghosh SJ, Thorat ND, Pawar SH (2015) Intracellular synthesis of silver nanoparticle by actinobacteria and its antimicrobial activity. Spectrochim Acta A Mol Biomol Spectrosc 136:1175–1180
Pace LE, Shulman LN (2016) Breast cancer in sub-Saharan Africa: challenges and opportunities to reduce mortality. Oncologist 21(6):739–744
Pantidos N, Horsfall LE (2014) Biological synthesis of metallic nanoparticles by Bacteria, Fungi and plants. J Nanomed Nanotechnol 5(5):233. https://doi.org/10.4172/2157-7439.1000233
Park TJ, Lee KG, Lee SY (2016) Advances in microbial biosynthesis of metal nanoparticles. Appl Microbiol Biotechnol 100(2):521–534
Paul D, Sinha SN (2014) Extracellular synthesis of silver nanoparticles using *Pseudomonas aeruginosa* KUPSB12 and its antibacterial activity. Jordan J Biol Sci 7(4):245–250

Peiris MK, Gunasekara CP, Jayaweera PM, Arachchi NDH, Fernando N (2017) Biosynthesized silver nanoparticles: are they effective antimicrobials? Rio de Janeiro 112(8):537–543

Petros RA, DeSimone JM (2010) Strategies in the design of nanoparticles for therapeutic applications. Nat Rev Drug Discov 9(8):615–627

Phanjom P, Ahmed G (2015) Biosynthesis of silver nanoparticles by *Aspergillus oryzae* (MTCC No. 1846) and its characterizations. Nanosci Nanotechnol 5(1):14–21

Prabhu S, Poulose EK (2012) Silver nanoparticles: mechanism of antimicrobial action, synthesis, medical applications, and toxicity effects. Int Nano Lett 2:32

Prasad R (2014) Synthesis of silver nanoparticles in photosynthetic plants. J Nanopart: 963961. https://doi.org/10.1155/2014/963961

Prasad R (2016) Advances and applications through fungal nanobiotechnology. Springer, Cham. ISBN: 978-3-319-42989-2

Prasad R (2017) Fungal nanotechnology: applications in agriculture, industry, and medicine. Springer International Publishing. ISBN 978-3-319-68423-9

Prasad R, Kumar V, Prasad KS (2014) Nanotechnology in sustainable agriculture: present concerns and future aspects. Afr J Biotechnol 13(6):705–713

Prasad R, Pandey R, Barman I (2016) Engineering tailored nanoparticles with microbes: quo vadis. WIREs Nanomed Nanobiotechnol 8:316–330. https://doi.org/10.1002/wnan.1363

Prasad R, Pandey R, Varma A, Barman I (2017a) Polymer based nanoparticles for drug delivery systems and cancer therapeutics. In: Natural Polymers for Drug Delivery (eds. Kharkwal H and Janaswamy S), CAB International, UK, 53–70

Prasad R, Bhattacharyya A, Nguyen QD (2017b) Nanotechnology in sustainable agriculture: recent developments, challenges, and perspectives. Front Microbiol 8:1014. https://doi.org/10.3389/fmicb.2017.01014

Prasad R, Kumar V, Kumar M and Shanquan W (2018a) Fungal Nanobionics: Principles and Applications. Springer Singapore (ISBN 978-981-10-8666-3) https://www.springer.com/gb/book/9789811086656

Prasad R, Jha A, Prasad K (2018b) Exploring the Realms of Nature for Nanosynthesis. Springer International Publishing (ISBN 978-3-319-99570-0) https://www.springer.com/978-3-319-99570-0

Prema P, Iniya PA, Immanuel G (2016) Microbial mediated synthesis, characterization, antibacterial and synergistic effect of gold nanoparticles using *Klebsiella pneumoniae* (MTCC-4030). R Soc Chem Adv 6:4601–4607

Priyadarshini E, Pradhan N, Sukla LB, Panda PK (2014) Controlled synthesis of gold nanoparticles using *Aspergillus terreus* IF0 and its antibacterial potential against Gram negative pathogenic bacteria. J Nanotechnol: 653198. https://doi.org/10.1155/2014/653198

Qu Y, You S, Zhang X, Pei X, Shen W, Li Z, Li S, Zhang Z (2018) Biosynthesis of gold nanoparticles using cell-free extracts of *Magnusiomyces ingens* LH-F1 for nitrophenols reduction. Bioprocess Biosyst Eng 41(3):359–367

Rahim KA, Mahmoud SY, Ali AM, Almaary KS, Mustafa AE, Husseiny S (2017) Extracellular biosynthesis of silver nanoparticles using *Rhizopus stolonifer* (mould). Saudi J Biol Sci 24(1):208–216

Rajam KS, Rani ME, Gumaseeli R, Munavar MH (2017) Extracellular synthesis of silver nanoparticle by the fungi *Emericella nidulans* EV$ and its application. Ind J Experim Biol 55:262–265

Rajeshkumar S, Malarkodi C, Paulkumar K, Vanaja M, Gnanajobitha G, Annadurai G (2013) Intracellular and extracellular biosynthesis of silver nanoparticles by using marine bacteria *Vibrio alginolyticus*. J Nanosci Nanotechnol 3:21–25

Rajput K, Raghuvanshi S, Bhatt A, Kumar Rai SK, Agrawal PK (2017) Review on synthesis silver nanoparticles. *Int J Curr Microbiol App Sci* 6(7):1513–1528

Ramesh V, Armash A (2015) Green synthesis of gold nanoparticles against pathogens and cancer cells. Int J Pharmacol Res 5(10):250–256

Ranjitha VR, Rai VR (2017) Actinomycetes mediated synthesis of gold nanoparticles from the culture supernatant of *Streptomyces griseoruber* with special reference to catalytic activity. 3 Biotech 7(5):299. https://doi.org/10.1007/s13205-017-0930-3

Rao Y, Inwati GK, Singh M (2017) Green synthesis of capped gold nanoparticles and their effect on gram-positive and gram-negative bacteria. Future Sci OA 3(4):2056–5623

Ronavari A, Igaz N, Gopisetty MK, Szerencses B, Kovacs D, Papp C, Vagvolgyi C, Boros IM, Konya Z, Kiricsi M, Pfeiffer I (2018) Biosynthesized silver and gold nanoparticles are potent antimycotics against opportunistic pathogenic yeasts and dermatophyte. Int J Nanomedicine 13:695–703

Sadowski Z, Maliszewska HI, Grochowalska B, Polowczyk I, Kozlecki T (2008) Synthesis of silver nanoparticles using microorganisms. Mater Sci 26:419–423

Sahay G, Alakhova DY, Kabanov AV (2010) Endocytosis of nanomedicines. J Control Release 145(3):182–195

Sakamoto JH, van de Ven Al, Godin B, Blanco E, Serda RE, Grattonic A, Ziemys A, Bouamrani A, Hut, Ranganathan SI, De Rosa E, Martinez JO, Smid CA, Buchanan RM, Lee SY, Srinivasan S, Landry M, Meyn A, Tasciottie E, Liu X, Decuzzi P, Ferrari M (2010) Enabling individualized therapy through nanotechnology. Pharm Res 62(2):57–89

Sanghi R, Verma P (2009) Biomimetic synthesis and characterisation of protein capped silver nanoparticles. Bioresour Technol 100(50):1–4

Sarangadharan S, Nallusamy S (2015) Biosynthesis and characterization of silver nanoparticles produced by *Bacillus licheniformis*. Int Pharm Med Biol Sci 4(4):236–239

Saravanan M, Vemu AK, Barik SK (2011) Rapid biosynthesis of silver nanoparticles from *Bacillus megaterium* (NCIM 2326) and their antibacterial activity on multi drug resistant clinical pathogens. Colloids Surf B: Biointerfaces 88:325–331

Saravanan M, Amelash T, Negash L, Gebreyesus A, Selvaraj A, Rayar V, Dheekonda K (2013) Extracellular biosynthesis and biomedical application of silver nanoparticles synthesized from baker's yeast. Int J Res Pharmaceut Biomed Sci 4(3):822–828

Sathiyanarayanan G, Kiran GS, Selvin J (2013) Synthesis of silver nanoparticles by polysaccharide bioflocculant produced from marine *Bacillus subtilis* MSBN17. Colloids Surf B Biointerfaces 102:13–20

Shah RK, Haider A, Das L (2017) Extracellular synthesis of ag nanoparticles using *Escherichia coli* and their antimicrobial efficacy. Int J Pharm Bio Sci 7(3):78–83

Shahverdi AR, Minaeian S, Shahverdi HR, Jamalifar H, Nohi AA (2007) Rapid synthesis of silver nanoparticles using culture supernatants of Enterobacteria: a novel biological approach. Process Biochem 42:919–923

Shanthi S, Jayaseelan BD, Velusamy P, Vijayakumar S, Chih CT, Vaseeharan B (2016) Biosynthesis of silver nanoparticles using a probiotic *Bacillus licheniformis* Dahb1 and their antibiofilm activity and toxicity effects in *Ceriodaphnia cornuta*. Micro Pathogen 93:70–77

Sheikhloo Z, Salouti M (2011) Intracellular biosynthesis of gold nanoparticles by the fungus *Penicillium Chrysogenum*. Int Nanosci Nanotechnol 7:102–105

Shivaji S, Madhu S, Singh S (2011) Extracellular synthesis of antibacterial silver nanoparticles using psychrophilic bacteria. Process Biochem 46(9):1800–1807

Siddiqi KS, Husen A, Rao RAK (2018) A review on biosynthesis of silver nanoparticles and their biocidal properties. J Nanobiotechnol 16(1):14. https://doi.org/10.1186/s12951-018-0334

Siegel RL, Miller KD, Fedewa SA, Ahnen DJ, Meester RG, Barzi A, Jemal A (2017) Colorectal cancer statistics. CA Cancer J Clin 67(3):177–193

Singh PK, Kundu S (2014) Biosynthesis of gold nanoparticles using bacteria. Proc Natl Acad Sci India – Sec B 84(2):331–336

Singh S, Kumar A, Karakoti A, Seal S, Self WT (2010) Unveiling the mechanism of uptake and sub-cellular distribution of cerium oxide nanoparticle. Mol BioSyst 6(10):1813–1820

Singh R, Wagh P, Wadhwani S, Gaidhani S, Kumbhar A, Bellare J, Chopade BA (2013) Synthesis, optimization, and characterization of silver nanoparticles from *Acinetobacter calcoaceticus* and their enhanced antibacterial activity when combined with antibiotics. Int J Nanomedicine 8:4277–4290

Singh N, Saha P, Rajkumar K, Abraham J (2014) Biosynthesis of silver and selenium nanoparticles by *Bacillus* sp. JAPSK2 and evaluation of antimicrobial activity. Pharm Lett 6(1):175–181

Singh T, Jyoti K, Patnaik A, Singh A, Chauhan R, Chandel SS (2017) Biosynthesis, characterization and antibacterial activity of silver nanoparticles using an endophytic fungal supernatant of *Raphanus sativus* journal of genetic engineering and. Biotechnology 15:31–39

Skalickova S, Baron M, Sochor J (2017) Nanoparticles biosynthesized by yeast: a review of their application. Kvasny Prum 63(6):290–292

Sowani H, Mohite P, Munot H et al (2016) Green synthesis of gold and silver nanoparticles by an actinomycete *Gordonia amicalis* HS-11: mechanistic aspects and biological application. Process Biochem 51(3):374–383

Sowbarnika R, Anhuradha S, Preetha B (2018) Enhanced antimicrobial effect of yeast mediated silver nanoparticles synthesized from baker's yeast. Int J Nanosci Nanotechnol 14(1):33–42

Srinath BS, Namratha K, Byrappa K (2017) Eco-friendly synthesis of gold nanoparticles by gold mine bacteria *Brevibacillus formosus* and their antibacterial and biocompatible studies. IOSR J Pharm 7(8):53–60

Stern ST, Adiseshaiah PP, Crist RM (2012) Autophagy and lysosomal dysfunction as emerging mechanisms of nanomaterial toxicity. Part Fibre Toxicol 9(1):20

Suman J, Neeraj S, Rahul J, Sushila K (2014) Microbial synthesis of silver nanoparticles by *Actinotalea* sp. MTCC 10637. Am J Phytomed Clin Therap 2(8):1016–1023

Sunkar S, Nachiyar CV (2012) Microbial synthesis and characterization of silver nanoparticles using the endophytic bacterium *Bacillus cereus*: a novel source in the benign synthesis. Global J Med Res 12:43–49

Suresh AK, Pelletier DA, Wang W, Broich ML, Moon JW, Gu B, Allison DP, Joy DC, Phelps TJ, Doktycz MJ (2011) Biofabrication of discrete spherical gold nanoparticles using the metal-reducing bacterium *Shewanella oneidensis*. Acta Biomater 7:2148–2152

Taran M, Rad M, Alavi M (2016) Characterization of ag nanoparticles biosynthesized by *Bacillus* sp. HAI4 in different conditions and their antibacterial effects. J Appl Pharm Sci 6(11):094–099

Thomas R, Janardhanan A, Varghese RT, Soniya EV, Mathew J, Radhakrishna EK (2014) Antibacterial properties of silver nanoparticles synthesized by marine *Ochrobactrum* sp. Braz J Microbiol 45(4):1221–1227

Truong NP, Whittaker MR, Mak CW, Davis TP (2015) The importance of nanoparticle shape in cancer drug delivery. Expert Opin Drug Deliv 12(1):129–142

Vadlapudi V, Kaladha DSVG (2014) Review: green synthesis of silver and gold nanoparticles. Middle-East J Sci Res 19(6):834–842

Vahabi K, Mansoori GA, Karimi S (2011) Biosynthesis of silver nanoparticles by fungus *Trichoderma reesei*. Insci J 1(1):65–79

Vala AK (2014) Exploration on green synthesis of gold nanoparticles by a marine-derived fungus *Aspergillus sydowii*. Environ Prog Sustain Energy. https://doi.org/10.1002/ep.11949

Verma VC, Singh SK, Solanki R, Prakash S (2011) Biofabrication of anisotropic gold nanotriangles using extract of endophytic *Aspergillus clavatus* as a dual functional reductant and stabilizer. Nanoscale Res Lett 6:16–22

Waghmare SR, Mulla NM, Marathe SR, Sonawane KD (2015) Silver nanoparticles using Candida utili. Biotechnology 5:33–38

Wicki A, Witzigmann D, Balasubramanian V, Huwyler J (2015) Nanomedicine in cancer therapy: challenges, opportunities, and clinical applications. J Control Release 200:138–157

Wypij M, Czarnecka J, Swiecimska M, Dahm H, Rai M, Golinska P (2018) Synthesis, characterization and evaluation of antimicrobial and cytotoxic activities of biogenic silver nanoparticles synthesized from *Streptomyces xinghaiensis* OF1 strain. World J Microbiol Biotechnol 34(23):1–13

Zahran MK, Mohamed AA, Mohamed FM, El-Rafie MH (2013) Optimization of biological synthesis of silver nanoparticles by some yeast fungi. Egypt J Chem 56(1):91–110

Zhang XF, Gurunathan S (2016) Combination of salinomycin and silver nanoparticles enhances apoptosis and autophagy in human ovarian cancer cells: an effective anticancer therapy. Int J Nanomedicine 11:3655–3675

Zhang X, Qu Y, Shen W, Wang J, Li H, Zhang Z, Li S, Zhou J (2016) Biogenic synthesis of gold nanoparticles by yeast *Magnusiomyces ingens* LH-F1 for catalytic reduction of nitrophenols. Colloids Surf A Physicochem Eng Asp 497:280–285

Zomorodian K, Pourshahid S, Sadatsharifi A, Mehryar P, Pakshir K, Rahimi MJ, Monfared AA (2016) Biosynthesis and characterization of silver nanoparticles by *Aspergillus* species. Biomed Res Int: 1–6, 5435397

Chapter 11
Antimicrobial Nanocomposites for Improving Indoor Air Quality

Disha Mishra and Puja Khare

Contents

11.1 Introduction: Indoor Air Pollution

Growing concerns about the significant health hazard caused by indoor air pollutants has directed research toward improving indoor air quality (IAQ). Indoor air pollution caused by various sources such as household activities, dust, electric equipment, heating activities, paints, groundwater supplies, and personal care products are the major sources of indoor air pollutants. More than 90% of the population spends time in an indoor air environment (Datta et al. 2017). Mostly, it is children and women who are vulnerable to indoor air pollution (Mannucci and Franchini 2017). The chemical or allergens of these sources often include nitrogen and sulfur oxides, ozone, pesticides, formaldehyde, infectious agents, and biological agents (microbial organisms, mold, dust mites) (Fernández et al. 2013). The level of threat caused by these allergens

D. Mishra · P. Khare (✉)
Agronomy and Soil Science Division, CSIR-Central Institute of Medicinal and Aromatic Plants, Lucknow, Uttar Pradesh, India

R. Prasad (ed.), *Microbial Nanobionics*, Nanotechnology in the Life Sciences,
https://doi.org/10.1007/978-3-030-16383-9_11

Table 11.1 Major sources and types of indoor air pollutants

S. No	Pollutant type	Source	Health effect
1.	Environmental tobacco smoke, carbon monoxide, nitrogen oxides, and other gases, organic chemicals, pesticides, formaldehyde	Heating appliances, combustion process, household activities and cleaning, pesticide usage, air fresheners, paints	Respiratory problems, dizziness, headache, lung cancer, mental dysfunction, allergic reactions, etc.
2.	Radon	Earth-derived buildings, groundwater supplies	Lung cancer
3.	Lead	Lead-based paint	Nerve disorders, anemia, damage to kidneys, growth retardation
4.	Formaldehyde	Pressed wood products, tobacco smoke	Allergic reactions, nausea, dizziness, coughing
5.	Volatile organic chemicals	Aerosol sprays, personal care products	Allergic reaction, damage to organs
6.	Respirable particles, asbestos	Wood-burning stoves, fireplaces, unvented kerosene space heaters gas-fired ranges, furnaces, water heaters	Respiratory infections, cancer, allergy
7.	Biological contaminants (molds, mildews, and fungi, bacteria, viruses, dust mites)	House dust, infected animals and humans, wet surfaces	Coughing, sensitivity, allergy, influenza, itching

varies according to temperature, humidity, and draught effects. The effect of these allergens on the population directly depends on the sensitivity of the person. Deleterious effects might be the result of single or multiple exposures to certain pollutants. The sources and health hazards of indoor pollutants are described in Table 11.1 and the contribution of each source is demonstrated in Fig. 11.1.

11.2 Indoor Air Quality

Growing concerns about the poor IAQ and serious etiological impacts have also drawn attention toward the term "Invironment" (Smith-Cavros and Eisenhauer 2014). This is new terminology, but actually refers the IAQ, which is not only related to comfort but is also affected by temperature, humidity, odor, and chemical and biological particles/organisms present in the air. A definition provided by the US Environmental Protection Agency (US EPA 2016) refers to the IAQ within and around buildings and structures, especially as it relates to the health and comfort of building occupants. Understanding and controlling common pollutants indoors can help reduce the risk of indoor health concerns (US EPA). Long-term exposure in an indoor environment is likely to exert detrimental health effects on humans. Therefore, it is essential to maintain IAQ so that the potential health hazards caused by different types of airborne pollutants can be minimized.

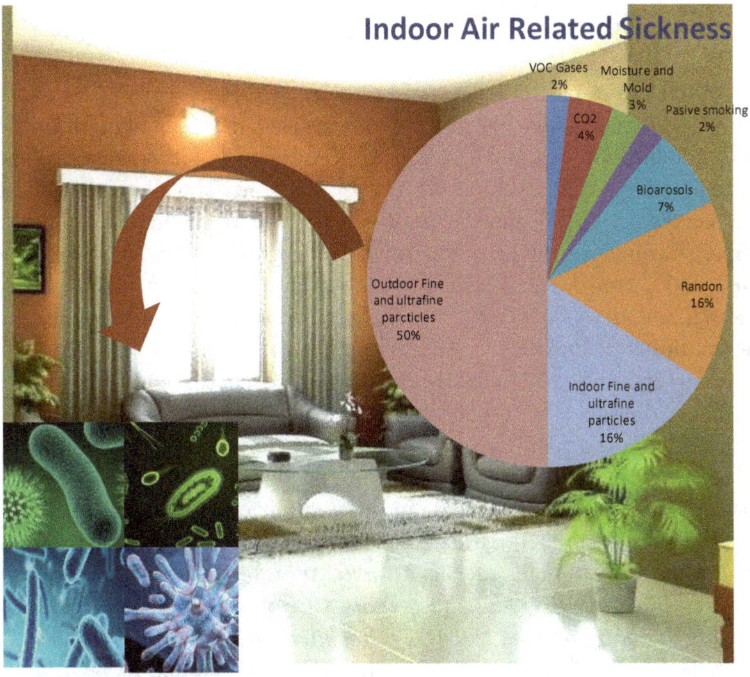

Fig. 11.1 The contribution of various sources for developing air-related sickness through microbes

11.3 Significance of Bioaerosols in Indoor Air

Aerosols are liquid or solid particles suspended in a gaseous medium with sizes ranging from 0.001 to 100 μm (Pöschl 2005). Aerosol particles of biological origin (cells, cell fractions or organic matter of animal, plant, and microbial origin) form a significant portion of atmospheric aerosols, sometimes reaching close to 50% of all aerosol particles (Despres et al. 2012). The air quality is majorly affected by particulates of bacteria, fungi, virus, actinomycetes, etc., and sometimes the droppings of insects and birds can also be the source of air pollution. These microorganisms build and grow after condensation due to improper ventilation. The dynamic nature and specific features of airborne particles have contributed to 3–30% of air pollution. Therefore, knowledge about the microbiology of IAQ is a serious concern in terms of bacterial and fungal loads in indoor air (Georgakopoulos et al. 2009). Exposure to several of these biological entities in addition to microbial fragments, such as cell wall fragments and flagella, and microbial metabolites, e.g., endotoxin, mycotoxins, and volatile organic compounds, may result in adverse health effects (Mandal and Brandl 2011).

11.4 Prevalence of Bioaerosols in Indoor Air

The living airborne particles (bacteria, viruses, and fungi) are generally termed bioaerosols or organic dust. Their presence in the air is the result of the dispersal from a site of colonization or growth. A high concentration of airborne bacteria is often associated with several health problems and infections. The allergenic or immunotoxic effects of various species such as *Bacillus*, *Streptomyces albus*, *Pantoea agglomerans*, *Pseudomonas chlororaphis*, *Arthrobacter globiformis*, *Thermoactinomyces vulgaris*, and *Corynebacterium* sp. are well documented (Ana et al. 2015). Members of *Klebsiella, Enterobacter*, and *Serratia* groups were able to grow well in environmental fluids (Pathak 2014). The occurrence of microbes in the varying indoor air environment such as *Micrococcus* sp., *Staphylococcus* sp., Streptococcaceae, Propionibacteriaceae, Corynebacteriaceae, Veillonellaceae, Prevotellaceae, Fusobacteriaceae, *Lactococcus*, *Firmicutes*, and *Actinobacteria* have been documented (Prussin and Marr 2015). The occurrence of aerosolized Gram-positive and Gram-negative bacteria was noticed in different air environments (Adhikari et al. 2014; Milanowski and Dutkiewicz 2002; Schwab et al. 2014). Apart from the fact that several species of molds such as *Absidia*, *Acrosperia*, *Alternaria*, *Arthrobotrys*, *Aspergillus*, *Bipolaris*, *Botrytis*, *Chaetomium*, *Cladosporium*, *Epicoccum*, *Fusarium*, *Humicola*, *Mucor*, *Phoma*, *Rhizopus*, *Tritirachium*, and *Verticillum* were also identified in house dust (Bhatia 2011). In addition, *Penicillium* and *Aspergillus* species are generally isolated from indoor environments and can cause allergic reactions (Vincent et al. 2017). Therefore, knowledge of the microbiological air quality is an important criterion that must be taken in to account for indoor workplaces and should be designed to provide a safe environment. The sampling methods and identification techniques of bioaerosols are described in Fig. 11.2. Table 11.2 shows the occurrence of various microbes in different air environments.

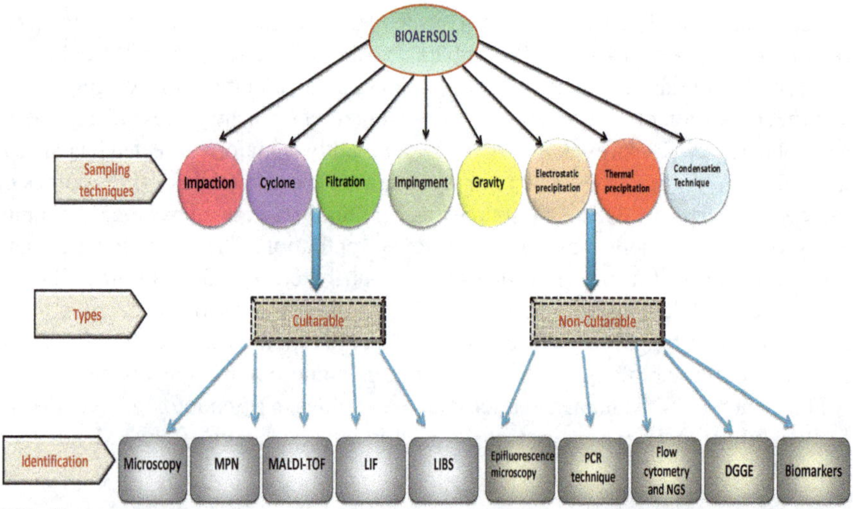

MPN: Most probable number counting assay, LIF: Laser induced fluorescence, MALDI-TOF: Matrix-assisted laser desorption/ionization-time of flight, NGS: Next generation sequencing, LIBS: Laser induced breakdown spectroscopy, DGGE: Denaturing gradient gel electrophoresis

Fig. 11.2 Sampling method, types, and identification techniques of bioaerosols

Table 11.2 Major microbial strains identified in the different air environments

Location	Sampling method	Identified genus		Identification method	Reference
		Bacteria	Fungi		
Office	Open face filters	*Proteobacteria*, *Firmicutes*, *Actinobacteria*, Lactobacillales Enterobacteriales	*Cladosporium*, *Aureobasidium*, *Phoma*, *Alternaria*, *Rhodotorula*, and *Penicillium*	Cultivation and DNA sequencing	Adams et al. (2015)
Hospital	Anderson sampler	*Bacillus*, *Micrococcus*, *Staphylococcus*, *Streptomyces*, *Corynebacterium*, *Brevibacterium*, *Pseudomonas*, *Enterobacter*, *Serratia*, *Klebsiella*	ND	Cultivation and microscopy and biochemical test	Soleimani et al. (2016)
Sludge dewatering house		Saccharibacteria, Planococcaceae, Cyanobacteria, *Pseudomonas*, *Zoogloea*, *Arcobacter*, *Flavobacterium*	ND	Cell enumeration and sequencing	Han et al. (2018)
School and day care	Air sampler	*Micrococcus, Paracoccus, Staphylococcus*, *Streptomyces,* *Pseudonocardia*, and *Nocardiopsis*	*Alternaria*, *Epicoccum*, *Curvularia*, *Cladosporium*, *Aspergillus, Penicillium*, *Hyphodontia*, and *Thanatephorus*	Cultivation and sequencing	Shin et al. (2015)
School dormitory and retirement home	Single-stage Andersen sampler	*Staphylococcus*, *Micrococcus*, *Bacillus*	*Penicillium, Cladosporium, Aspergillus*	Cultivation and enumeration	Faridi et al. (2015)
University campus	Biostage sampler	Alphaproteobacteria, Actinobacteria, Bacilli, Betaproteobacteria, Cocci, Deinococci, Flavobacteria, Gammaproteobacteria	*Cladosporium, Aspergillus, Penicillium,* and *Alternaria*	Cultivation and enumeration	Priyamvada et al. (2018)
Library	Buck bio-culture pump	*Bacillus*, *Streptococcus*	*Aspergillus*, *Rhizopus*	Microscopy	Ghosh et al. (2013)

11.5 Health Hazards of Bioaerosols and Control Mechanisms

Microbiological pollution is a very serious issue in the current era because it contains bacteria, viruses, fungi, actinomycetes, and spores, which are extremely hazardous to the living organism. They are airborne, can easily be transmitted up to long distances, and can cause a variety of diseases such as allergic rhinitis, asthma, chronic obstructive pulmonary disease, influenza, and severe acute respiratory syndrome (Soriano et al. 2017). The metrological factors, size, and pathogenic nature of these bioaerosols are the key determining factors of their lethal effects (Brągoszewska and Pastuszka 2018; Jones and Harrison 2004). The spread of bioaerosols among humans is mainly due to inhalation, direct contact or through contaminated sources. The physical processes that govern the assembly of indoor microbial communities are shown in Fig. 11.3.

Therefore, bioaerosols can be easily prevented by opting for appropriate detection methods and control mechanisms. The immediate control mechanism includes inactivation, removal or collection. Thermal energy, UV irradiation technique, and air ion emission techniques are often used as effective methods of eradicating bioaerosols from indoor environments (Grinshpun et al. 2007). In addition, air filters are now being employed to control the growth of microorganisms by absorbing air dust or reversing the air flow. A group of researchers have developed antimicrobial air filters because they are more effective than conventional air filters (Joe et al. 2016; Wang C et al. 2016; Wang S et al. 2016; Wang Z et al. 2016a). The air filtration technology

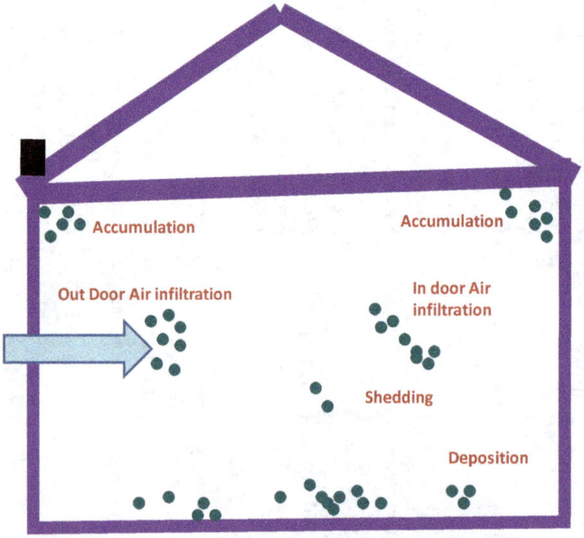

Fig. 11.3 Physical processes that govern the assembly of indoor microbial communities (Peccia and Kwan 2016)

often includes utilization of inorganic air filters, which mainly use inorganic nanoparticles (NPs). After air filtration, deposits on the air filters can cause decreased filter efficacy and filter deterioration. The microbial cells also proliferate on these air filters and release volatile organic compounds that cause damage to filters and to human health too. Therefore, despite being part of the remediation techniques for improvement of IAQ, antimicrobial air filters based on inorganic species have become a source of pollution themselves. Various antimicrobial air filters have been developed, which are based on Cu, TiO_2, Ag NPs and carbon nanotubes for the removal of bioaerosols (Hans et al. 2013; Ko et al. 2014; Li et al. 2014). However, the toxicity of these inorganic species is documented to cause damage to DNA or mammalian cells. Moreover, long-term exposure to these NPs can result in respiratory dysfunction. Fig. 11.2 shows the type, sampling technique, and identification method for bioaerosols.

Recently, plant extracts and essential oils have been recognized for utilization in air filters because they exert sufficient antimicrobial efficacy. The active phytochemicals present in natural products are known to have antimicrobial potential against a wide range of bacteria and fungi (Ahmad and Beg 2001; Dahiya and Purkayastha 2012). The development of drug-resistant features inside the bacterial cells has led to the development of novel antimicrobial agents. The nanosized filter membrane is now an efficient method of the development of an effective membrane surface with sufficient antimicrobial activity owing to the increased surface area (Wang and Pan 2015). Also, the release of various phytochemicals was markedly slow after loading onto the nano surfaces (Martínez-Ballesta et al. 2018); therefore, these types of antimicrobial filters are able to release phytochemicals into the air for a long duration, thereby sustaining antimicrobial activity.

11.6 Herbal Antimicrobial Agents

Herbs have been proven to be effective antimicrobial agents since ancient times and being of herbal origin they have provided an interesting solution for replacing chemicals (Dhiman et al. 2016). This concern has brought major attention to aromatic plants. The aroma-based plants are well known to have certain bioactive compounds in their essential oil or extract, which are primarily utilized to retard microbial growth in various mediums. This essential oil/extract is the combination of various phytochemicals such as mono- and sesquiterpenes, alkaloids, saponins, carbohydrates, phenols, alcohols, ethers, aldehydes, and ketone, which play an important role as antimicrobial agents (Foroughi et al. 2016). In addition, the interaction effect and synergistic effects of these secondary metabolites have contributed to the enhancement of antimicrobial activity. They have also proved to be better antimicrobial agents against various pathogenic strains. The major benefit regarding the use of these herbal constituents is that they play different antimicrobial mechanisms owing to the presence of various chemicals; therefore, microbial strains were not easily able to generate resistance against these. Ambrosio et al. (2017) have determined the antimicrobial activity of around 28 essential oils, including *Cymbopogon citratus*, *Lippia alba*, *Cordia verbenacea*, *Eucalyptus* sp., *Zingiber officinale*, and *Melaleuca alternifolia*, against *Salmonella enteritidis*, *Escherichia coli*, *Staphylococcus aureus*, *Listeria innocua*, and *Enterococcus faecalis* (Ambrosio et al. 2017). The beneficial bacteria evaluated were *Bacillus subtilis*, *Lactobacillus plantarum*, and *Lactobacillus rhamnosus*, and it was found that selected essential oils had shown inhibitory action against certain strains. Cinnamon essential oils were tested against pathogenic bacteria *E. coli* and *Staphylococcus* and showed antibacterial activity in 60% and 79.4% respectively over 7 hours (Zhang et al. 2016). The phytochemical alkaloids, such as saponins, flavone, and phenolic of *Oliveria decumbens* essential oil, were tested against infectious strains and the results showed remarkable antibacterial activity (Behbahani et al. 2018). Generally, the mechanism of these phytochemicals for inactivating microorganisms is based on their interaction with the microbial membrane. These compounds are known to penetrate through the microbial membrane and cause the leakage of ions and cytoplasmic content, thus leading to cellular breakdown (Fig. 11.4). Apart from the essential oil, the herbal extract of various plants such as *Thymus vulgaris*, *Rosmarinus officinalis*, *Salvia officinalis*, *Punica granatum*, *Syzygium cumini*, and *Psidium guajava* are known to have phytochemicals with proven antimicrobial potential (Nascimento et al. 2000). Considering the potential of these medicinal and aromatic plants, the antimicrobial potential of a wide range of these plants should also be tried in the reduction of the microbial load in the air environment.

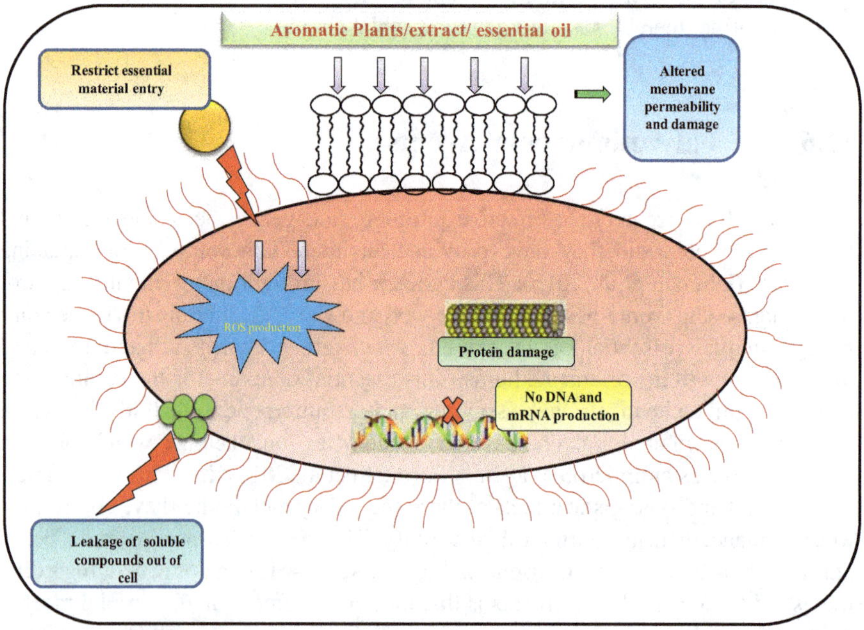

Fig. 11.4 Mechanism of the antimicrobial activity of essential oil/extract on the bacterial cells

11.7 Utilization of Herbs for the Construction of Antimicrobial Air Quality

The rising threat of infectious diseases as a result of the degraded IAQ is due to airborne microorganisms. Multiple growing microorganisms release certain volatile organic compounds into the air. Recently, numerous antimicrobial herbal extracts and essentials oil have been proposed as alternative antimicrobial materials rather than inorganic materials because they are less toxic. However, the volatile nature and degradation potential of these phytochemicals have limited their commercial application. For this, a suitable matrix is required that can control their volatility and also help to enhance their antimicrobial potential at the same as sustained release. Therefore, apart from using traditional methods, this technology would be helpful in the continuous elimination of microbes from indoor air environments and is also simple and cost-effective. The antimicrobial efficiency of these filters depends on the surface area, contact time, exposure time, and concentration of loaded agents. Table 11.3 summarizes the use of herbal constituents for antimicrobial purposes.

The loading of *Sophora flavescens* extract onto pristine air filters has resulted in the inactivation of 80% of *S. epidermidis* strains within 1 h and this increased to 98% after 10 h. Some authors have also shown that the deposition of larger amounts of antimicrobial substances is related to greater removal efficiency (Choi et al. 2015; Kang et al. 2016). Jung et al. (2011) found that the relative microbial survival of *Staphylococcus epidermidis*, *Bacillus subtilis*, and *Escherichia coli* decreased with increasing levels of NPs of the extract of the same plant on polyurethane resin fiber filters (Jung et al. 2011). Filters coated with *Euscaphis japonica* NPs were found to show sufficient antibacterial efficiency (82%) with regard to *S. epidermidis* than *Micrococcus luteus* strains (Hwang et al. 2015). Grape seed extract and propolis were sprayed onto polyethylene terephthalate fibers to create nanosized antimicrobial filters and the inactivation of *S. aureus* was found in up to 80%, together with the finding that the release time of the extract from the surface also affected the inactivation rate (Han et al. 2015). Protein-functionalized bacterial nanocellulose was also tested for its potential to trap pollutants and the data showed a low-pressure drop due to the removal of pollutants with high efficiency (Fan et al. 2018). The antibacterial activity of the gaseous phase of essential oils was found to be suitable for application in ventilation systems (Fan et al. 2018). Souzandeh et al. (2017) also tested the filtration efficiency of protein- and gelatin-crosslinked nanofibers against *Shewanella oneidensis* and *S. aureus* and suggested that a positively charged group and nanosized fiber morphology were able to kill the microorganism and carry out multifunctional air filtration (Souzandeh et al. 2017). Thus, the treatments of various nanosized surfaces with naturally occurring compounds proved to be effective antimicrobial agents and may be utilized as air-filtering materials because of their specific surface area.

Table 11.3 Encapsulation of bioactive compounds of herbal origin

Encapsulating matrix	Bioactive compound	Fabrication method	Tested strains	References
PVA/CNC	Basil leaves extract	Solvent casting	*Listeria monocytogenes* and *Bacillus cereus*	Singh et al. (2018)
CNF/soy protein	Clove essential oil	Homogenization and casting	*B. cereus*, *Escherichia coli*, *Salmonella enteritidis* and *Staphylococcus aureus*	Ortiz et al. (2018)
Sodium alginate/ montmorillonite	*Origanum majorana*	Homogenization	*E. coli*, *L. monocytogenes*, *B. cereus*, and *S. aureus*	Alboofetileh et al. (2018)
Chitosan	Lime oil	Nanoprecipitation	*Shigella dysenteriae*	Sotelo-Boyás et al. (2017)
Polyvinyl alcohol/β -CD	Cinnamon essential oil	Electrospinning	*L. monocytogenes*, *Salmonella enteritidis*, *Aspergillus niger* and *Penicillium*	Feng et al. (2017)
Chitosan/cotton	*Azadirachta indica*	Ball milling	*E. coli* and *S. aureus*	Yadav and Balasubramanian (2016)
Chitosan	Cardamom	Ionic gelation	*E. coli* and *S. aureus*	Liu et al. (2016)
Agar/cellulose	Savory essential oil	Homogenization	*L. monocytogenes*, *S. aureus*, *Bacillus cereus*, *E. coli*	Maryam et al. (2015)
Chitosan	*Senna auriculata* and *Achyranthes aspera* extract	Coacervation method	*E. coli* and *S. aureus*	Chandrasekar et al. (2014)
Sodium alginate/ chitosan	*Ocimum* extract	Cation-induced controlled jellification	*E. coli* and *S. aureus*	Rajendran et al. (2013)

CNC cellulose nanocrystals, *CNF* cellulose nanofibers, *PVA* polyvinyl alcohol, *CD* cyclodextrin

11.8 Utilization of Nanomaterials in Air Filtration

In the development of high-performance antimicrobial air filters, natural biodegradable polymers offer a better opportunity than materials of organic/inorganic origin. These green air filters of polymeric origin exhibit newer and different properties than the individual constituents with active functionalities. These polymers may be of synthetically derived origin or based on biomass. Both types of materials are successfully utilized for fabrication of the composite surface of nano scales. Nanostructured polymers are basically nanosized materials, which could be nanofibers or nanospheres. The outstanding properties of polymeric material, such as its light weight, durability, chemical stability, and cost-effectiveness, make it a feasible material for the production of air fibers. Apart from this nanosized surface area, high porosity allows large amounts of dust, pollen, or microorganisms to be absorbed. The key requirement for the development of effective filters is dependent on the selection of polymers that can easily manage filtration efficiency, pressure drop, and life span of filters. Polymers such as silk fibroin, chitosan, gelatin, cellulose, and lignin have been used to produce nanofibers (Ago et al. 2012; Elsabee et al. 2012; Lai et al. 2014; Maleknia and Majdi 2014). Different methods have been proposed to generate polymeric nanomaterials that can promote the physical absorption of the drug onto the fiber surface, and the establishment of covalent bonds between the polymer and the antibacterial agent. The generated matrix promotes the sustained release profiles of encapsulated or loaded antibacterial agents, which could be the main strategy of interest. Therefore, the development of polymeric nanomaterial-based antimicrobial filters will be helpful in killing microbes in the air. A large number of techniques are available for the construction of nanocomposites such as sol-gel, electrospinning, solution casting, intercalation, spinning, and in situ polymerization. Many characterization methods, for example, scanning and imaging techniques (scanning electron microscopy, transmission electron microscopy, atomic force microscopy), identification of functional groups (Fourier transform infrared, Raman spectroscopy), or thermal and mechanical performance testing (thermogravimetry/differential scanning calorimetry, dynamic mechanical analysis), are commonly employed to assure their physical and chemical priopencities.

11.9 Conclusion and Future Prospects

In conclusion, herb-based air filter/fresheners could be able to create favorable IAQ where the risk of the emission of bioaerosols could be minimized. However, data on IAQ in this context are in their infancy. However, preliminary results revealed that nanocomposite surfaces loaded with natural aromatics, such as essential oils or natural herbs, are able to substantially reduce the microbial load for longer periods. For the commercialization and industrial application of such products, a long-term study on IAQ with different sources is required.

References

Adams RI, Bhangar S, Pasut W, Arens EA, Taylor JW, Lindow SE, Nazaroff WW, Bruns TD (2015) Chamber bioaerosol study: outdoor air and human occupants as sources of indoor airborne microbes. PLoS One 10:e0128022

Adhikari A, Kettleson EM, Vesper S, Kumar S, Popham DL, Schaffer C, Indugula R, Chatterjee K, Allam KK, Grinshpun SA (2014) Dustborne and airborne Gram-positive and Gram-negative bacteria in high versus low ERMI homes. Sci Total Environ 482:92–99

Ago M, Okajima K, Jakes JE, Park S, Rojas OJ (2012) Lignin-based electrospun nanofibers reinforced with cellulose nanocrystals. Biomacromolecules 13:918–926

Ahmad I, Beg AZ (2001) Antimicrobial and phytochemical studies on 45 Indian medicinal plants against multi-drug resistant human pathogens. J Ethnopharmacol 74:113–123

Alboofetileh M, Rezaei M, Hosseini H, Abdollahi M (2018) Morphological, physico-mechanical, and antimicrobial properties of sodium alginate-montmorillonite nanocomposite films incorporated with marjoram essential oil. J Food Process Preserv 42:e13596

Ambrosio CM, de Alencar SM, de Sousa RL, Moreno AM, Da Gloria EM (2017) Antimicrobial activity of several essential oils on pathogenic and beneficial bacteria. Ind Crop Prod 97:128–136

Ana GR, Morakinyo OM, Fakunle GA (2015) Indoor air quality and risk factors associated with respiratory conditions in Nigeria. Current Air Quality Issues. IntechOpen, 2015

Behbahani BA, Yazdi FT, Vasiee A, Mortazavi SA (2018) Oliveria decumbens essential oil: chemical compositions and antimicrobial activity against the growth of some clinical and standard strains causing infection. Microb Pathog 114:449–452

Bhatia L (2011) Impact of bioaerosols on indoor air quality – a growing concern. Adv Bioresearch 2:120–123

Brągoszewska E, Pastuszka JS (2018) Influence of meteorological factors on the level and characteristics of culturable bacteria in the air in Gliwice, Upper Silesia (Poland). Aerobiologia 34:241–255

Chandrasekar S, Vijayakumar S, Rajendran R (2014) Application of chitosan and herbal nanocomposites to develop antibacterial medical textile. Biomed Aging Pathol 4:59–64

Choi J, Yang BJ, Bae G-N, Jung JH (2015) Herbal extract incorporated nanofiber fabricated by an electrospinning technique and its application to antimicrobial air filtration. ACS Appl Mater Interfaces 7:25313–25320

Dahiya P, Purkayastha S (2012) Phytochemical screening and antimicrobial activity of some medicinal plants against multi-drug resistant Bacteria from clinical isolates. Indian J Pharm Sci 74:443–450

Datta A, Suresh R, Gupta A, Singh D, Kulshrestha P (2017) Indoor air quality of non-residential urban buildings in Delhi, India. Int J Sust Built Env 6:412–420

Després VR, et al. (2012) Primary biological aerosol particles in the atmosphere: a review. Tellus B: Chem Phys Meteorol 64(1):15598

Dhiman R, Aggarwal N, Aneja KR, Kaur M (2016) In vitro antimicrobial activity of spices and medicinal herbs against selected microbes associated with juices. Int J Microbiol 2016:9015802

Elsabee MZ, Naguib HF, Morsi RE (2012) Chitosan based nanofibers, review. Mater Sci Eng C 32:1711–1726

Fan X, et al. (2018) A nanoprotein-functionalized hierarchical composite air filter. ACS Sust Chem Eng 6(9):11606–11613

Faridi S, Hassanvand MS, Naddafi K, Yunesian M, Nabizadeh R, Sowlat MH, Kashani H, Gholampour A, Niazi S, Zare A (2015) Indoor/outdoor relationships of bioaerosol concentrations in a retirement home and a school dormitory. Environ Sci Pollut Res 22:8190–8200

Feng K, Wen P, Yang H, Li N, Lou WY, Zong MH, Wu H (2017) Enhancement of the antimicrobial activity of cinnamon essential oil-loaded electrospun nanofilm by the incorporation of lysozyme. RSC Adv 7:1572–1580

Fernández LC, Alvarez RF, González-Barcala FJ, Portal JAR (2013) Indoor air contaminants and their impact on respiratory pathologies. Archivos de Bronconeumología (English Edition) 49:22–27

Foroughi A, Pournaghi P, Zhaleh M, Zangeneh A, Zangeneh MM, Moradi R (2016) Antibacterial activity and phytochemical screening of essential oil of *Foeniculum vulgare*. Int J Pharm Clin Res 8:1505–1509

Georgakopoulos D, Després V, Fröhlich-Nowoisky J, Psenner R, Ariya P, Pósfai M, Ahern H, Moffett B, Hill T (2009) Microbiology and atmospheric processes: biological, physical and chemical characterization of aerosol particles. Biogeosciences 6:721–737

Ghosh B, Lal H, Kushwaha R, Hazarika N, Srivastava A, Jain V (2013) Estimation of bioaerosol in indoor environment in the university library of Delhi. Sustain Environ Res 23:199–207

Grinshpun SA, Adhikari A, Honda T, Kim KY, Toivola M, Ramchander Rao KS, Reponen T (2007) Control of aerosol contaminants in indoor air: combining the particle concentration reduction with microbial inactivation. Environ Sci Technol 41:606–612

Han B, Kang J-S, Kim H-J, Woo C-G, Kim Y-J (2015) Investigation of antimicrobial activity of grapefruit seed extract and its application to air filters with comparison to propolis and shiitake. Aerosol Air Qual Res 15:1035–1044

Han Y, Wang Y, Li L, Xu G, Liu J, Yang K (2018) Bacterial population and chemicals in bioaerosols from indoor environment: sludge dewatering houses in nine municipal wastewater treatment plants. Sci Total Environ 618:469–478

Hans M, Erbe A, Mathews S, Chen Y, Solioz M, Mücklich F (2013) Role of copper oxides in contact killing of bacteria. Langmuir 29:16160–16166

Hwang GB, Heo KJ, Yun JH, Lee JE, Lee HJ, Nho CW, Bae G-N, Jung JH (2015) Antimicrobial air filters using natural Euscaphis japonica nanoparticles. PLoS One 10:e0126481

Joe YH, Park DH, Hwang J (2016) Evaluation of Ag nanoparticle coated air filter against aerosolized virus: anti-viral efficiency with dust loading. J Hazard Mater 301:547–553

Jones AM, Harrison RM (2004) The effects of meteorological factors on atmospheric bioaerosol concentrations – a review. Sci Total Environ 326:151–180

Jung JH, Hwang GB, Park SY, Lee JE, Nho CW, Lee BU, Bae G-N (2011) Antimicrobial air filtration using airborne *Sophora flavescens* natural-product nanoparticles. Aerosol Sci Technol 45:1510–1518

Kang JS, Kim H, Choi J, Yi H, Seo SC, Bae G-N, Jung JH (2016) Antimicrobial air filter fabrication using a continuous high-throughput aerosol-based process. Aerosol Air Qual Res 16:2059–2066

Ko Y-S, Joe YH, Seo M, Lim K, Hwang J, Woo K (2014) Prompt and synergistic antibacterial activity of silver nanoparticle-decorated silica hybrid particles on air filtration. J Mater Chem B 2:6714–6722

Lai G-J, Shalumon K, Chen S-H, Chen J-P (2014) Composite chitosan/silk fibroin nanofibers for modulation of osteogenic differentiation and proliferation of human mesenchymal stem cells. Carbohydr Polym 111:288–297

Li P, Wang C, Zhang Y, Wei F (2014) Air filtration in the free molecular flow regime: a review of high-efficiency particulate air filters based on carbon nanotubes. Small 10:4543–4561

Liu H, Xu H, Zhang C, Gao M, Gao X, Ma C, Lv L, Gao D, Deng S, Wang C (2016) Emodin-loaded PLGA-TPGS nanoparticles combined with heparin sodium-loaded PLGA-TPGS nanoparticles to enhance chemotherapeutic efficacy against liver cancer. Pharm Res 33:2828–2843

Maleknia L, Majdi ZR (2014) Electrospinning of gelatin nanofiber for biomedical application. Orient J Chem 30:2043–2048

Mandal J, Brandl H (2011) Bioaerosols in indoor environment – a review with special reference to residential and occupational locations. Open Environ Biological Monit J 4:83–96

Mannucci PM, Franchini M (2017) Health effects of ambient air pollution in developing countries. Int J Environ Res Public Health 14:1048

Martínez-Ballesta M, Gil-Izquierdo Á, García-Viguera C, Domínguez-Perles R (2018) Nanoparticles and controlled delivery for bioactive compounds: outlining challenges for new "smart-foods" for health. Foods 7:72

Maryam I, Huzaifa U, Hindatu H, Zubaida S (2015) Nanoencapsulation of essential oils with enhanced antimicrobial activity: a new way of combating antimicrobial resistance. J Pharmacog Phytochem 4:165

Milanowski J, Dutkiewicz J (2002) Exposure to airborne microorganisms in furniture factories. Ann Agric Environ Med 9:85–90

Nascimento GG, Locatelli J, Freitas PC, Silva GL (2000) Antibacterial activity of plant extracts and phytochemicals on antibiotic-resistant bacteria. Braz J Microbiol 31:247–256

Ortiz CM, Salgado PR, Dufresne A, Mauri AN (2018) Microfibrillated cellulose addition improved the physicochemical and bioactive properties of biodegradable films based on soy protein and clove essential oil. Food Hydrocoll 79:416–427

Pathak AK (2014) Aero bacteriology of concentrated animal feeding operations: a review. Cibtech J Zool 5(1):40–47

Peccia J, Kwan SE (2016) Buildings, beneficial microbes, and health. Trends Microbiol 24:595–597

Pöschl U (2005) Atmospheric aerosols: composition, transformation, climate and health effects. Angew Chem Int Ed 44:7520–7540

Priyamvada H, Priyanka C, Singh RK, Akila M, Ravikrishna R, Gunthe SS (2018) Assessment of PM and bioaerosols at diverse indoor environments in a southern tropical Indian region. Build Environ 137:215–225

Prussin AJ, Marr LC (2015) Sources of airborne microorganisms in the built environment. Microbiome 3:78

Rajendran R, Radhai R, Kotresh T, Csiszar E (2013) Development of antimicrobial cotton fabrics using herb loaded nanoparticles. Carbohydr Polym 91:613–617

Schwab F, Gastmeier P, Meyer E (2014) The warmer the weather, the more gram-negative bacteria – impact of temperature on clinical isolates in intensive care units. PLoS One 9:e91105

Shin S-K, Kim J, Ha S-M, Oh H-S, Chun J, Sohn J, Yi H (2015) Metagenomic insights into the bioaerosols in the indoor and outdoor environments of childcare facilities. PLoS One 10:e0126960

Singh S, Gaikwad KK, Lee YS (2018) Antimicrobial and antioxidant properties of polyvinyl alcohol bio composite films containing seaweed extracted cellulose nano-crystal and basil leaves extract. Int J Biol Macromol 107:1879–1887

Smith-Cavros E, Eisenhauer E (2014) Overtown: neighbourhood, change, challenge and "invironment". Local Environ 19:384–401

Soleimani Z, Goudarzi G, Sorooshian A, Marzouni MB, Maleki H (2016) Impact of Middle Eastern dust storms on indoor and outdoor composition of bioaerosol. Atmos Environ 138:135–143

Soriano JB, Abajobir AA, Abate KH, Abera SF, Agrawal A, Ahmed MB, Aichour AN, Aichour I, Aichour MTE, Alam K (2017) Global, regional, and national deaths, prevalence, disability-adjusted life years, and years lived with disability for chronic obstructive pulmonary disease and asthma, 1990–2015: a systematic analysis for the Global Burden of Disease Study 2015. Lancet Respir Med 5:691–706

Sotelo-Boyás M, Correa-Pacheco Z, Bautista-Baños S, Corona-Rangel M (2017) Physicochemical characterization of chitosan nanoparticles and nanocapsules incorporated with lime essential oil and their antibacterial activity against food-borne pathogens. LWT Food Sci Technol 77:15–20

Souzandeh H, Molki B, Zheng M, Beyenal H, Scudiero L, Wang Y, Zhong W-H (2017) Cross-linked protein nanofilter with antibacterial properties for multifunctional air filtration. ACS Appl Mater Interfaces 9:22846–22855

U.S. EPA (2016) Greenhouse gas emissions. https://www3.epa.gov/climatechange/ghgemissions/ Accessed 10 January 2016.

Vincent M, Percier P, De Prins S, Huygen K, Potemberg G, Muraille E, Romano M, Michel O, Denis O (2017) Investigation of inflammatory and allergic responses to common mold species: results from in vitro experiments, from a mouse model of asthma, and from a group of asthmatic patients. Indoor Air 27:933–945

Wang Z, Pan Z (2015) Preparation of hierarchical structured nano-sized/porous poly (lactic acid) composite fibrous membranes for air filtration. Appl Surf Sci 356(2015):1168–1179

Wang Z, Pan Z, Wang J, Zhao R (2016a) A novel hierarchical structured poly (lactic acid)/titania fibrous membrane with excellent antibacterial activity and air filtration performance. J Nanomater 2016:39

Wang C, Wu S, Jian M, Xie J, Xu L, Yang X, Zheng Q, Zhang Y (2016b) Silk nanofibers as high efficient and lightweight air filter. Nano Res 9:2590–2597

Wang S, Zhao X, Yin X, Yu J, Ding B (2016c) Electret polyvinylidene fluoride nanofibers hybridized by polytetrafluoroethylene nanoparticles for high-efficiency air filtration. ACS Appl Mater Interfaces 8:23985–23994

Yadav R, Balasubramanian K (2016) Bioabsorbable engineered nanobiomaterials for antibacterial therapy, engineering of nanobiomaterials. Elsevier, Amsterdam, pp 77–117

Zhang Y, Liu X, Wang Y, Jiang P, Quek S (2016) Antibacterial activity and mechanism of cinnamon essential oil against *Escherichia coli* and *Staphylococcus aureus*. Food Control 59:282–289

Chapter 12
Microbial Photosynthetic Reaction Centers and Functional Nanohybrids

Anjana K. Vala and Bharti P. Dave

Contents

12.1 Introduction

Nanotechnology, the cutting edge technology of today, though considered one of the new divisions of modern science, had been in practice since even during ancient times. However, with publication of the first paper on nanotechnology by Eric Drexler in 1981, conceptual revolution started (Drexler 1981; Prathna et al. 2010; Vala et al. 2016). With advancement in analytical techniques, nanotechnology has been now considered as the "Future of all technologies" (Prathna et al. 2010). A range of nanomaterials with diverse applications have been reported. Yang et al. (2011) could increase efficiency of tandem solar cell from 5.22% to 6.24% using gold nanoparticles. Weak localization of light by gold bionanofluids has been observed by Dave et al. (2015). Guldi et al. (2006) has reviewed carbon nanostructures as integrative constituents for electron donor-acceptor ensembles. Role of microorganisms in recent years has also been noteworthy. In addition to diverse ecologically important roles played by microorganisms, recently, they have been

A. K. Vala (✉) · B. P. Dave
Department of Life Sciences, Maharaja Krishnakumarsinhji Bhavnagar University, Bhavnagar, Gujarat, India
e-mail: akv@mkbhavuni.edu.in

R. Prasad (ed.), *Microbial Nanobionics*, Nanotechnology in the Life Sciences,
https://doi.org/10.1007/978-3-030-16383-9_12

observed as promising candidates for generating electricity, either as microbial fuel cell (MFC) or photosynthetic microbial fuel cell (PMFC). In case of MFC, bacteria act as anode, and during wastewater treatment, organic wastes are consumed and oxidized leading to generation of electrons (Siegert et al. 2019). Whereas, in case of PMFC, photosynthetic microbes like cyanobacteria act as the anode for harnessing solar energy and consuming carbon dioxide ultimately leading to generation of electrons through photosynthetic process (Strik et al. 2008; Singh et al. 2018).

Creation of functional bionanohybrid complexes has led to the emergence of new discipline "nanobionics." Nanobionics assures to provide the complexes (with unique properties and having diverse potentials) that are considered as materials for the future (Shoseyov and Levy 2008; Darder et al. 2007; Giraldo et al. 2014; Nagy et al. 2014; Szabó et al. 2015; Hajdu et al. 2017).

Wallace et al. (2012) in a review highlighted the fabrication of nanostructured electrodes having relevance in the field of medical bionics. Giraldo et al. (2014) reported nanomaterials as promising tools for engineering plant functions; however, understanding about their absorption, transport, and distribution within photosynthetic organisms has not yet been completely understood.

12.2 Photosynthetic Reaction Center (RC)

Photosynthetic organisms possess very well-organized fine-tunable arrangements with structural and functional hierarchy which are the sites for transformation of light energy into chemical energy. The light-harvesting complexes absorb the incident photons, and the excitation energy is funneled into reaction center (RC) protein complexes having chlorophyll molecules that are redox active. Once the primary charge separations occur in the RCs, vectorial transport of charges (electrons and protons) in the photosynthetic membrane takes place. Due to properties that RCs possess, their use in solar energy-converting and integrated optoelectronic systems is practicable (Szabó et al. 2015).

The photosynthetic reaction center (RC) is the site where the first step of harnessing the light energy takes place. Hence, RC is one of the most important proteins (Nagy et al. 2010). The size of RC is about 10 nm, and it works also in nano-efficiency as one charge separation is initiated by one photon (Dorogi et al. 2006; Wraight and Clayton 1974; Nagy et al. 2010). Figure 12.1 illustrates the involvement of RC in separation of charge and stabilization processes in the photosynthetic membrane.

High reducing power for carbon reduction is generated by an electron driven across the membrane harnessing the free energy of light (Nagy et al. 2010). RCs are one of the characteristic features of photosynthetic organisms like plants, cyanobacteria, and photosynthetic bacteria. Despite significant differences in their structure and functions, common tasks performed in all cases include electron excitation (a pigment connected to the protein is excited by light), charge separation inside the protein, stabilization of the primer charge-pair by consecutive redox processes inside the protein, and protonation and deprotonation of certain amino acids

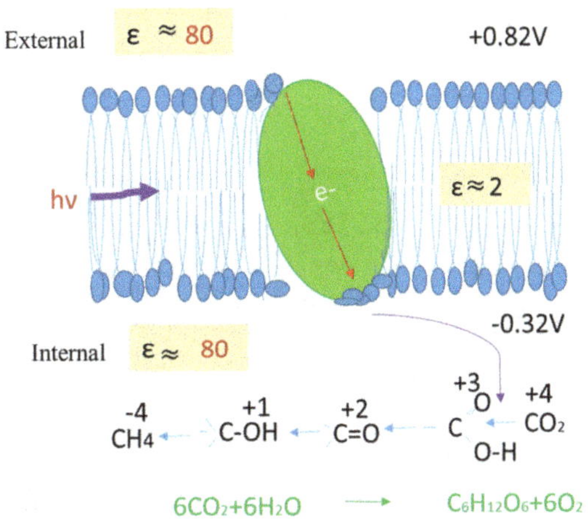

Fig. 12.1 RC proteins in charge separation and stabilization in photosynthetic membrane. Harnessing the free energy of light, an electron is driven across the membrane, and high reducing power for carbon reduction is created. (Adopted from Nagy et al. 2010)

that are linked to charge separation and stabilization. Conformational changes of amino acids and hydrogen bonding realignment accompany the charge movements (Nagy et al. 2010). Photosynthetic bacteria possess simple phototransduction machinery, with one reaction center. While in purple bacteria it is pheophytin-quinone, in green sulfur bacteria, it's the Fe-S type (passing electron through a quinone to an Fe-S center). Cyanobacteria, like plants, have two photosystems arranged in tandem.

12.2.1 RCs and Nanohybrids

RCs are gaining attention from research communities taking into consideration their academic, technical, agricultural, and ecological importance. RC could be an excellent model protein for several processes. Recently, RCs are viewed as quite useful in nanosystems, for instance, in imaging technologies or in integrated optoelectronic devices as circuit part. Initially during 1960s, *Rhodobacter sphaeroides* and *Blastochloris viridis* (formerly *Rhodopseudomonas viridis*) purple bacteria were the only source from which reaction centers could be retrieved. However, purification of RCs from a number of species is now possible (Nagy et al. 2010). Looking to the technical importance, these proteins with almost 100% photon energy transformation efficiency also possess technical properties which could be useful for other applications as well, e.g., characteristic light absorption in the near-infrared range (700–1000 nm) makes them important in developing security devices. Figure 12.2 shows a few of the bio-nanocomposite materials containing RCs. RCs when bound to transition metal oxides, carbon nanotubes, or other electrode surfaces could be useful in alternative fuel production and in CO_2 reduction by using sunshine (Dorogi et al. 2006; Lu et al. 2005; Das et al. 2004; Lebedev et al. 2006; Nagy et al. 2010). RCs also could be applied for reducing CO_2 level from the atmosphere and hence tackling the issue of greenhouse effect (Nagy et al. 2010).

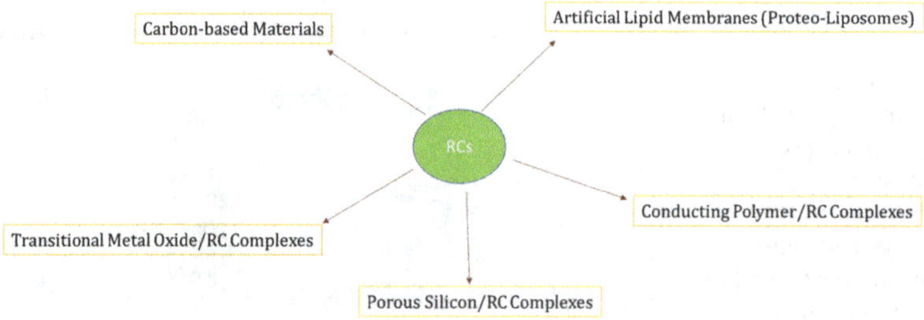

Fig. 12.2 Applications of RCs in various nanostructures

Hollander et al. (2011) harnessed bacterial photosynthetic reaction center pigment-protein complexes (RCs) interfaced with a metal electrode in order to maximize photocurrent generation. Better surface coverage of RCs surrounded by light-harvesting 1 complex (LH1) and role of tetraheme cytochrome as a connecting wire led to higher photocurrents. Cytochrome-c (cyt-c) introduction also led to increased current. This could be possibly due to intercalation amid the RCs or RC-LH1 complexes and the electrode to facilitate electron transfer. Under ambient conditions, generation of high currents could be achieved consecutively for many hours or days.

Native and untreated photosynthetic membranes of *R. sphaeroides* adsorbed onto a gold surface have been functionally characterized (Magis et al. 2010). Gold adsorbed membranes generated a steady high photocurrent for many minutes. The activity could be retained for 3 days (Magis et al. 2010).

Friebe et al. (2016) developed a biohybrid photocathode consisting of bacterial reaction center-light-harvesting 1 (RC-LH1) complexes self-assembled on a nanostructured silver substrate. This system yielded maximum photocurrent 416 $\mu A\ cm^{-2}$ under 4 suns, the highest reported so far using a bare metal electrode. On the rough silver substrate, a 2.5-fold increase in light absorption per RC-LH1 complex was observed. An increase in protein stability under continuous illumination was attributed to nanostructuring of the silver substrate. Simplicity in construction, enhanced protein loading capability, and stability besides more effective use of light made the hybrid material an outstanding candidate for developing plasmon-enhanced biosensors and biophotovoltaic devices.

While interconnection of proteins and protein complexes with electrodes remains a challenge, Kamran et al. (2014) employed Langmuir-Blodgett technology to produce a densely packed monolayer of protein complexes. The monolayer positioned onto a gold electrode produced the highest light-induced photocurrents per protein complex. The authors claim the work to be a first report revealing retention of intrinsic quantum efficiency of primary photosynthesis even outside biological cell.

Magyar et al. (2011) purified photosynthetic reaction center proteins (RCs) from purple bacterium *Rhodobacter sphaeroides* R-26 and bound it to single-walled carbon nanotubes (SWNTs). The authors observed accumulation of positive and negative charges upon binding of SWNT to RC and slow reorganization of protein structure after excitation. Stability of the photochemical activity of SWNT/RC complexes even in dried form for several weeks was a noteworthy observation. As per polarographic analysis, after light excitation, there was an electronic interaction between the RCs and SWNTs. Distinct properties of the SWNT/protein complexes make them potentially useful for applications in various fields like microelectronics, analytics, or energy conversion and storage.

Hajdu et al. (2018) purified photosynthetic reaction center proteins (RCs) from purple bacteria *Rhodobacter sphaeroides* R-26 and immobilized them on carboxyl- and amine-functionalized multiwalled carbon nanotubes (CNTs). The effect of carbon nanotubes on the $^1O_2$ production in RC/MWCNT photoactive hybrid materials was characterized. CNTs were observed as efficient quenchers of $^1O_2$ physical energy transfer or by chemical reaction, resonance, or oxidation, respectively.

Hajdu et al. (2012) immobilized photosynthetic reaction center (RC) purified from the purple bacterium *R. sphaeroides* R-26 on the surface of porous silicon microcavities (PSiMc) using two different binding methods, viz., covalent binding and non-covalent attachment, via a specific peptide interface ("peptide binding"). Binding affinity of the RCs to the peptide-coated PSi was observed to be nearly double. Activity of the immobilized RC was examined using various analytical techniques, and the immobilized RC was observed to be functional. Functional integrity of the bio-nanomaterial has been considered as a promising aspect in developing applications of silicon-based electronics and biological redox system.

Hajdu et al. (2011) carried out structural, optical, and electrochemical studies of RC bio-nanocomposite prepared using reaction centers from *R. sphaeroides* R-26 and different carrier matrices like borosilicate glass, conducting indium tin oxide (ITO), non-functionalized and NH_2-functionalized multiwalled carbon nanotubes (MWNTs), and conductive artificial materials. Binding of photosynthetic RCs could be achieved, and there was only partial loss of photochemical activity of the RC. An electrostatic interaction existed between the RCs and the carrier material if the carrier matrix was a conductive material. Hajdu et al. (2017) have reviewed characteristics of photosynthetic RC proteins and their potential applications for preparing functional molecular bionanohybrid resources as materials for future.

In natural photosynthesis, only one reaction center combined with a number of light-harvesting units is involved. Harnessing metal nanoparticles, dendrimers, oligopeptides, nano-carbon materials, porphyrin nanotubes, etc., it could be possible to combine multiple reaction centers with light-harvesting units. Development of efficient light-to-energy conversion systems and artificial photosynthesis for solar fuel production could be possible using such materials (Fukuzumi and Ohkubo 2012).

Oda et al. (2006) extracted the photosynthetic light-harvesting complex LH2 protein from the intracytoplasmic membranes from anoxygenic thermophilic purple photosynthetic bacterium *Thermochromatium tepidum* and purified it. The LH2 complex having a cylindrical structure of 7.3 nm diameter, containing 27 bacteriochlorophyll a and 9 carotenoid molecules, was introduced into a folded-sheet silica mesoporous material (FSM) having nanometer-size pores of honeycomb-like hexagonal cylindrical structure inside. Even after adsorption to FSM, the LH2 exhibited almost intact absorption bands of bacteriochlorophylls and was observed to be fully active in capturing and transferring excitation energy. Higher circular symmetry was suggested as increased heat stability of the exciton-type absorption band of bacteriochlorophylls (B850) was exhibited by the LH2 complex inside the FSM. The authors proposed the environment inside the hydrophobic silica nanopores as a new matrix for the membrane proteins to exhibit their functions. New probes and reaction systems could be constructed using the silica-membrane protein adduct.

Further, Oda et al. (2010) purified photosynthetic reaction center pigment-protein complex from a thermophilic purple photosynthetic bacterium *Thermochromatium tepidum* and incorporated it into silica nanopores. Even after incorporation into the silica nanopore, the RC protein retained full activity and performed photosynthetic charge separation reaction and electron transfer reactions when the pore diameter fitted to the size of RC protein complex. ATR-FTIR data revealed that the overall RC protein structure also remained unchanged. The authors suggested a possibility of preparation of silica mesoscopic material conjugates with multiple enzymes (e.g., hydrogenases) that can utilize reducing power from the RC. Useful materials can be produced harnessing such conjugates.

While there is a great interest in fabricating versatile photosynthetic reaction center (RC)-functionalized nanocomposite films, Lu et al. (2007) have reviewed different types of RC-functionalized films and considered self-assembled monolayers (SAMs), protein-embedded gel films, and protein-entrapped nanoporous films more practical looking to convenience in protein film preparation and control over orientation of the close packed proteins layers and keeping a mild environment for preserving the protein surface microstructure, etc.

Due to unique properties of reaction centers (RCs), harnessing them in solar energy-converting and integrated optoelectronic systems could be possible. Hence, there has been a growing interest towards using them in molecular devices. It has been reviewed that RCs bound to different carrier matrices largely retained their photophysical and photochemical activities in the nanosystems (Szabó et al. 2015). Upon employing various physical and chemical methods for binding of RCs to different nanomaterials, it is usually observed that the P+(QAQB)− charge pair is stabilized after the attachment of the RCs to the nanostructures, followed by slow reorganization of the protein structure. Electric conductivity measurement indicated an electronic interaction between the protein and inorganic carrier matrices. This could be important in developing biohybrid devices for biosensor and/or optoelectronic applications (Szabó et al. 2015).

Liposomes have also been observed to be important in reconstitution of photoreaction units (PRUs) involving RC and light-harvesting (LH) members (Crielaard et al. 1989; Ajiki et al. 1998; Goc et al. 1996). Immobilized PRU liposomes could be useful as a photo-electrochemical conversion layer for biosensors or photocells (Crielaard et al. 1989; Ajiki et al., 1998; Lu et al. 2007). Goc et al. (1996) reported construction of a PRU by harnessing RC and either Bchl a or chlorophyll (Chl) a molecules inserted together in liposome membrane, wherein without aid of polypeptides, the pigments incorporated directly into the lipid bilayer and worked as LH antennae for RC. Improved efficiency in photobleaching of RC at 860 nm and photoelectric conversion was revealed by the system with reconstructed antennae. Donor-acceptor energy transfer between the membrane-embedded pigments and RC occurred as indicated by quenching of fluorescence emission upon addition of RC to liposomes.

12.3 Conclusion

Explorations on biohybrid, especially bionanohybrid materials, have been an interesting area of research. RC proteins, vital components of light energy-converting systems, due to their unique properties, are not only fundamental to sustaining life on earth but are also important to nanobionic technologies. Recently, RC proteins have been observed to play significant role in synthesizing bionanohybrids with diverse potentials and, hence, gained importance in developing materials possible for real application in the future.

References

Ajiki S, Sugino H, Toyotama H, Hara M, Miyake J (1998) Reconstitution and immobilization of photo-reaction units from photosynthetic bacterium *Rhodopseudomonas viridis*. Mater Sci Eng C 6:285–290

Crielaard W, Hellingwerf KJ, Konings WN (1989) Reconstitution of electrochromically active pigment-protein complexes from *Rhodobacter-sphaeroides* into liposomes. Biochim Biophys Acta 973(2):205–211

Darder M, Aranda P, Ruiz-Hitzky E (2007) Bionanocomposites: a new concept of ecological, bioinspired and functional hybrid materials. Adv Mater 19:1309–1319

Das R, Kiley PJ, Segal M, Norville J, Yu A, Wang L, Trammell SA, Reddick LE, Kumar R, Stellacci F, Lebedev N, Schnur J, Bruce BD, Zhang S, Baldo M (2004) Integration of photosynthetic protein molecular complexes in solid-state electronic devices. Nano Lett 4:1079–1083

Dave V, Vala AK, Patel RJ (2015) Observation of weak localization of light in gold nanofluids synthesized using the marine derived fungus *Aspergillus niger*. RSC Adv 5:16780–16784

Dorogi M, Bálint Z, Cs M, Vileno B, Milas M, Hernádi K, Forró L, Gy V, Nagy L (2006) Stabilization effect of single walled carbon nanotubes on the functioning of photosynthetic reaction centres. J Phys Chem B 110:21473–21479

Drexler KE (1981) Molecular engineering: An approach to the development of general capabilities for molecular manipulation. Proc Natl Acad Sci U S A 78:5275–5278

Friebe VM, Delgado JD, Swainsbury DJK, Gruber JM, Chanaewa A, van Grondelle R, von Hauff E, Millo D, Jones MR, Frese RN (2016) Plasmon-enhanced photocurrent of photosynthetic pigment proteins on nanoporous silver. Adv Funct Mater 26:285–292

Fukuzumi S, Ohkubo K (2012) Assemblies of artificial photosynthetic reaction centres. J Mater Chem 22:4575–4587

Giraldo JP, Landry MP, Faltermeier SM, McNicholas TP, Iverson NM, Boghossian AA, Reuel NF, Hilmer AJ, Sen F, Brew JA, Strano MS (2014) Plant nanobionics approach to augment photosynthesis and biochemical sensing. Nat Mater 13:400–408

Goc J, Hara M, Tateishi T, Miyake J (1996) Reconstructed light-harvesting system for photosynthetic reaction centres. J Photochem Photobiol A Chem 93:137–144

Guldi DM, Aminur Rahman GM, Sgobba V, Ehli C (2006) Multifunctional molecular carbon materials-from fullerenes to carbon nanotubes. Chem Soc Rev 35:471–487

Hajdu K, Szabó T, Magyar M, Bencsik G, Németh Z, Nagy K, Forró L, Váró G, Hernádi K, Nagy L (2011) Photosynthetic reaction center protein in nanostructures. Phys Status Solidi B 248(11):2700–2703

Hajdu K, Gergely C, Martin M, Zimányi L, Agarwal V, Palestino G, Hernádi K, Németh Z, Nagy L (2012) Light-harvesting bio-nanomaterial using porous silicon and photosynthetic reaction center. Nanoscale Res Lett 7:400

Hajdu K, Szabó T, Sarrai AE, Rinyu L, Nagy L (2017) Functional nanohybrid materials from photosynthetic reaction center proteins. Int J Photoenergy: 9128291. https://doi.org/10.1155/2017/9128291

Hajdu K, Rehman AU, Vass I, Nagy L (2018) Detection of singlet oxygen formation inside photoactive biohybrid composite material. Materials 11:28. https://doi.org/10.3390/ma11010028

Hollander MJ, Magis JG, Fuchsenberger P, Aartsma TJ, Jones MR, Frese RN (2011) Enhanced photocurrent generation by photosynthetic bacterial reaction centers through molecular relays, light-harvesting complexes, and direct protein-gold interactions. Langmuir 27(16):10282–10294

Kamran M, Delgado JD, Friebe V, Aartsma TJ, Frese RN (2014) Photosynthetic protein complexes as bio-photovoltaic building blocks retaining a high internal quantum efficiency. Biomacromolecules 15(8):2833–2838

Lebedev N, Trammell SA, Spano A, Lukashev E, Griva I, Schnur J (2006) Conductive wiring of immobilized photosynthetic reaction centre to electrode by cytochrome C. J Am Chem Soc 128:12044–12045

Lu Y, Liu YY, Xu J, Chunhe Xu C, Liu B, Kong J (2005) Bio-nanocomposite photoelectrode composed of the bacteria photosynthetic reaction centre entrapped on a nanocrystalline TiO_2 matrix. Sensors 5:258–265

Lu Y, Xu J, Liu B, Kong J (2007) Photosynthetic reaction center functionalized nano-composite films: Effective strategies for probing and exploiting the photo-induced electron transfer of photosensitive membrane protein. Biosens Bioelectron 22:1173–1185

Magis GJ, Hollander MJ, Onderwaater WG, Olsen JD, Hunter CN, Aartsma TJ, Frese RN (2010) Light harvesting, energy transfer and electron cycling of a native photosynthetic membrane adsorbed onto a gold surface. Biochimica et Biophysica Acta (BBA): Biomem 1798(3):637–645

Magyar M, Hajdu K, Szabó T, Hernádi K, Dombi A, Horváth E, Magrez A, Forró L, Nagy L (2011) Long term stabilization of reaction center protein photochemistry by carbon nanotubes. Phys Status Solidi B 248(11):2454–2457

Nagy et al (2010) Photosynthetic reaction centres – from basic research to application. Not Sci Biol 2(2):7–13

Nagy L, Magyar M, Szabo T et al (2014) Photosynthetic machineries in nano-systems. Curr Protein Pept Sci 15:363–373

Oda I, Hirata K, Watanabe S, Shibata Y, Kajino T, Fukushima Y, Iwai S, Itoh S (2006) Function of membrane protein in silica nanopores: incorporation of photosynthetic light-harvesting protein LH2 into FSM. J Phys Chem B 110:1114–1120

Oda I, Iwaki M, Fujita D, Tsutsui Y, Ishizaka S, Dewa M, Nango M, Kajino T, Fukushima Y, Itoh S (2010) Photosynthetic electron transfer from reaction center pigment-protein complex in silica nanopores. Langmuir 26(16):13399–13406

Prathna TC, Mathew L, Chandrasekaran N, Raichur AM, Mukherjee A (2010) Biomimetic synthesis of nanoparticles: science, technology & applicability. In: Mukherjee A (ed) Biomimetics, learning from nature. https://doi.org/10.5772/8776

Shoseyov O, Levy I (2008) Nanobiotechnology: bioinspired devices and materials of the future. Humana Press Inc., Totowa

Siegert M, Sonawane JM, Ezugwu CI, Prasad R (2019) Economic assessment of nanomaterials in bio-electrical water treatment. In: Prasad R, Thirugnanasanbandham K (eds) Advanced research in nanosciences for water technology. Springer International Publishing AG, Cham, pp 5–23

Singh K V, Juyal V D, Chakravorty A, Upadhyay N, Thapliyal D (2018) Approaches in microbial fuel cell array arrangements for conventional DC battery charger. In: 2nd international conference on Inventive Systems and Control (ICISC), Coimbatore, 2018, pp. 418–422

Strik DPBTB, Terlouw H, Hamelers HVM, Buisman CJN (2008) Renewable sustainable biocatalyzed electricity production in a photosynthetic algal microbial fuel cell (PAMFC). Appl Microbiol Biotechnol 81:659–668

Szabó T, Magyar M, Hajdu K, Dorogi M, Nyerki E, Tóth T, Lingvay M, Garab G, Hernádi K, Nagy L (2015) Structural and functional hierarchy in photosynthetic energy conversion-from molecules to nanostructures. Nanoscale Res Lett 10:458–470

Vala AK, Trivedi HB, Dave BP (2016) Marine-derived fungi: potential candidates for fungal nano-biotechnology. In: Prasad R (ed) Advances and applications through fungal nanobiotechnology. Springer International Publishing, Cham, pp 47–69
Wallace GG, Higgins MJ, Moulton SE, Wang C (2012) Nanobionics: the impact of nanotechnology on implantable medical bionic devices. Nanoscale 4:4327–4347
Wraight CA, Clayton R (1974) The absolute quantum efficiency of bacteriochlorophyll photo-oxidation in reaction centres of *Rhodopseudomonas sphaeroides*. Biochem Biophys Acta 333:246–260
Yang J, You J, Chen CC, Hsu WC, Tan H, Zhang XW, Hong Z, Yang Y (2011) Plasmonic polymer tandem solar cell. ACS Nano 5:6210–6217

Chapter 13
Nanomaterials in Microbial Fuel Cells and Related Applications

Theivasanthi Thirugnanasambandan

Contents

Abbreviations

2D	Two dimension
3D	Three dimension
AC	Activated carbon

T. Thirugnanasambandan (✉)
International Research Centre, Kalasalingam Academy of Research and Education (Deemed University), Krishnankoil, TN, India

R. Prasad (ed.), *Microbial Nanobionics*, Nanotechnology in the Life Sciences,
https://doi.org/10.1007/978-3-030-16383-9_13

Am^{-2}	Ampere per meter square (unit for current density)
Am^{-3}	Ampere per cubic meter (unit for current density)
ANB	Acid navy blue r dye
ARB	Anode-respiring bacteria
Au	Gold
BECs	Bioelectrochemical cells
bioMEMS	Biomedical microelectromechanical systems
BOD	Biochemical oxygen demand
CE	Coulomb efficiency
CeO_2	Ceric oxide (or) ceric dioxide (or) ceria (or) cerium oxide (or) cerium dioxide (or) cerium(IV) oxide
CFE	Carbon felt electrode
CMC	Carboxymethyl cellulose
CNTs	Carbon nanotubes
COD	Chemical oxygen demand
CPPEs	Carbon paste paper electrodes
Cr(III)	Trivalent chromium
Cr(VI)	Hexavalent chromium
CS	Chitosan
e^-	Electrons
EDX	Energy dispersive X-Ray analyzer
Fe-AAPyr	Iron-aminoantipyrine
GNS	Graphene nanosheet
GO	Graphene oxide
H^+ ions	Protons
HRTEM	Transmission electron microscope
kΩ	Kiloohm (unit for resistance)
MFC	Microbial fuel cells
MnO_2	Manganese oxide (or) manganese dioxide (or) manganese(IV) oxide
MnOOH	Hydroxy-oxido-oxomanganese
mV	Millivolt (unit for potential)
MWCNTs	Multiwall carbon nanotubes
mW/m^2	Milliwatts per square meter (unit for power density)
Ni	Nickel
nm	Nanometer
nW/cm^2	Nano-watts per square centimeter (unit for power density)
ORR	Oxygen reduction reaction
PA	Phosphoric acid
PANI	Polyaniline
Pd	Palladium
PDMS	Polydimethylsiloxane
PEM	Proton-exchange membrane
pH	Power of hydrogen
PMMA	Polymethyl methacrylate

POE	Poly(oxyethylene)
Pt	Platinum
PTFE	Polytetrafluoroethylene
PVA	Poly(vinyl alcohol)
PVAc-g-PVDF	Polyvinyl acetate-polyvinylidene fluoride coated cotton fabric
rGO	Reduced graphene oxide
SCOD	Soluble chemical oxygen demand
SEM	Scanning electron microscope
STEM	Scanning transmission electron microscope
TCOD	Total chemical oxygen demand

13.1 Introduction: Microbial Fuel Cells (MFCs)

A fuel cell is an energy-generating device that converts chemical energy into electrical energy. It consists of an electrolytic cell with the electrolyte and two electrodes (cathode and anode) which are connected together with a load resistance. MFC is an electrochemical fuel cell. It is used for sustainable energy generation by wastewater treatment with the help of electrochemically active microorganisms. The microorganisms are fed with organic substances which are called as substrates. They form biofilms over the anode for efficient electron transfer between the substrate and the anode. MFC has potential applications in alternative energy, wastewater treatment, environmental protection, and bio-sensing (for oxygen and pollutants) (Siegert et al. 2019).

Biofilm is a sticky matrix which has conducting nanowires. The microorganisms oxidize the substrate, and electron transfer may occur directly or with the help of mediators. The electrons are used to reduce oxygen into water at the cathode. Zhou et al. reported that the bacterial biofilm produced at anode acts as catalyst. It converts the chemical energy of the organic molecule into electrons during oxygen reduction at cathode and forms water (Zhou et al. 2013). Mustakeem reported the biofilm formation in MFC. It is shown in Fig. 13.1 (Mustakeem 2015). MFCs are classified into various types like single-chamber MFC or two-chamber MFC (with two separate compartments for anode and cathode). In a two-chamber MFC, the chambers are separated by a PEM which allows the protons to permeate through it to participate in redox reactions. They can be used to power small electronic devices like biosensors.

In MFC, organic materials and waste materials are the source of biofuels. The ions (anion and cation) in the water and the water-soluble ions of these materials serve as electrolytes. Microorganisms serve as biocatalysts (Khan et al. 2017). When these materials are catalyzed by the microorganisms, electrons and protons are transferred and moved (Moqsud et al. 2015). Initially, electrons (e^-) go to the anode electrode. Next, they move to the cathode electrode through the external circuit. In the case of protons (H^+ ions), they go to the cathode electrode by the electrolyte. In the cathode, they react with oxygen. Hence, MFC generates electrical power

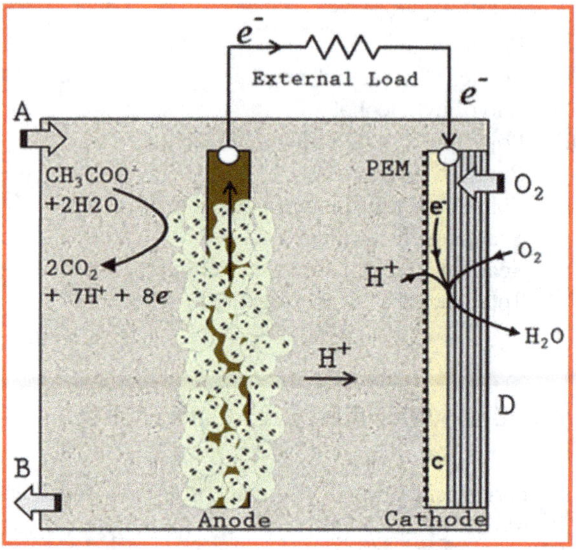

Fig. 13.1 Schematic diagram shows the bacterial biofilm formed at the anode of a typical single-chamber MFC. (**a**) Effluent inlet. (**b**) Effluent outlet. (**c**) Catalyst layer. (**d**) Diffusion layers at air-cathode. (From Mustakeem, 2015)

(Ersan et al. 2010; Zhao et al. 2013; Khudzari et al. 2016; Khan et al. 2017; Siegert et al. 2019). Figure 13.2 shows the bioelectricity production from organic and waste materials.

In the report of Buzea et al. (2007), nanomaterials are described as, in principle, materials of which a single unit is sized (in at least one dimension) between 1 and 1000 nanometers (10^{-9} meter) but usually is 1 and 100 nm. Nanomaterial synthesis has been developed in support of microfabrication research. They often have unique optical, electronic, or mechanical properties (Hubler and Osuagwu 2010). Nanomaterials are being commercialized slowly (Eldridge 2014) and emerging as commodities (McGovern 2010). These materials are applied in various fields, industries, and products including healthcare, electronics, and cosmetics.

Since MFC is working with microorganisms, there is a need for biocompatible materials as electrodes. So, carbon materials are found to be suitable candidates as electrode materials for MFC. Nanotechnology offers many efficient materials like graphene, carbon nanotubes, and their composites with conducting polymers like polyaniline, polypyrrole, and other conducting metal oxides like iron oxide and manganese oxide. Using novel nanomaterials, increasing the power density of MFC is a trend for practical applications. Various instruments are used to study the reaction mechanism in a microbial fuel cell.

Confocal laser scanning microscopy is used to view the growth and thickness of the biofilms on electrodes. Cyclic voltammetry is applied to study the extracellular electron transfer. But it can be used only at low scan rates since the reactions involve

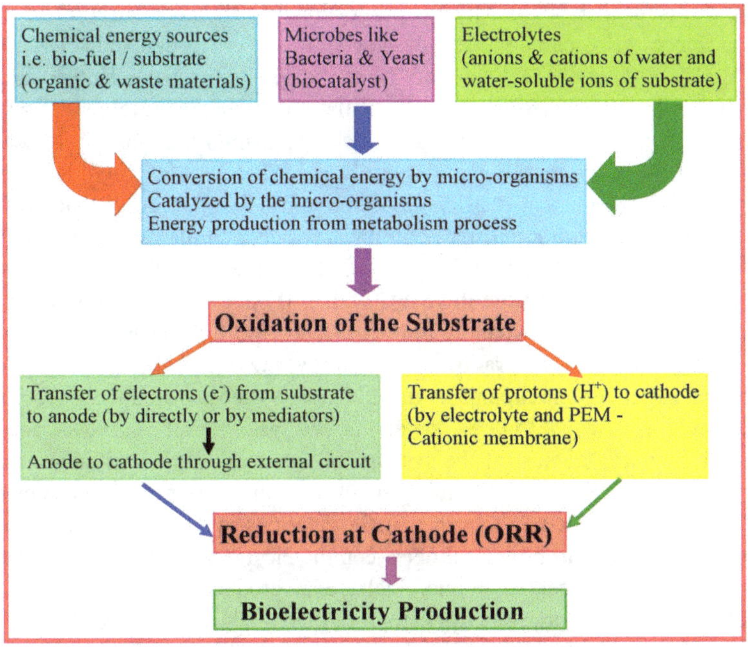

Fig. 13.2 Flow chart of the production of bioelectricity from organic and waste materials by microbial activities

living bacteria (Omkar et al., 2017). Bacteria produce power indefinitely as long as there are enough food sources to nurture the bacteria. MFCs are applied in the production of hydrogen fuel cells and desalination of seawater and provide sustainable energy sources for remote areas (Reuben 2018).

Generally, in the MFC, wastes are utilized as a source of substrate due to the dual advantage, i.e., cheap electricity generation and wastewater treatment. Some of the common wastes are domestic wastewater, vegetable wastes, agrowastes, azo dye wastes, and wastewater of various industries like food industries, breweries, cellulose/starch industries, organic industries, and chemical (chromium, selenium, and nitrate) industries. *Shewanella oneidensis*, *Geobacter sulfurreducens*, *Klebsiella pneumoniae*, *Rhodoferax ferrireducens*, *Escherichia coli*, *Rhodopseudomonas palustris*, *Desulfovibrio desulfuricans*, *Acidithiobacillus ferrooxidans*, *Clostridium cellulolyticum*, *Enterobacter cloacae*, *Trichococcus pasteurii*, *Streptomyces enissocaesilis* KNU (K strains), *Nocardiopsis* sp. KNU (S strain), *Pseudomonas* species, *Parabacteroides*, *Proteiniphilum*, and *Catonella* are some of the bacteria utilized in the MFC. Apart from bacteria, microorganism like yeast (*Saccharomyces cerevisiae*, *Candida melibiosica*, *Hansenula anomala*, *Hansenula polymorpha*, *Arxula adeninivorans*, and *Kluyveromyces marxianus*) is utilized as biocatalysts.

The power output of the MFC is proportional to the concentration of organic wastes. Hence, MFC can be used as a biochemical oxygen demand sensor (Chang et al. 2005; Kim et al. 2003). Apart from the substrate, many factors including microorganisms alter performance of the MFC. Some higher-level power output of the MFC (using various substrates and microorganisms) is reported: as per Wang et al. report, MFC using rice straw hydrolysate substrate (400 mg/mL concentration) generates the current density 137.6 mA/cm^2 (Wang et al. 2014); MFC with *Pseudomonas* sp. microorganism, peptone substrate, and methylene blue mediator generates the power density 979 mW/m^2 (Daniel et al. 2009); current output 2236 mW/m^2 is obtained from the MFC with 200 ppm concentration of acid navy blue r (ANB) dye waste (Khan et al. 2015); and 2900 mW/m^2 is the output of the MFC with 25 mg/ L concentration of selenium waste (Catal et al. 2009).

Utilizations of aerobic organisms in MFC and wastewater treatment require oxygen. Hence, it is essential to flow oxygen into the system for the survival of bacteria. It leads to increase the cost of the system but anaerobic organisms such as anode-respiring bacteria (ARB) reduces the cost substantially. Anaerobic organisms do not need oxygen. Also, ARB is able to transfer electrons in between the electrodes. It helps to the electrons to flow freely between the electrodes. Alternative form of the MFC, i.e., electrolytic cell using microbial agents, produces hydrogen at the cathode instead of electricity (Waste-management-world.com 2013). Growing of aerobic bacteria in cathode chamber (improves the oxygen reduction rate and inhibits proton transfer) and growing of anaerobic organisms like ARB in anode chamber (supports for electrons flow) will improve the MFC performance.

13.2 Energy Needs of the World

As per the International Energy Agency (IEA), current power demand is 12 billion tonne oil equivalent. It is expected to increase up to 18 billion tonne oil equivalent in the coming decades (Chu and Majumdar 2012). Enormous energy is found in organic matter such as carbohydrates. They are present in the municipal waste. Fuel cells are coming under alternative energy sources having a negligible CO_2 emission. Also MFCs do not need any external energy input. This is an additional advantage for the extensive applications of MFCs particularly where electrical amenities less or no (Chandrasekhar et al. 2018).

Biomass and household waste are identified as resources for the bioenergy. Reuse, recovery, and recycling of wastewater can offer a sustainable wastewater management system in freshwater-scarce regions. Microbial fuel cell is the solution for both water and energy needs (Gajda et al. 2016). The world shortage of potash, nitrogen, and phosphorus suggests the usage of treated wastewater for farming. Water and wastewater treatment and distribution consume 1.4% of the total national electricity consumption. Figure 13.3 shows the energy production from nonrenewable energy resources causes environmental pollution and MFC produces clean energy.

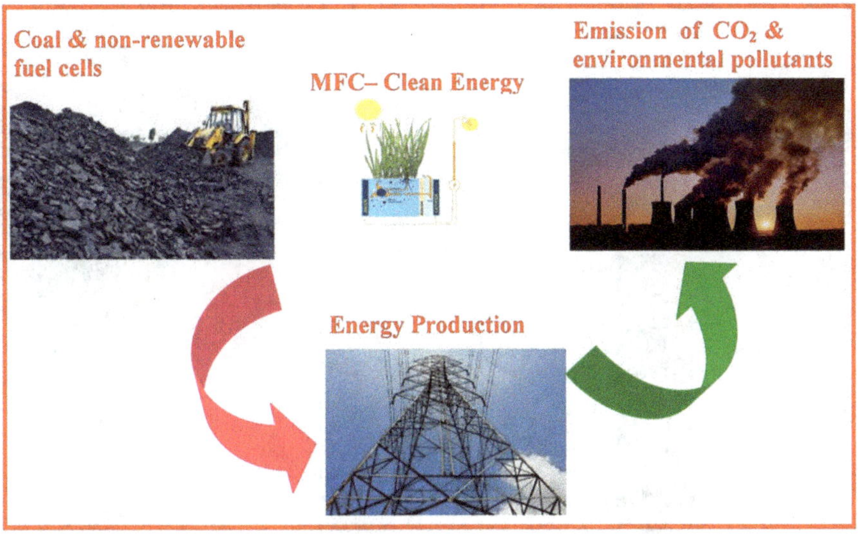

Fig. 13.3 Diagram shows the utilization of nonrenewable energy resources in energy production to meet the energy demand. This leads to depletion of the resources and causes environmental pollution

13.3 Construction of MFC

A two-chamber MFC usually consists of two chambers in "H" shape separated by a tube with proton-exchange membrane. The anode should be highly conductive, biocompatible, and chemically stable under the conditions present inside the fuel cell. Metals like copper are avoided because of corrosion which has toxic effect on microorganisms. The cathode ought to have good reduction performance, good re-oxidation, and high long-term stability. The salt or agarose bridges used in MFC have high internal resistance. In a MFC aerobic respiration and fermentation should be avoided, and anaerobic respiration must be accelerated (igem.org 2018). Figure 13.4a, b shows the schematic working model of dual-chambered MFC and single-chambered MFC, respectively (Chaturvedi and Verma 2016).

13.3.1 Biocathode of MFC

The higher cost, non-environmental friendly, and complexities in the catalyst production are the factors leading to the biocathode development. Microorganisms serve as the catalyst in these biocathodes. The biofilm created on the cathode surface catalyzes the reduction reaction (Harnisch and Schröder 2010). Figure 13.5 illustrates a biocathode with biofilm on the surface. The biofilm catalyzes the reduction of chemical active species such as nitrate and oxygen (Mustakeem 2015).

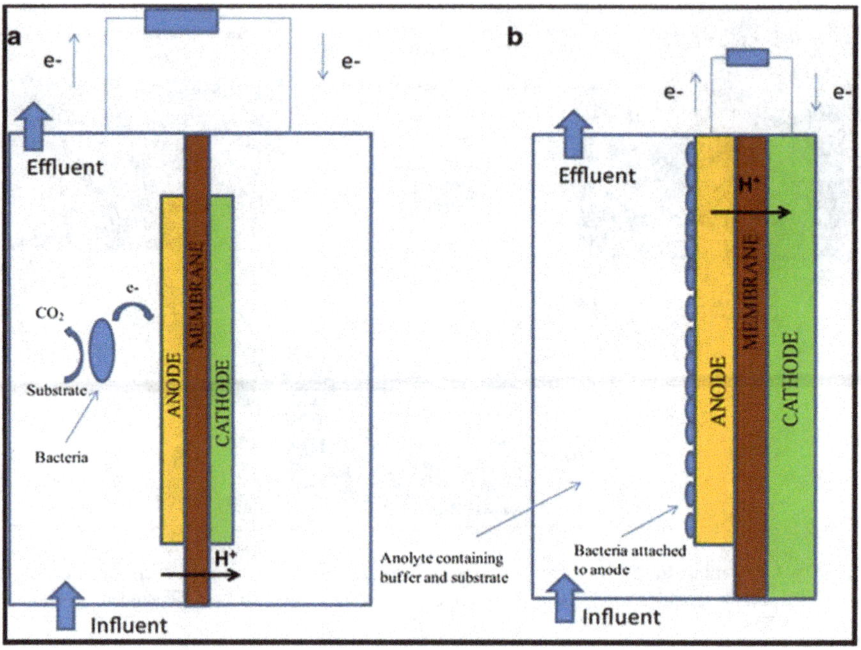

Fig. 13.4 MFC working model. (**a**) Dual-chambered MFC and (**b**) single-chambered MFC. (From Chaturvedi and Verma 2016)

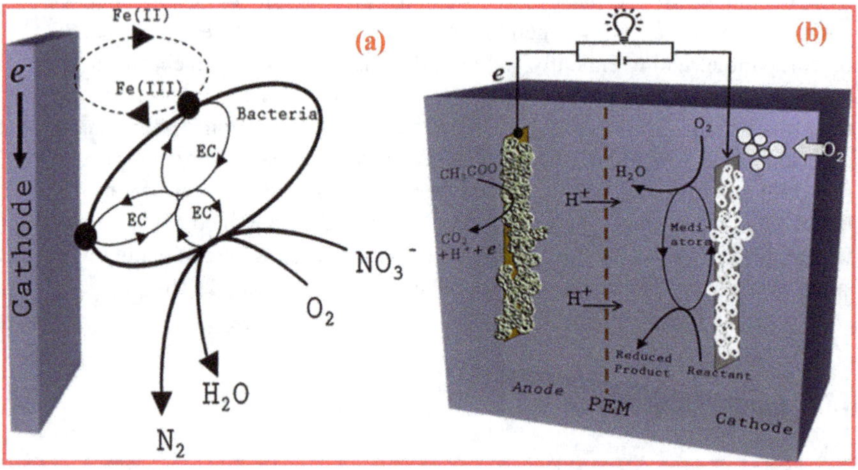

Fig. 13.5 MFC biocathode: (**a**) Oxygen and nitrate are reduced by direct electron transfer or by mediator. Transition metal mediators transfer the electrons to oxygen. (**b**) Reduction of oxygen and chemical reactants. This MFC will generate power and will drive biochemical synthesis reactions. (From Mustakeem 2015)

Biocathodes are categorized as aerobic and anaerobic biocathodes. Oxygen acts as a terminal electron acceptor, and hydrogen peroxide acts as an intermediate in aerobic biocathodes. Transition metals like iron and manganese serve as electron mediators in between the electrode and oxygen. Electrons are transferred from the cathode to the terminal electron acceptor in this method (Park and Zeikus 2003). In anaerobic biocathodes oxygen is not present. In this type, nitrates and sulfates can be utilized as terminal electron acceptors instead of oxygen.

13.3.2 Proton-Exchange Membrane (PEM)

Commercially available membranes (like Nafion) are highly conductive. These membranes have high ion-exchange capacity and proton conductivity due to their structural properties. However Nafion has problems of oxygen leakage from the cathode to the anode, substrate losses and biofouling, conductivity at low water content, and poor mechanical strength at high temperatures. Proton-exchange membrane is synthesized by solution casting method utilizing the materials such as poly(oxyethylene), poly(vinyl alcohol), chitosan, and phosphoric acid. Water absorption capacity is directly proportional to the concentration of PVA. A higher percentage of water uptake means higher ion-exchange capacity of the membrane (Dharmadhikari et al. 2018).

Membranes made of natural materials such as eggshell membrane are used in MFC. Sensor devices with those membranes possess low internal resistances and high sensitivity (Chouler et al. 2017). The membrane must transfer protons from the anode to the cathode and should prevent the transfer of oxygen and substrates. Oxygen penetration to the anode part makes the anode aerobic. Also, it reduces power density and wastewater treatment efficiency. Penetration of oxygen into the cathode substrate leads to the reduction of power out and microorganisms' efficiency in chemical oxygen demand (COD) removal (Ghasemi et al. 2015).

MFC uses expensive proton-exchange membrane for separating anode and cathode. Recently some inexpensive clay is studied as an ion-exchange medium. They are known as cation-exchange membranes. They have structurally stronger properties. Both proton-exchange membrane and cation-exchange membrane reduce oxygen diffusion in the anode chamber of the cell. But proton-exchange membrane cannot be reused and is difficult to adapt to desired shapes. Clay mixed with kaolin and bentonite and finally fired designed a system. This kind of clay system produced an open-circuit voltage up to 1.5 volts. Figure 13.6 shows the diagram of double-chamber MFC designed with cation-exchange membrane (Reuben 2018). Similarly an inexpensive nanocomposite membrane based on cotton fabric is designed. A highest proton conductivity of PVAc-g-PVDF-coated cotton fabric of $(1.5 \pm 0.2) \times 10^{-2}$ S/cm at 25 °C and lowest glucose permeability of $(12 \pm 1) \times 10^{-6}$ cm^2/s are obtained (Zhang et al. 2017a).

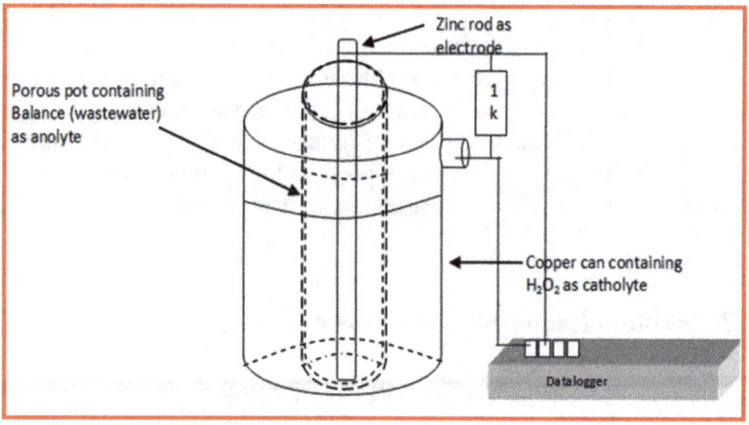

Fig. 13.6 A simple double-chamber MFC design. It has porous pot anode chamber and copper cathode chamber. Porous pot is cation-exchange membrane. (From Reuben 2018)

13.3.3 Microorganisms as Biocatalysts

Microorganisms produce bacterial flagellar-like structures that are made up of several proteins which possess metallic-like conductivity. This type of extracellular electrically conductive protein nanofilaments are identified in *Geobacter sulfurreducens* called as microbial nanowires. These wires are also formed in organisms like *Rhodopseudomonas palustris*, *Desulfovibrio desulfuricans*, and *Acidithiobacillus ferrooxidans*. The electrical conductivity of these microbial nanowires is comparable to carbon nanotubes and some organic conductors. Proteins contain several aromatic amino acids which are responsible for electrical conduction. So, these microorganisms found applications in the field of bioremediation of heavy metals like uranium, arsenic, and chromium and in bioelectronics since these wires are having enough mechanical strength (Young's modulus ~1 GPa) for construction of electronic devices (Sure et al. 2016).

Microorganisms can transfer electrons from the substrate to the anode without the help of mediators or exoelectrogens. They produce specific proteins or genes for their inevitable performance toward electricity generation in MFCs. The mechanisms involved in this process are biofilm formation, metabolism, and electron transfer. The expression of certain genes for outer membrane multiheme cytochromes, redox-active compounds, and conductive pili participate in the exoelectrogenic activity of microorganism genus such as *Geobacter* and *Shewanella* (Kumar et al. 2015).

The anode-respiring bacteria (ARB) can thrive in oxygen-free environments, and they are called as anaerobic organisms. These anode-respiring bacteria are able to transfer electrons to the negative terminal, thus generating useable current in the process. They attach themselves to the anode of a battery. They consume an organic substrate for food and release excess electrons as part of their metabolic pathway. The biofilm formed is a living matrix of protein and sugar. Geobacter is an ARB

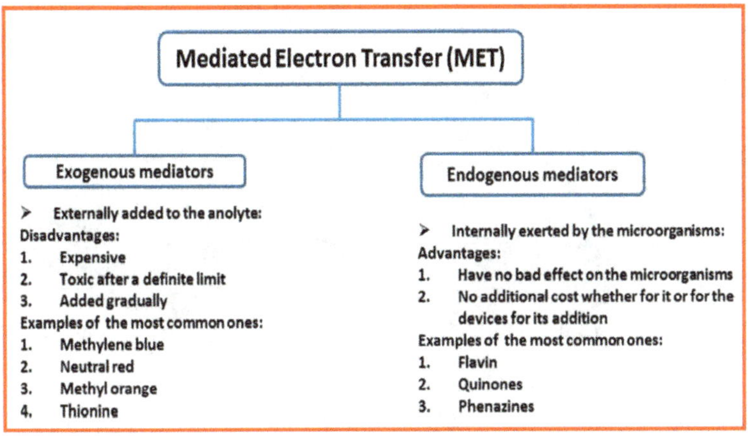

Fig. 13.7 External and internal mediators in MFCs. (From Sayed and Abdelkareem 2017)

which is useful in the production of high current densities. ARB like only one compound, i.e., acetate, which is a fatty acid (Waste-management-world.com 2013).

Direct transfer of electrons takes place from bacteria to anode. Sometimes mediators like thionine, methyl viologen, and humic acid are used for electron transfer. These mediators are more expensive and toxic to the microorganisms. Bacteria like *Shewanella oneidensis* and *Geobacter sulfurreducens* produce electrically conductive appendages called bacterial nanowires. These nanowires facilitate direct transfer of electrons to the anode, thus increasing the efficiency and reducing the cost (Lal Deeksha 2013). *Escherichia coli* are often used for power generation because they can grow fast and are quite robust regarding cultivation conditions. Electrons are transported through nanowires called pili, which possess delocalized electronic states to function as protein wires with metallic-like conductivity (Omkar et al. 2017).

Internal mediators have advantages more than the external mediators. They are cheap and have no toxic effect on the microorganism. Figure 13.7 shows some examples for the internal and external mediators with their advantages and disadvantages. Several yeast strains like *Saccharomyces cerevisiae*, *Candida melibiosica*, *Hansenula anomala*, *Hansenula polymorpha*, *Arxula adeninivorans*, and *Kluyveromyces marxianus* have been studied as biocatalysts in MFC. Figure 13.8 shows the electron transfer during the metabolism of the organic materials (cellular respiration process) in the yeast cell (Sayed and Abdelkareem 2017).

13.4 Functions and Mechanisms of MFC

13.4.1 Mimicking and Resemblance of MFC

MFC mimics the interactions of microbes found in nature. Bacteria like *Listeria monocytogenes* deliver electrons. The mechanism involved in the energy production from wastes by MFC somewhat resembles the metabolism process by the gut flora

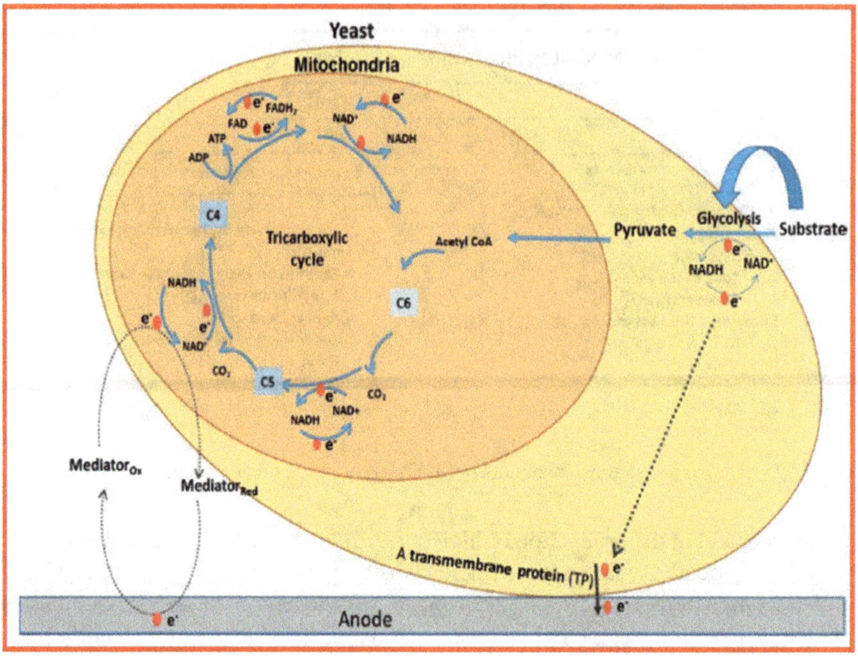

Fig. 13.8 Illustration for the possibilities of electrons' origin and transfer of yeast cells into MFC. (From Sayed and Abdelkareem 2017)

of humans and other animals (including insects). Microbes generate energy from glucose by cellular respiration process. In aerobic cellular respiration, glucose is broken down into carbon dioxide and water using oxygen. In the case of anaerobic cellular respiration, nitrate (NO_3) or other molecules are utilized instead of oxygen.

Extracellular electron transfer (EET) are the microbial-bioelectrochemical processes which transfer electrons. Food-borne pathogen *L. monocytogenes* is a fermentative gram-positive bacterium which decays organic matter. It uses flavin-based mechanism to deliver electrons into the surrounding environments. The characterization of a flavin-based EET mechanism establishes new avenues for the electrochemical activities and bioenergetic applications (Light et al. 2018). *L. monocytogenes* can be utilized to improve MFC.

13.4.2 Electron Transfer Mechanism

The electron transfer rate has controls on the output power of MFC. Mustakeem (2015) reported that the extracellular transport of electrons to electrodes takes place in three mechanisms, i.e., direct electron transfer, electron transfer through mediators, and electron transfer through nanowires. Apart from these mechanisms, another mechanism of electron transport is suggested by Schaetzle and co-workers. They have reported that the electron transport is done by the oxidation of the excreted catabolites (Schaetzle et al. 2009).

Some bacteria genera like *Geobacter* and *Shewanella* deliver electrons from the oxidative metabolic pathways to the external environment. They are called as exoelectrogens (Reguera et al. 2005). They directly transfer the electrons to the electrode surface (Lies et al. 2005). Also, they transfer electron through mediators, i.e., conductive appendages or nanowires (El-Naggar et al. 2010; Gorby et al. 2006). The electronic conductivity of these nanowires is higher than the synthetic metallic nanostructure (Malvankar et al. 2011). Few bacteria genera like *Shewanella* and *Pseudomonas* secrete some chemicals (like flavins). These chemicals transfer electrons from the bacteria to electrodes (Yang et al. 2012; Schroder 2007).

13.4.3 Oxygen Reduction Reaction (ORR)

The performance of MFC is influenced by the kinetics of the electrode reactions. The performance of the electrodes depends on the materials from which the electrodes are made. A wide range of materials have been tested to improve the performance of MFCs. Carbon-based nanomaterials, composite materials, and various transition metal oxides have emerged as promising materials for both anode and cathode constructions. These materials have the potential as alternatives to conventional expensive metals like platinum particularly for the oxygen reduction reaction (Mustakeem 2015).

Electrode materials play a key role in MFC. Carbon material like graphite can be used as both anode and cathode (Mashkour and Rahimnejad, 2015). The electrode material selection is vital for the performance of MFC. It determines the various factors of the MFC including bacterial adhesion, electron transfer, and electrochemical efficiency (Logan 2010). The biocompatibility of the electrode increases the property of bacterial adhesion. However, carbon-based materials have poor catalytic activity. An additional catalyst is required to increase the ORR of these materials (Mustakeem 2015).

A low-level ORR at a neutral pH and low temperature reduces the MFC performance. In MFC, the ORR takes place at the three-phase interface of air (gas), electrolyte (liquid), and electrode (solid). The cathode has three layers, viz., diffusion layer, conducting support material, and catalyst. Though the identical materials are utilized to make anode and cathode, a robust cathode should have mechanical strength, catalytic property, and electronic and ionic conductivity properties at higher level. Figure 13.9a shows the ORR at the interface of gas, liquid, and solid. Figure 13.9b shows ORR at the three-phase interface of the cathode layers, i.e., catalyst layer, electrode base material, and oxygen diffusion layer (Mustakeem 2015).

A catalyst increases the ORR rate by decreasing the activation energy barrier (Lim and Wilcox 2012). In acidic medium, ORR occurs in two different electron pathways. The 4-electron and the 2-electron pathways are expressed in Eqs. 1 and 2 (Erable et al. 2012).

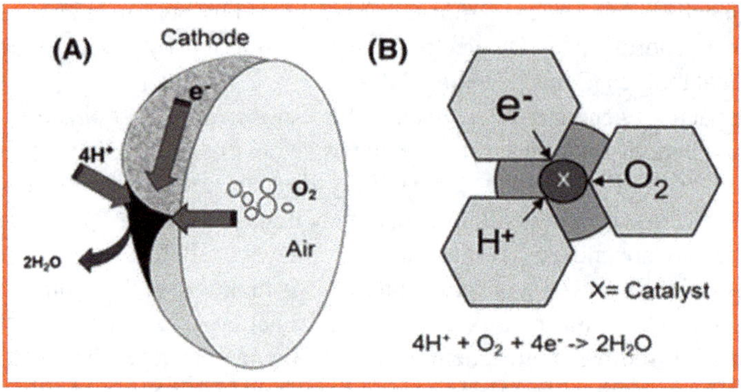

Fig. 13.9 Schematic diagram shows the cathode reaction. The reaction occurs at the triple-phase interface of air, solution, and catalyst. (**a**) Oxygen reduction reaction (ORR) at the interface of air, electrolyte, and electrode produces water as product. (**b**) ORR at three-phase interface of cathode layers. (From Mustakeem 2015)

$$O_2 + 4H^+ + 4e^- \rightarrow 2H_2O \qquad (13.1)$$

$$O_2 + 2H^+ + 2e^- \rightarrow H_2O_2$$

$$H_2O_2 + 2H^+ + 2e^- \rightarrow 2H_2O \qquad (13.2)$$

Various literatures report that MnO_2-catalyzed cathode increases the ORR. The output power density of the MnO_2 cathode is much higher than the cathode materials such as carbon black, carbon nanotubes (CNTs), graphite, and stainless steel. MnO_2 is a *two-dimensional* (*2D*) material. Zhang et al. confirm that the MnO_2 deposited over glassy carbon using polyvinylidene fluorine (PVDF) as binder performs similar to a platinum catalyst. In the ORR, initially, MnO_2 is reduced to MnOOH by electrons. Oxidation occurs in the next step. The current density of ORR from MnO_2 deposited cathode is ten times higher than the glassy carbon cathode. It indicates the increased ORR catalytic property of MnO_2 (Zhang et al. 2009).

13.5 Nanomaterials in MFC

13.5.1 Carbon Nanomaterials in MFC

Carbon nanomaterials are highly conductive and mechanically stable with larger surface area and higher electrochemical catalytic activity (Ghasemi et al. 2013). Carbon nanomaterials can work both as anode and cathode. As the anode, carbon materials can promote microbial colonization and accelerate the formation of biofilms, thus increasing the power density. As the cathode, carbon-based materials can function as catalysts for the oxygen reduction reaction even reaching the

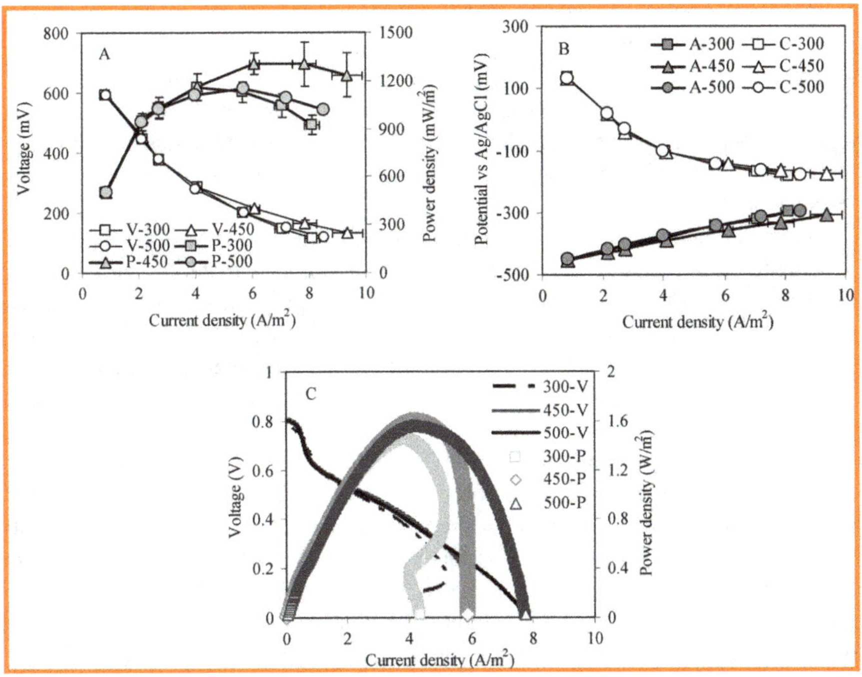

Fig. 13.10 Carbon brush anodes of MFCs treated with different temperatures (300 °C, 450 °C, and 500 °C) by variable external resistance method. (**a**) Power densities, (**b**) electrode potentials, (**c**) power curves by linear sweep voltammetry. (From Yang Qiao et al. 2017)

performance of Pt catalysts (Li et al. 2017). The performances of different carbon materials like carbon felt, carbon cloth, and carbon paper are analyzed and compared. The power output depends not only on the material but also on the surface features. Based on this, carbon felt is superior in treating winery wastewater which is a highly organic-loaded waste to the other two materials (Penteado et al. 2017).

Pretreatment and modification of the anode improve the performance of the anode. The carbon brush is heated at various temperatures (300 °C, 450 °C, and 500 °C), and the performance of the MFC is analyzed. Figure 13.10a, b, and c show the results related to power density, electrode potential, and power curves of the preheated carbon brush anode. Figure 13.10a shows the maximum power densities obtained from the reactor with the 300 °C and 500 °C anode are 1160 ± 82 and 1149 ± 11 mW/m^2, respectively. The 450 °C anode generates higher-power density, i.e., 1305 ± 67 mW/m^2, at 6.1 ± 0.2 A/m^2 current density. Figure 13.10b indicates that the cathode potentials remain consistent but variations occur in the anode potentials of different preheating anodes like power densities. As per Fig. 13.10c power density curves (results obtained from the linear sweep voltammetry analysis), the maximum power densities are 1450 (300 °C), 1607 (450 °C), and 1561 (500 °C) mW/m^2 (Yang Qiao et al. 2017).

Carbon nanomaterials are also prepared by carbonization of the biological materials at high temperatures (Yuan et al. 2013). Nitrogen-enriched carbon nanomaterials have potentials to prepare anode materials of MFC. The maximum power density obtained was found to be 1600 ± 50 mW m^{-2} when nitrogen-doped carbon nanotubes (NCNTs) are used as cathodic catalysts. Further the bamboo-shaped and vertically aligned NCNTs had lower internal resistance and higher cathode potentials (Feng et al. 2011).

Graphene-based composite materials are used as cathode in MFCs. As per report of Valipour et al., $RGO_{HI\text{-}AcOH}$ (highly conductive graphene material) cathode gives a power density of 1683 ± 23 mW/m^2, and RGO/Ni (graphene/nickel) nanoparticle composite cathode gives a power density of 1015 ± 28 mW/m^2. The catalytic activity of $RGOHI_{HI\text{-}AcOH}$ is mainly due to the high surface area and degree of graphitization. The double loading of catalyst offers stable power generation and long-term operation of MFCs (Valipour et al. 2016).

In MFCs, low hydrophilicity property of graphene adversely affects the performance of the graphene-modified anodes (G anodes). To elevate the hydrophilicity and performance of the graphene anode, different amounts (0.15 mg·mg^{-1} to 0.2 mg·mg^{-1} and 0.25 mg·mg^{-1}) of graphene oxide (GO) are doped. In this case, the contact angle decreases considerably. The G anode (doped with GO 0.2 mg·mg^{-1}) produces the static water contact angle (θ_c) value as 64.6 ± 2.75°. It exhibits the optimal performance. MFC with this anode generates the maximum power density (P_{max}) 1100.18 mW·m^{-2} which is 1.51 times higher than the bare graphene anode. The results are shown in Fig. 13.11a and b. Also, it has COD removal efficiency, and coulomb efficiency (CE) higher than the other MFCs with GO-doped anode as well as G anode. It has COD removal efficiency 82.78 ± 0.45% and CE 33.76 ± 0.43%, but the MFC with G anode exhibits 79.69 ± 0.65% and 30.24 ± 0.46%, respectively (Yang Na et al. 2016b).

When carbon nanotubes are mixed with wastewater (10 mg L^{-1} to 200 mg L^{-1} concentration), it produces positive effect on power generation. It stimulates the power generation due to the increased conductivity of the MWCNTs. Also COD removal efficiency is also enhanced from 74.2% to 84.7% (Miran et al. 2016). When compared to the conventional carbon cloth electrode, the intertwined CNT-textile fibers have larger biofilm. This film facilitates electron transfer from exoelectrogens to the CNT-textile anode (Xie et al. 2010). Graphene and CNT are coated on carbon cloth and stainless steel mesh electrodes to improve their performance (Tsai et al. 2015; Hsu et al. 2017).

High-strength wastewaters such as landfill leachate contains dissolved extracts and suspended matter. When it is treated with MFC, ammonia is separated from the leachate. Activated carbon performs well in this treatment as anode. The percentages of ammonia removal of activated carbon, zinc, and black carbon are found to be 96.6%, 66.6%, and 92.8%, respectively (Alabiad et al. 2017). In MFC, graphite nanofibers with diameter 6–10 μm also work as an efficient anode. Their surface area to volume ratio is 15,000 m^2/m^3. The activated carbon air-cathode is an alternative for Pt-based cathode. While utilizing it as cathode, it works almost equal to the platinum catalyst and performs well for long duration.

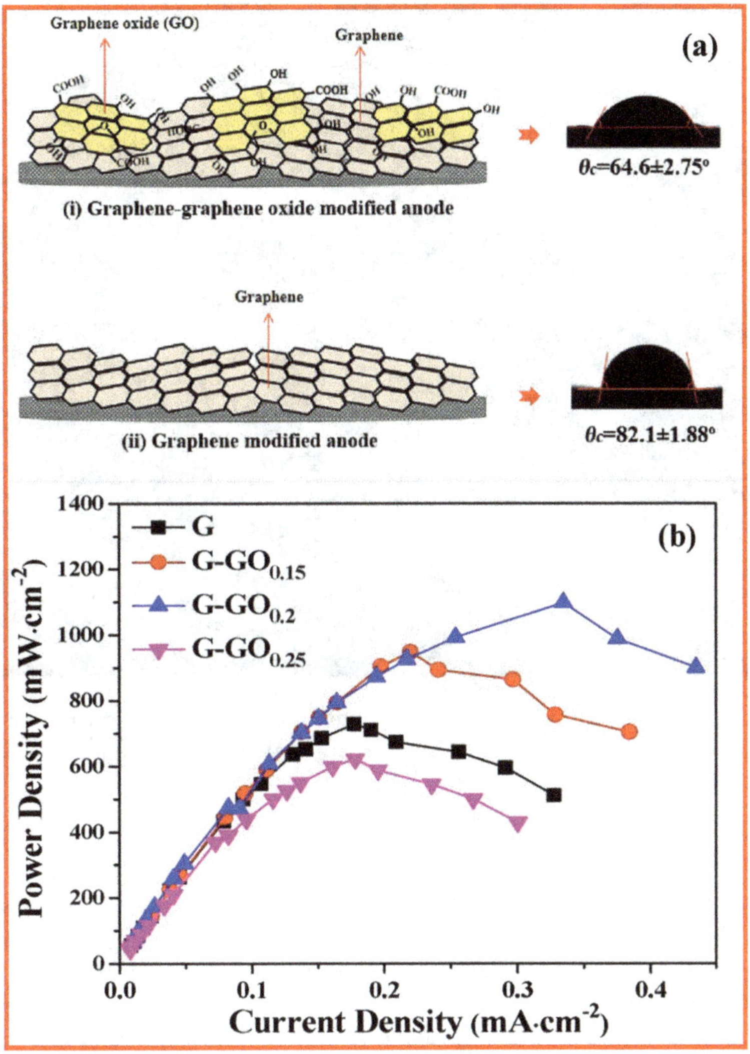

Fig. 13.11 Graphene oxide (GO) doped graphene (G) anode. (**a**) Schematic diagram of G-GO anode and G anode with static water contact angle (θ_c) values. (**b**) Power densities and current densities of the MFCs with G and G-GO anodes. MFC with G-GO anode (0.2 mg·mg^{-1} GO doped) generates power density higher than the MFCs with G anode and other doped anodes. (From Yang Na et al. 2016b)

It is observed that coating of nanomaterials like multilayer graphene (MG) and CNTs on the stainless steel mesh (SSM) electrode improves the power density and reduces the internal resistance of a MFC system. The surface modification of anode or cathode with CNTs and graphene increases the power generation by approximately 3–7 and 1.5–4.5 times, respectively. When comparing to the MFC with an untreated

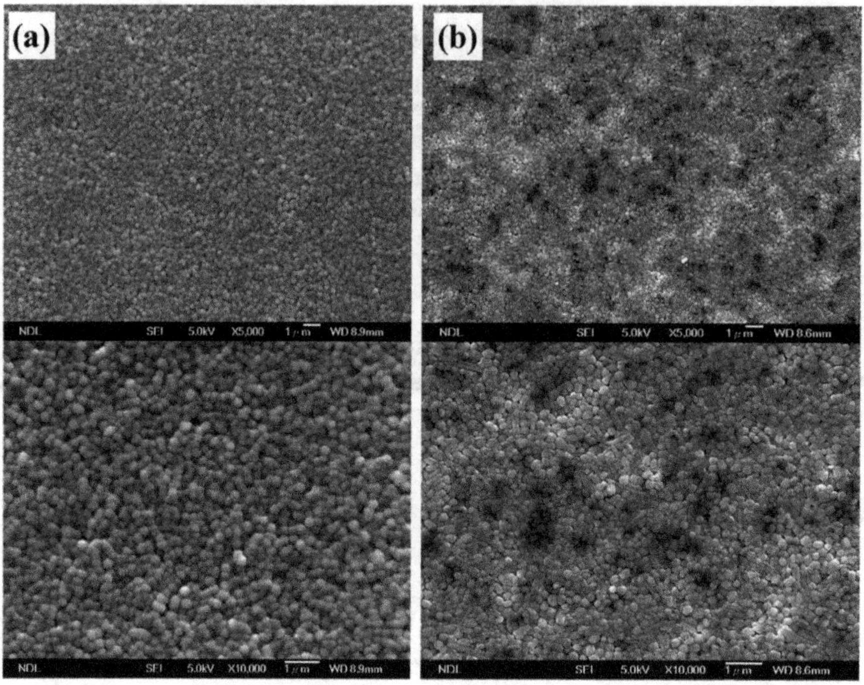

Fig. 13.12 SEM images (5000× and 10,000× magnification) of cathodes covered with treated PTFE. (**a**) PTFE mixed with multilayer graphene. (**b**) PTFE mixed with MWCNTs. (From Hsu et al. 2017)

anode, the internal resistances of MFCs with CNTs and MG-modified anodes are in the reduced level, i.e., 18 and 30%, respectively. SSM cathode surface modification is done by the coating of MG and MWCNTs mixed PTFE solution. SEM images of the MG and MWCNTs mixed PTFE are shown in Fig. 13.12a and Fig. 13.12b, respectively. It is observed from these figures that the surface roughness of the MWCNTs mixed PTFE is more than the MG mixed PTFE (Hsu et al. 2017).

Carbon cloth electrodes are modified with carbon nanotube and graphene to investigate the performance of single-chamber MFC. In this investigation, *Escherichia coli* HB101 is used as catalyst in an air-cathode MFC. It is observed from the results that the carbon cloth electrodes modified with both materials, i.e., CNT and graphene, improve the power density of MFCs. The internal resistance of normal electrodes decreases dramatically from 377 kΩ to 5.6 kΩ (while using both electrodes modified by graphene with a cathodic catalyst). Among the all electrodes, graphene-modified electrode exhibits superior performance. When comparing to the modified cathode, the modified anode exhibits greater performance (Tsai et al. 2015).

The single-chamber MFC is fabricated using polymethyl methacrylate (PMMA) which is shown in Figure 13.13a. The cathode electrode is fixed to the air side, and the anode electrode is fixed opposite side. The reactor volume is 75 mL and the

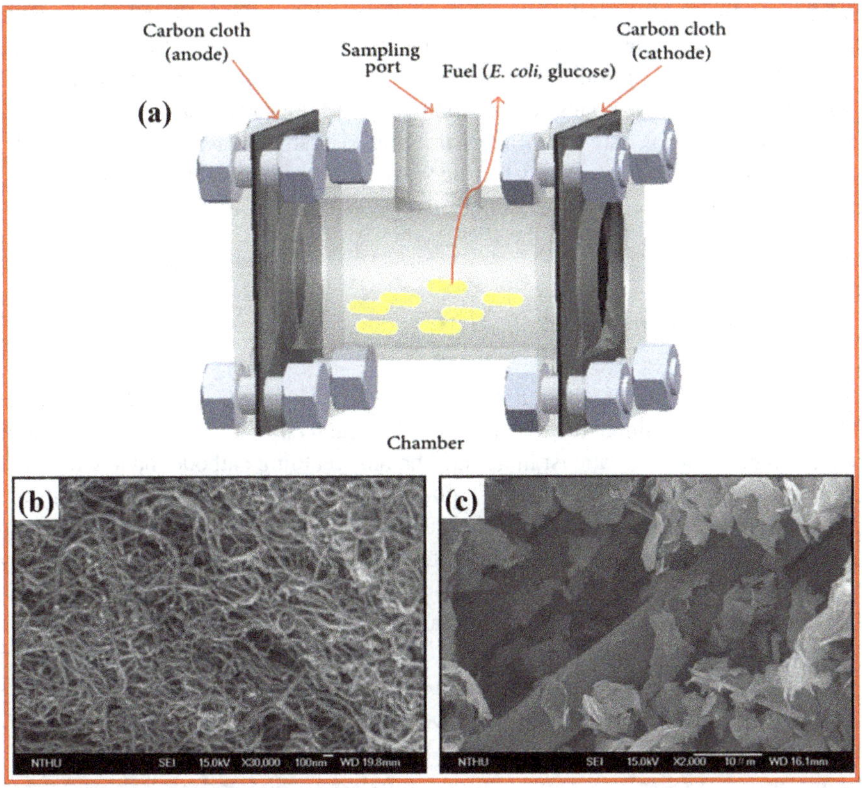

Fig. 13.13 Schematic diagram of air-cathode MFC: (**a**) The single-chamber MFC has PMMA chamber as the air-cathode, modified or unmodified carbon cloth as electrodes, glucose as fuel, and *Escherichia coli* as anode's catalyst. Copper wires are used to connect with external resistance. (**b**) and (**c**) SEM images of MWCNTs and graphene coated on the carbon cloth, respectively. (From Tsai et al. 2015)

surface area of electrode is 12.57 cm^2. SEM images of MWCNTs and graphene coated on the carbon cloth are shown in Fig. 13.13b, c, respectively. Both materials are uniformly distributed on the carbon cloth which improves the specific surface area of the electrode. Specific surface area of graphene-coated electrode is two times more than the CNT-coated electrode (Tsai et al. 2015). This improved specific surface area leads to the superior performance of graphene-coated electrode.

Metals and carbon felt are used as electrodes. Metals rapidly corrode, and carbon felt is porous and prone to clogging. So, the carbon felt is replaced with paper coated with carbon paste. Lamberg and Bren reported about the carbon paste paper electrodes (CPPEs). These electrodes are fabricated by coating of carbon paste on a paper strip followed by polyaniline coating. The carbon paste is made by the mixture of graphite powder and mineral oil. The CPPEs are assessed as anodes in bioelectrochemical cells (BECs). In this assessment, *Shewanella oneidensis* MR-1 bacteria is utilized to donate the electrons through extracellular electron transfer.

When comparing to the polyaniline-modified carbon felt electrode (CFE), the BEC using the CPPE anode works better. It generates a current density (maximum value of 2.2 A m^{-2}) after 24 hours of the inoculation. It is two times more than the BEC with CFE anode. It generates current continuously 4 days without the need for additional fuel (lactate). It is confirmed from this assessment that CPPE is a simple, low-cost, and promising new bioelectrode material for microbial fuel cells (Lamberg and Bren 2016).

13.5.2 Nanocomposites in MFC

To improve the cathode kinetics of MFC, activated carbon (AC), graphene nanosheet (GNS), and iron-aminoantipyrine (Fe-AAPyr), catalyst materials are integrated as an alternative cathode catalyst material. The air-breathing cathode made with GNS and Fe-AAPyr materials generates higher power. The power generated from the MFC utilizing these cathode materials is enumerated in Table.13.1. Figure 13.14a, b shows the SEM images of Fe-AAPyr material and graphene nanosheet, respectively (Kodali et al. 2018).

Anode is prepared by using materials like conductive polymers and carbon nanomaterial composites. Qiao Yan and co-workers tested a composite of carbon nanotube and polyaniline in the electrochemical impedance spectroscopy. In this test, *Escherichia coli* bacteria are utilized as the microbial catalyst with 20 wt. %

Table 13.1 Details of power generated by different cathode catalyst materials

Cathode catalyst material	Power generated (μW cm^{-2})
GNS and Fe-AAPyr	235 ± 1
Fe-AAPyr	217 ± 1
GNS	150 ± 5
AC	104

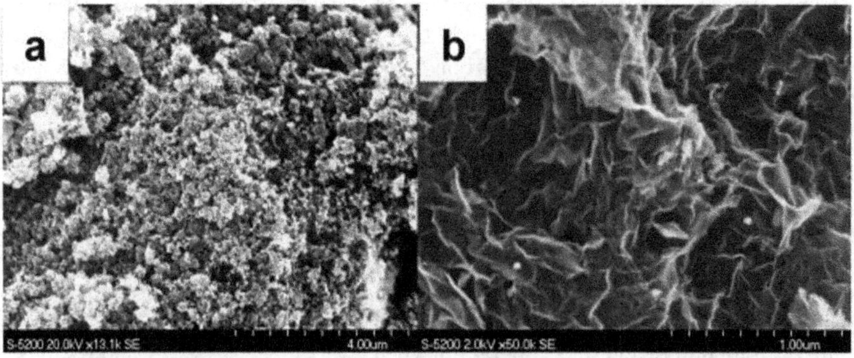

Fig. 13.14 SEM images of cathode catalyst materials. (**a**) Iron-aminoantipyrine (Fe-AAPyr) catalyst (**b**) Graphene nanosheets. (From Kodali et al. 2018)

CNT composite anode. The obtained power density (maximum) is 42 mW m^{-2} (Qiao et al. 2007).

Deng et al. explain the interaction between the exoelectrogens and their induced GO reduction in the high-performance MFC application. During the in situ preparation of graphene/exoelectrogen composite biofilm electrode, graphene oxide is reduced by *E. coli* bacteria. This reduction provides a rough surface and three-dimensional structure to the electrode which leads to larger specific surface area for microbes to settle. Graphene-modified anode increases the height of biofilm thickness largely. The increased level of the total protein in microbial reduction of graphene oxide (mrGO)-modified anode demonstrates the attachment of more bacteria on the anode surface. After five electricity production cycles, the power density of the mrGO-modified MFC reaches a maximum value of 1140.63 mW m^{-2} (Deng et al. 2017).

Two-dimensional porous anodes have small pore sizes. So, bacteria clog on the surface and are inaccessible to the interior of the anode. This seriously limits the anode efficiency. The problems associated with the 3D structures include low specific surface area (due to lack of microscopic or nanoscopic structures), small pore sizes for bacteria penetration, poor conductivity, and disruption of bacterial membrane by sharp nanomaterials. While using the novel 3D macroporous anode, these problems are avoided. This anode is designed by the freestanding, flexible, conductive, and monolithic graphene foam decorated with the conductive polymer, i.e., PANI. The 3D graphene is synthesized by chemical vapor deposition with nickel foam as the substrate and using ethanol as the carbon source. The pore size of graphene foam is much larger than the size of bacteria (1–2 μm). Hence, bacteria can easily diffuse inside and colonize. The MFC equipped with carbon cloth anode generates the power density ~110 mW/m^2 at 6 h. MFC (with 3D graphene/PANI foam anode) generates the same power density at 6 h, but at 24 h, the power density is higher than at 6 h, i.e.,190 mW/m^2 (Yong et al. 2012).

13.5.3 Biogenic Inorganic Nanoparticles

Nanomaterials are having more surface area which allows a very good adhesion of microorganisms at the anode. Some biogenic inorganic nanoparticles facilitate extracellular electron transfer in MFCs. Jiang et al. studied the performance of MFC with iron sulfide nanoparticles and *Shewanella bacteria.* In this study, it is observed that iron sulfide nanoparticles are in intimate contact with the cell membrane by uniform coating. The charge transport occurs in the presence of live *Shewanella.* It improves the electron transfer at cell/electrode interface and the cellular networks which leads to the enhanced current output (Jiang et al. 2014). Nano-CeO_2 is utilized for the modification of the carbon felt anode in the MFC. The modified anode obtains the higher closed circuit voltage resulting from the lower anode potential. The MFC with nano-CeO_2-modified carbon felt anode generates maximum power density 2.94 W m^{-2} with lower internal resistance 77.1 Ω (Yin et al. 2016).

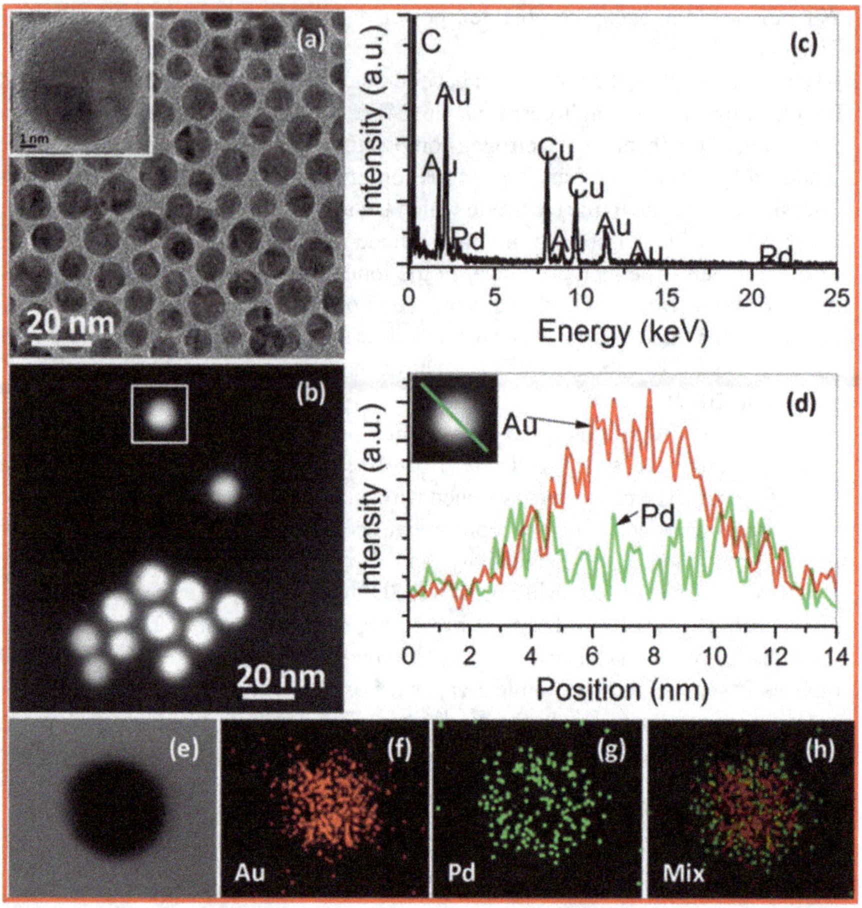

Fig. 13.15 Images of Au-Pd core-shell prepared in oleylamine at elevated temperature. (**a**) TEM image. Insert HRTEM image of a single Au-Pd particle. (**b**) STEM image. (**c**) EDX analysis. (**d**) Elemental profiles in STEM mode. (**e–h**) Nanoscale element mappings of (**e**) the formation of core-shell Au-Pd structure and (**f–h**) distributions of Au, Pd, and bimetallic nanoparticles. (From Yang Gaixiu et al. 2016a)

In MFCs, the electrodes should be immersed in electrolytes in neutral pH. In such conditions, Au core- and Pd shell-type bimetallic nanoparticles exhibit better ORR due to the enhanced catalytic properties. The lattice strain produced in between the core and shell regions leads to the enhanced catalytic properties. The maximum power density produced in the membrane less single-chamber MFC using wastewater and Au-Pd core-shell cathode catalysts is ca. 16 W m^{-3}. It is stable for more than 150 days. Optimization of Au core size and Pd shell thickness enhances the core-shell properties. Au-Pd core-shell analyses results and images related to the single-chamber MFC are shown in Figs.13.15 and 13.16, respectively (Yang Gaixiu et al. 2016a).

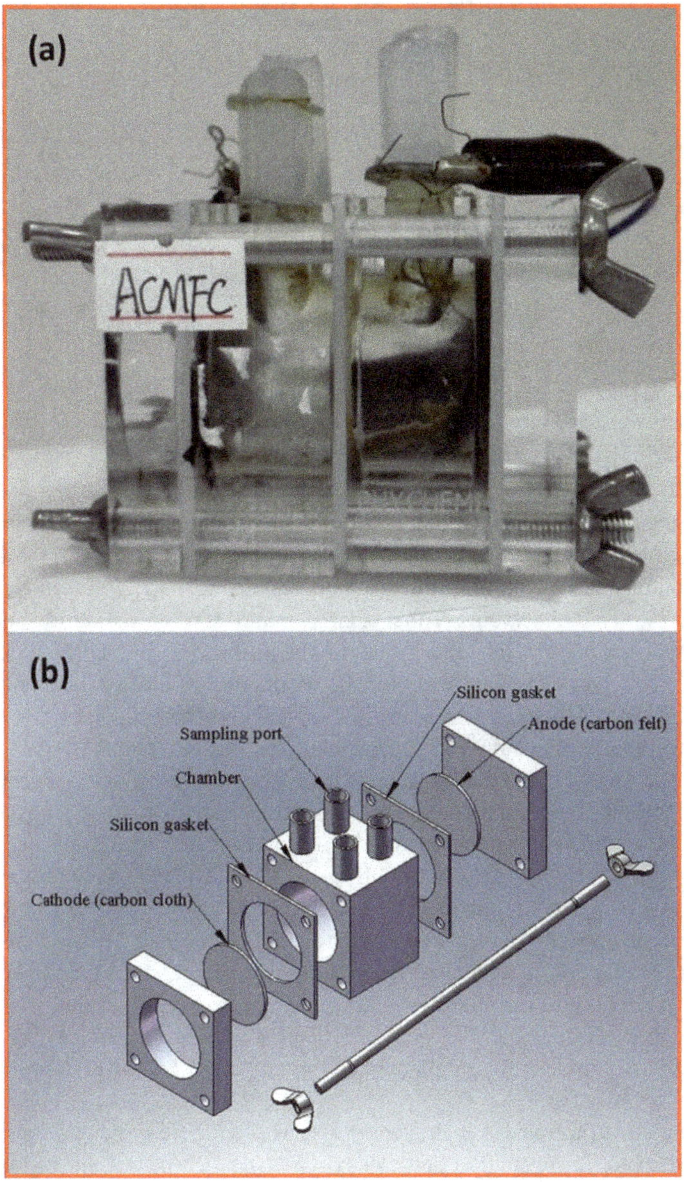

Fig. 13.16 Construction of single-chamber MFC to assess the ORR catalytic properties of the bimetallic core-shell Au-Pd nanoparticles. (**a**) The practical construction. (**b**) Schematic illustration shows the parts of single-chamber MFC (carbon cloth cathode, carbon felt anode, silicon gasket, sample chamber, and sampling ports). (From Yang Gaixiu et al. 2016a)

The performance of MFC using core-shell Au-Pd nanoparticle cathode is compared with the MFC using hollow Pt nanostructure air-cathode catalyst. When comparing the performance of both cathode catalysts, the core-shell Au-Pd cathode shows better performance. It has higher ORR catalytic performance. The maximum

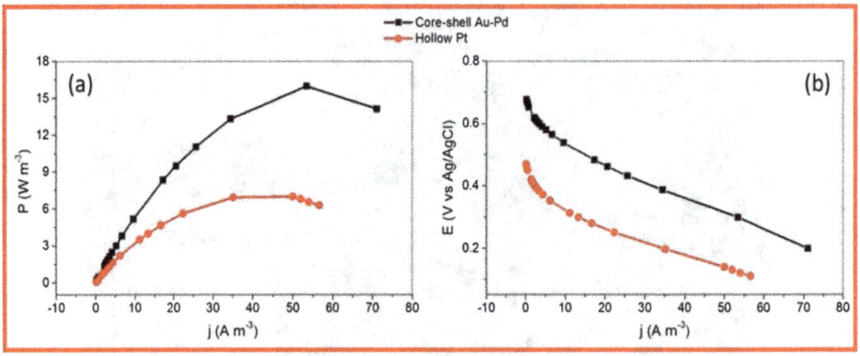

Fig. 13.17 Performance graph of MFC using core-shell Au-Pd nanoparticles cathode and MFC using hollow Pt nanostructures air-cathode catalyst. (**a**) Power density comparison of the both cathodes. (**b**) Electrode polarization (voltage) graph. (From Yang Gaixiu et al. 2016a)

power density and cathode voltages of core-shell Au-Pd nanoparticles are higher than the hollow structured Pt cathode. The maximum power density of the core-shell Au-Pd cathode and the hollow structured Pt cathode are 16.0 W m^{-3} and 7.1 W m^{-3}, respectively. The observed results clearly manifest that the core-shell Au-Pd nanoparticle cathode is an alternative to the Pt-based cathode catalysts. These nanoparticles can be used in MFC for the production of electricity without losing power efficiency and stability. Figure 13.17a, b shows the power density and polarization curves of both cathodes, respectively (Yang Gaixiu et al. 2016a).

MFC is a renewable and clean energy-generating system. Bacterial growth is the most important thing in this system. Composts rich in organic substances are commonly used for such growth. In addition, non-inert (metal) and inert (usually carbon-based) electrodes are used in MFC. In this study, zinc anode is evaluated for MFC, and it is observed that zinc is a corrosion-resistant material. Compost mixed with MFC having zinc anode and graphite cathode produces high-power density (5.33 W/m^2). According to the measurements and calculations of this study, zinc has promising electrode technology with good electrochemical and biochemical performances (Nurettin 2017). Various metal oxide nanoparticles like titanium dioxide and iron oxide along with conducting polymer like polyaniline are utilized as bioanodes in MFC.

Anodic electron transfer is the main process of electricity generation in MFCs. The anodic biofilm formation and electron transfer can be accelerated by adding biosurfactants to the anolyte. Rhamnolipid biosurfactant with the quantity of 40 mg/L, 80 mg/L, and 120 mg/L is added to the anolyte. It increases the abiotic capacitance from 15.12 F/m^2 (control) to 16.54 F/m^2, 18.00 F/m^2, and 19.39 F/m^2, respectively. After 7th day of the inoculation, anodic biofilm formation is facilitated for dosing 40 mg/L of rhamnolipids, with anodic biofilm coverage from 0.43% to 42.51% and the power density from 6.92 ± 1.18 W/m^3 to 9.93 ± 0.88 W/m^3. Rhamnolipid concentration with 80 mg/L and 120 mg/L blocks the electron transfer. This analysis reveals that rhamnolipids facilitate the enrichment of exoelectrogen (Zhang et al. 2017b).

Nanostructured hexagonal klockmannite copper selenide (CuSe) grown on the hybrid material exhibits the superior ORR catalytic performance. It also has more positive onset potential, higher current density, smaller Tafel slope, and excellent stability. The hybrid material is the mixture of reduced graphene oxide (rGO) and carbon nanotubes. MFC equipped with CuSe@rGO-CNTs cathode exhibits larger energy output than the MFC with carbon-supported platinum (Pt/C) catalyst (Tan et al. 2016).

13.6 Microfabrication in MFC

Microfabrication technology is employed in semiconductor manufacturing industries for thin film deposition, photolithography, etching, microelectromechanical system, and lab-on-a-chip toward batch fabrication with low expense and precisely controlled geometry. Miniaturized MFC in lab-on-a-chip devices find applications in biosensors for toxic chemical detection (Ren and Chae 2015). Micro-sized MFCs are miniature energy scavengers. They can be useful power sources for lab-on-a-chip applications and integrated onto chips for low-power electronic devices or sensors (Rojas and Hussain 2015).

Miniaturized energy-harvesting devices can be built using advanced microfabrication techniques. Micro-sized MFC is constructed using single-chamber MFC concept with an air-cathode. In this system, proton-exchange membrane is removed. MFCs are fabricated as system-on-chip functionality. This paves the way for MFC applications in sensors, watches, and mobile phones. Ni, Au, and MWCNT as anode are designed with air-cathode (Mink and Hussain 2013).

The micro-sized MFC has advantages like utilization of less electrode area and less liquid fuel volume. Justine E Mink et al. (2014) fabricate a mobile and inexpensive micro-sized MFC that using human saliva as fuel. This 25 μl MFC has graphene as an anode for efficient current generation and an air-cathode for enabling the use of the oxygen present in air. This system makes the entire operation completely mobile without using any laboratory chemicals. It produces the higher current densities of 1190 Am^{-3}. The graphene anode generated 40 times power more than carbon cloth anode. Also, test results (using acetate as organic material instead of saliva) demonstrate a linear relationship between the organic loading and current. Findings of this report lead to the applications of saliva-powered fuel cell technology for lab-on-a-chip devices or portable point-of-care diagnostic devices (Mink et al. 2014).

Micro-sized MFC can be utilized as an ovulation predictor based on the conductivity of a woman's saliva. It is observed that before the 5 days of ovulation period, a sharp decrease in the conductivity of saliva occurs. It is caused by a high level of estrogen and low level of electrolyte concentration in saliva (Huang et al. 1997). This micro-sized MFC analyzes the conductivity changes of saliva which help to identify the fertility period and to maintain woman's health. Also, it helps for better family planning in a noninvasive method.

MFC can be used as a potential power source for implanted bioMEMS devices. The MFC is biocatalyzed by *Saccharomyces cerevisiae*. This microorganism converts chemical energy stored in glucose of the blood stream. The MFC has 0.2 μm thickness gold evaporated polydimethylsiloxane (PDMS) anode and cathode separated by a Nafion 117 proton-exchange membrane. MFC with this type of micropillar structure shows excellent performance than silicon micro-machined MFCs, i.e., 4.9 times higher average current density and 40.5 times higher average power density. The MFC uses 15 μL of human plasma containing 4.2 mM glucose. It produces maximum open-circuit potential (OCP) of 488.1 mV, maximum current density of 30.2 $\mu A/cm^2$, and a maximum power density of 401.2 nW/cm^2. During continuous operation for 60 minutes, it produces an average OCP of 297.4 mV, average current density of 4.3 $\mu A/cm^2$, and average power density of 42.4 nW/cm^2 at 1 k Ω load. The coulombic efficiency of electron conversion from blood glucose is 14.7% (Chiao 2008).

13.7 Wastewater Treatment in MFC

MFC technology utilizes wastewater effectively to generate energy (Logan and Regan 2006a, b). Agrowaste materials produced during various agricultural operations are rich in COD. Some of them are useful in bioelectricity generation as well as wastewater treatment. MFC can be utilized for Cr(VI) wastewater treatment. It is observed from the reports that the MFC system that has mixed cultures of bacteria yields output better than the MFC system with single-culture bacteria. Also, reports indicate that cassava mill wastewater has potential to generate electricity from MFCs. During the wastewater treatment process using microorganisms, clean energy, i.e., hydrogen production, is also possible.

Wastewater from paper industries contains water-insoluble materials such as cellulose. The cellulosic waste materials are the attractive source of energy for electricity production in MFCs. However, this process requires anaerobes that can degrade cellulose and transfer electrons to the electrode (exoelectrogens). MFC with two-chamber system avoids oxygen contamination of the anode. Single-chamber MFC with air-cathode produces higher-power densities than aqueous catholyte MFC due to less internal resistance. Also, it avoids energy input for the cathodic reaction. While examining the changes in the bacterial consortium in a single-chamber, air-cathode MFC fed cellulose, it is observed that the main genera developed after extended operation of the MFC are *Parabacteroides*, *Proteiniphilum*, *Clostridium*, and *Catonella*. These results confirm that different bacteria evolve in single-chamber air-cathode MFC than the two-chamber reactors. Details of bacteria abundance in MFC are in Fig. 13.18. Polarization and power density curves obtained from the MFC are illustrated in Fig. 13.19 (Toczyłowska-Maminska et al. 2018).

The amount of energy needed for the treatment of wastewater is very high in the present situations. MFC can be a very good solution because the energy generated is sufficient for the system to run. Yue Dong et al. reported that energy self-sufficiency is essential for the sustainable wastewater treatment. They combine a microbial fuel cell and an intermittently aerated biological filter (MFC-IABF) to

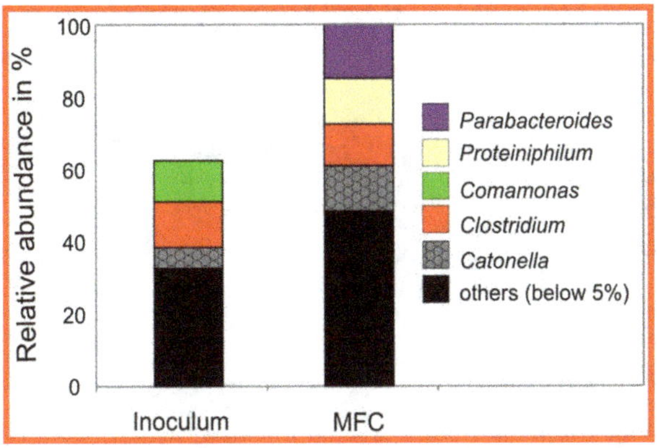

Fig. 13.18 Relative abundance of dominating bacteria genera in fresh inoculum and in single-chamber, air-cathode MFC fed cellulose system (after the operation). (From Toczyłowska-Maminska et al. 2018)

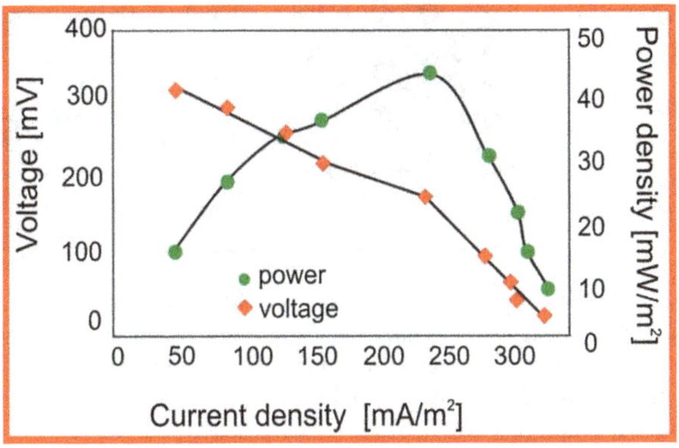

Fig. 13.19 Polarization and power density curves of single-chamber air-cathode cellulose-fed MFC. After 1 month of the MFC operation, the maximum current produced is 331 mA/m^2 (R = 100 Ω), and the maximum power produced is 44 mW/m^2 (R = 1000 Ω). (From Toczyłowska-Maminska et al. 2018)

treat the effluent with self-sufficient energy. Schematic diagram of the MFC-IABF system and its wastewater treatment performance are shown in Figs. 13.20 and 13.21, respectively (Dong Yue et al. 2015).

In the combined MFC-IABF system, synthetic wastewater (COD = 1000 mg L^{-1}) is fed continuously for more than 3 months using a capacitor-based circuit. This system is operated at room temperature. As the output of this work, the MFC generates electricity and supplies to IABF along with COD removal. It is observed that the MFC produces energy (0.27 kWh m^{-3}) which is sufficient to the pumping system (0.014 kWh m^{-3}) and aeration system (0.22 kWh m^{-3}). The IABF works in the intermittent aeration mode (aeration rate 1000 ± 80 mL h^{-1}), removes the

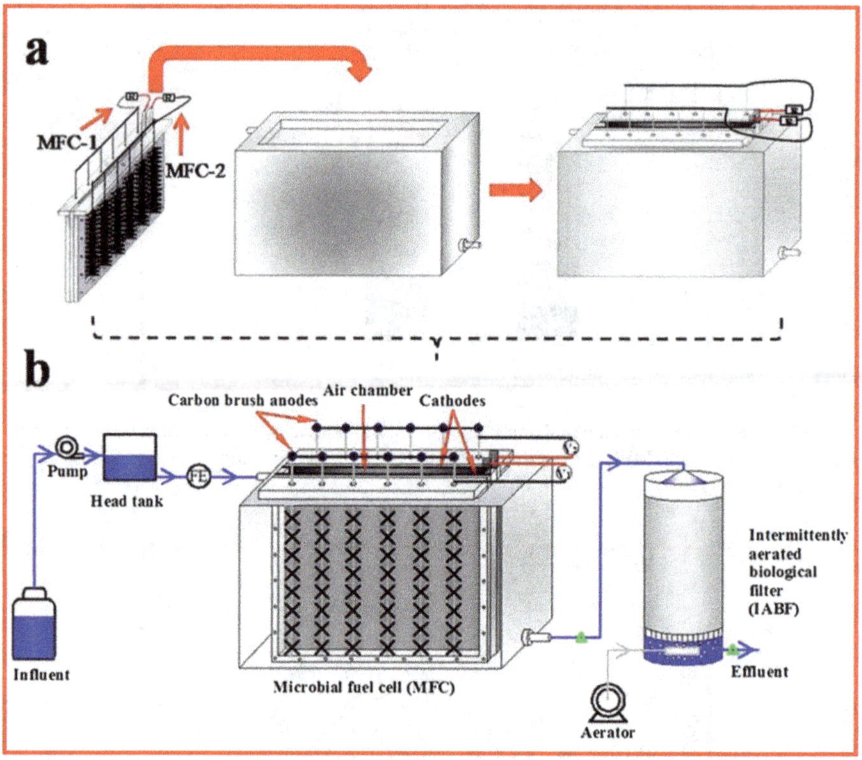

Fig. 13.20 Schematic diagram of (**a**) MFC reactor for electricity generation and COD removal. (**b**) Two-stage combined MFC-IABF system. This combined system operates with self-sufficient energy and treats the wastewater more efficiently than MFC reactor. Air is flowed inside the IABF system by the aerator. (From Dong Yue et al. 2015)

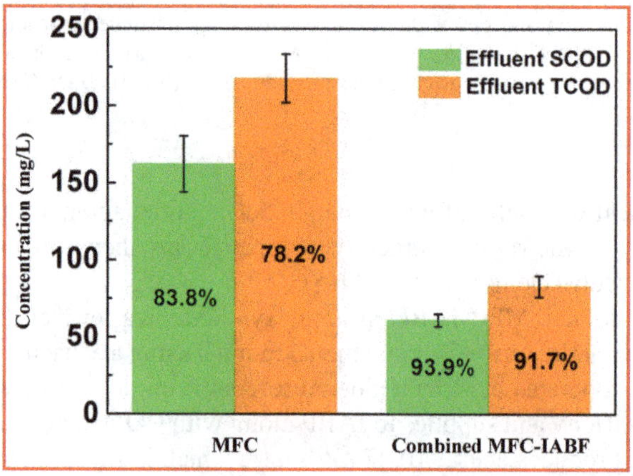

Fig. 13.21 Wastewater treatment performance of MFC and two-stage combined MFC-IABF system. The value inside the bars indicates the SCOD and TCOD removal rate (influent COD concentration is 1000 mg/L). (From Dong Yue et al. 2015)

wastes, and improves the water quality (HRT = 7.2 h). This combined system removes SCOD 93.9% and TCOD 91.7% (effluent SCOD = 61 mg L^{-1} and TCOD = 82.8 mg L^{-1}). These results confirm that the combined MFC-IABF system operates in an energy self-sufficient manner and treats the wastewater efficiently (Dong Yue et al. 2015).

Ren et al. have obtained the maximum power density 143 mW/m^2 with 1 g/L carboxymethyl cellulose (CMC) from a dual-chambered MFC. In this process, a binary culture of cellulose-degrading bacteria *Clostridium cellulolyticum* and electrochemically active bacteria *Geobacter sulfurreducens* is utilized (Ren et al. 2008). Rezaei et al. have enriched the cellulose-degrading bacteria *Enterobacter cloacae* strain FR in MFC with wastewater. This MFC produces maximum power density of 4.9 mW/m^2 using cellulose 4 g/L (Rezaei et al. 2009). Sedky et al. have utilized cellulose as substrate in a MFC. In this setup, cellulose-degrading bacteria *Nocardiopsis* sp. KNU and *Streptomyces enissocaesilis* KNU are employed for cellulose degradation in anode. The cathode has 50 mM ferricyanide. The maximum power density produced from this culture is 188 mW/m^2 consuming 1 g/L cellulose (Hassan et al. 2012).

A high amount of starch-rich wastewater is released during the starch production from cassava. It has high chemical oxygen demand (COD), biochemical oxygen demand (BOD), total solids, and cyanoglycosides. These cyanoglycosides form cyanide. Many wastewaters have cyanide concentration up to 200 mg/L. Hence, proper treatment of the cassava wastewater is essential prior to its release into the environment. Cassava wastewater sludge has a high organic content of 16,000 mg/L. MFC utilizes cassava wastewater, removes COD approximately 88% within 120 h, and generates maximum power 1771 mW/m^2 (Kaewkannetra et al. 2009). Prasertsung et al. (2012) have identified that increasing the pH of anode chamber in a MFC increases the production of electricity. The generated maximum power density was 22.19 W/m^3 at pH 9 from a single-chambered MFC that was used with cassava mill wastewater. COD of the wastewater was 1086 mg/L, and the initial pH was 5.0 (Prasertsung et al. 2012).

Chromium has industrial applications like leather tanning, metallurgy, electroplating, and wood preservatives. It exists in the aqueous solution either as hexavalent chromium [Cr(VI)] or trivalent chromium [Cr(III)]. Cr(VI) is considered as more hazardous material because of its mutagenic and carcinogenic properties (Humphries et al. 2004). Hence, Cr(VI) wastewater treatment can be coupled with electricity generation using MFC. In this method, Cr(VI) is reduced in the cathode of an MFC by the microorganisms *Trichococcus pasteurii* and *Pseudomonas aeruginosa*. Acetate and bicarbonate are added into the anode and cathode compartments, respectively. Cr(VI) is reduced by microbial activity utilizing the electrons and protons generated from the oxidation of acetate. It generates current and power density of 123.4 mA/m^2 and 55.5 mW/m^2, respectively (Tandukar et al. 2009).

A more efficient biocathode is designed with reticulated vitreous carbon (RVC) and carbon nanotube (CNT) to use in MFC for Cr(VI) removal application. It is prepared by the electrophoretic deposition of CNT on RVC. The material RVC is cheap and commercially available. It is an open-pore foam carbon material that is

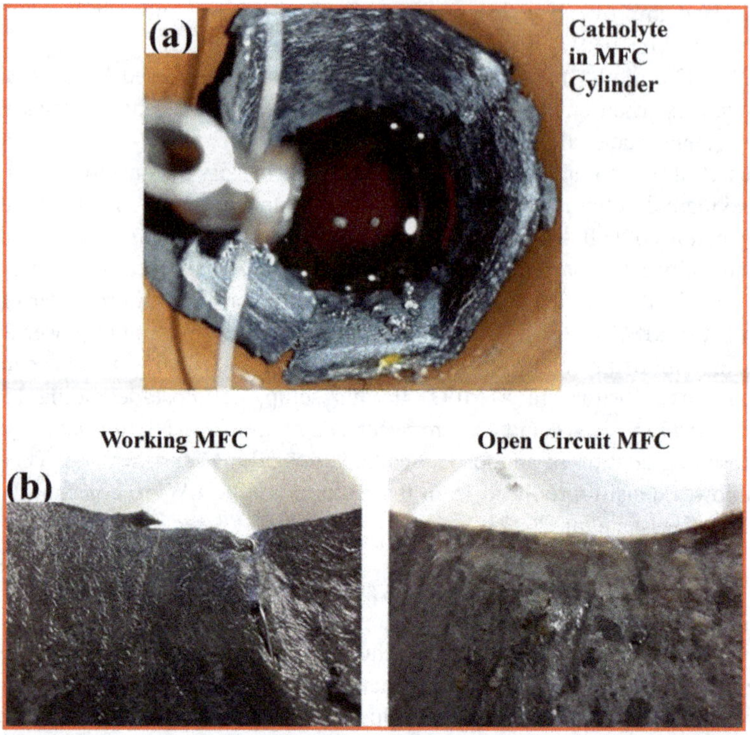

Fig. 13.22 Illustration of catholyte production from wastewater treatment using MFC. (**a**) Photograph shows the formation of droplets and accumulation catholyte solution inside the MFC cylinder. (**b**) Image shows the gas diffusion side of the cathode of working MFC (loaded) and open-circuit MFC (control). Biofilm growth is present only on the open-circuit MFC, i.e., that do not produce electricity. This indicates that the catholyte inhibits the growth of microorganisms and prevents the biofouling of cathode as well as the membrane. (From Gajda et al. 2016)

used in MFCs. The approximate power density of the MFC using RVC-CNT electrode is 132.1 ± 2.8 mWm^{-2}. This device removes 80.9% of Cr(VI) within 48 h of the operation (Fei et al. 2017).

MFC can be used for the production of catholyte within the reactor and for recovering of nitrogen from wastewater (shown in the Fig. 13.22). An electrolyzed basic solution (pH > 11) is produced from the cathode chamber, and the production rate is largely proportional to electrical current generation. This catholyte possess bactericidal properties. The bactericidal effect is confirmed using bacterial kill curves constructed by exposing a bioluminescent *Escherichia coli* target. The catholyte solution has cleaning properties. It reduces the microbial populations and limits undesired biofilm formation. Hence, it can be utilized as a washing agent in waterless urinals to improve sanitation. The demonstrated self-driven MFC system leads to the development of bioprocesses for sustainable wastewater treatment (Gajda et al. 2016).

13.8 Challenges in Microbial Fuel cells

MFCs have advantages like production of low-cost electricity from waste materials all-round the year (particularly in places where energy production plants are not available) and alternate method for bioremediation. However, they face some challenges. Utilizations of waste material in MFCs produce power density lower than pure carbon sources such as glucose. It is not possible for the utilization of pure sources routinely because it will increase the cost. These challenges hinder the commercialization of MFCs (Chaturvedi and Verma 2016). Also, the economically noncompetitive and high-cost components of MFCs are the major barriers for commercialization.

Some other drawbacks of MFCs are less surface area of the anode, inefficient electron transfer, proton mass transfer, poor oxygen reduction rate, low-power generation, lack of modularity, short life span of biochemical substrate, and very slow performance in wastewater treatment. The anode material should have more surface area and more affinity for microorganisms. The electron transfer mechanism of the anode materials is not understood well. Utilizations of materials like platinum as electrode increases the cost of MFC to higher end. Microorganisms are also one of the great challenges of MFC. The disadvantage of using mixed culture in a MFC is that it may contain pathogenic bacteria. Also, some bacteria may be sensitive to different kinds of stress. As per the report of Joseph Miceli, ARB helps to flow the electrons freely in between the electrodes (Waste-management-world.com 2013).

Materials such as nickel foam, stainless steel wool, platinum-coated stainless steel mesh, and molybdenum disulfide-coated stainless steel mesh electrodes are used as alternative to commercially non-viable electrode material like platinum (Ma Xiaoli et al. 2017). To bring the MFC technology out of the laboratory (for energy production at larger scales), pilot-scale tests are performed. These are good indicators that commercialization of this technology is possible (Logan 2010).

The power output is low in MFCs because of their high internal resistance. Also, it is lower than chemical fuel cells. The internal resistance is due to the proton mass transfer as well as the poor oxygen reduction rate at the cathode. Aerobic bacteria have higher affinity for oxygen than the abiotic cathode materials. Oxygen reduction kinetics and performance of the MFC is improved while using a cathode with aerobic bacteria (including the bacteria developing the corroding biofilm). Addition of the inorganic compounds into the anode chamber as nutrients and development of proton-specific membrane lead to inhibit proton transfer. MFCs should be optimized for its reactor configuration and electrolyte to reduce the internal resistance and to harvest the entire microbial catalytic potential (Kim et al. 2007).

To demonstrate MFC as power generator to supply power to the electronic systems, there is a possibility to use solid-phase organic matter at the anode. Sediment MFCs are designed from marine sediment to utilize as the source of bacteria and organic matter. The synthetic solid anolyte (SSA) is made by dissolving of agar, carbonaceous, and nitrogen sources into diluted seawater. This long-lasting portable SSA-MFC (shown in Fig. 13.23) overcomes problems like hydraulic pump system and biochemical substrate replacement (to sustain bacteria metabolism). It generates

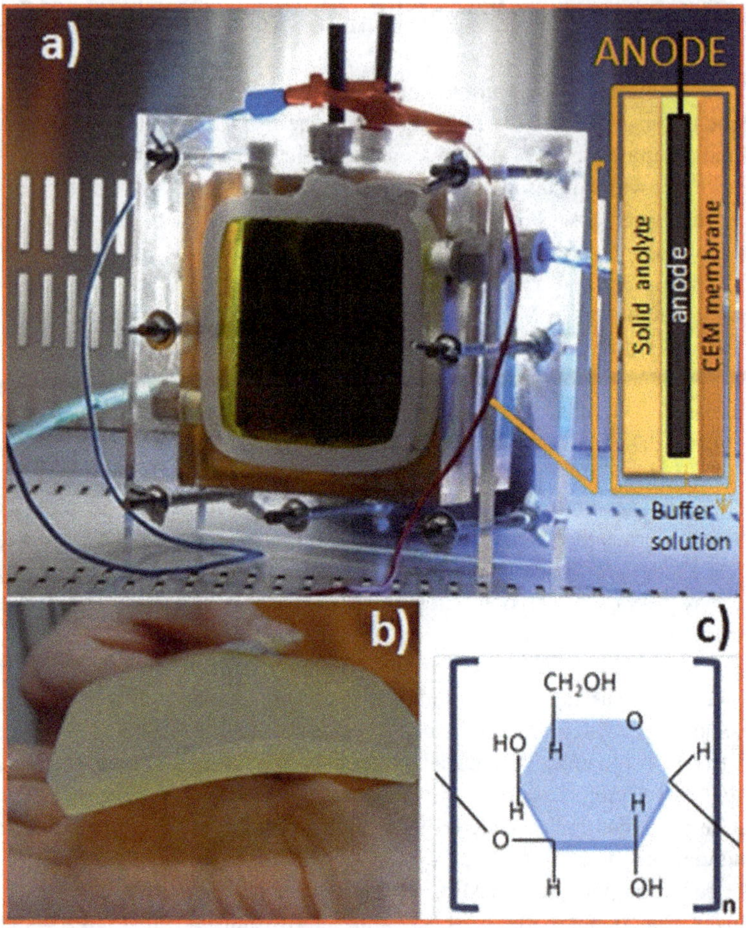

Fig. 13.23 Photograph of SSA-MFC. (**a**) Two-chamber MFC. The inset shows the SSA in the anode chamber. The anode is sandwiched between the SSA and CEM membrane and immersed in the buffer solution. (**b**) Photograph of the agar-based SSA embedded with energy storage system for bacteria metabolism. (**c**) Molecular structure of agar. (From Tommasi et al. 2016)

the maximum power density 60 mW/m^2 on the 7th day of the operation. It has 4 months lifetime without the need of electrolyte replacement and human assistance (Tommasi et al. 2016).

13.9 Influencing Factors of MFC Output

It is analyzed from the various reports that the output of the MFC depends on factors like substrates, concentration of waste, electrodes, pH, chamber construction, proton-exchange membrane, microorganism species, and their culture types. Some of the factors influence the performance of MFC. They are enumerated below.

In order to enhance the power generation of MFC:

- Increasing the pH of anode chamber (low ORR occurs at neutral pH)
- Increasing the electrons flow
- Increasing the ORR (catalyst increases the ORR rate).
- Increasing the concentration of organic waste materials (biofuels)
- Reducing the proton transfer
- Reducing internal resistance of MFC (by coating of nanomaterials like graphene and carbon nanotubes)
- Addition of the inorganic nutrients into the anode chamber
- Development of proton-specific membrane

The factors mentioned below help to improve the overall performance of MFC:

- Single-chamber air-cathode MFC supports for bacteria development better than the two-chamber MFC.
- Utilization of mixed bacteria cultures instead of single-culture bacteria increases the MFC output.
- Growing of anaerobic bacteria in anode chamber. It helps for electrons flow which leads to enhance the MFC performance.
- Utilization of cathode with aerobic bacteria inhibits proton transfer. It leads to improve the oxygen reduction kinetics as well as MFC performance.
- Biocompatible electrode increases the adhesion of bacteria which improves the MFC performance.
- Temperature maintenance is essential (low temperature reduces the MFC performance).

13.10 Conclusion

MFC is a novel wastewater treatment device with energy recovery from the waste. It mimics the interactions of microbes present in the nature. It converts chemical energy into electricity using microorganisms. The energy production mechanism of the MFC resembles the metabolism process by the gut flora. MFC has great potential in alternative energy source, wastewater treatment, environmental protection, bioremediation, and biosensor for oxygen and pollutants. In addition to the various applications, MFCs have certain drawbacks.

The drawbacks hinder in the practical applications of MFCs. Hence, extensive optimization is required for the efficient and wide applications of MFCs and to obtain the maximum microbial potential from them. Production of low-power density is the major drawback. It is rectified by using various nanomaterials and potent microorganisms. Some of the microorganisms have been identified that can transfer the electrons efficiently. Also, limited surface area of the electrodes prevents the microorganisms to adhere. Utilizations of nanomaterials increase the surface area of the electrodes.

Nanomaterials such as carbon-based materials, transition metal oxides, biogenic inorganic nanoparticles, and nanocomposites are widely used in MFCs due to their properties like high conductivity, biocompatibility, and chemical stability. Apart from these factors, they are utilized as alternative materials to reduce the high cost of components. Many methods have been identified to improve the performance of MFCs and to reduce the cost for feasible implementation of MFCs in commercial applications. Although MFCs face some challenges, there are good scopes and future prospects for them.

Acknowledgment The author expresses thanks to her husband Mr. G. Sankar for his assistances in this work. Also, she acknowledges the assistances of the International Research Center, Kalasalingam Academy of Research and Education (Deemed University), Krishnankoil – 626 126 (India).

References

Alabiad I, Ali UFM, Zakarya IA, Ibrahim N, Radzi RW, Zulkurnai NZ, Azmi NH (2017) Ammonia removal via microbial fuel cell (MFC) dynamic reactor. In: IOP conference series: materials science and engineering, vol. 206, no. 1. IOP Publishing, p 012079

Buzea C, Pacheco II, Robbie K (2007) Nanomaterials and nanoparticles: sources and toxicity. Biointerphases 2(4):MR17–MR71

Catal T, Bermek H, Liu H (2009) Removal of selenite from wastewater using microbial fuel cells. Biotechnol Lett 31(8):1211–1216

Chandrasekhar K, Kadier A, Kumar G, Nastro RA, Jeevitha V (2018) Challenges in microbial fuel cell and future scope. In: Microbial fuel cell. Springer, Cham, pp 483–499

Chang IS, Moon H, Jang JK, Kim BH (2005) Improvement of a microbial fuel cell performance as a BOD sensor using respiratory inhibitors. Biosens Bioelectron 20(9):1856–1859

Chaturvedi V, Verma P (2016) Microbial fuel cell: a green approach for the utilization of waste for the generation of bioelectricity. Bioresour and Bioprocess 3(1):38

Chiao M (2008) A microfabricated PDMS microbial fuel cell. J Microelectromech Syst 17(6):1329–1341

Chouler J, Bentley I, Vaz F, O'Fee A, Cameron PJ, Di Lorenzo M (2017) Exploring the use of cost-effective membrane materials for Microbial Fuel Cell based sensors. Electrochim Acta 231:319–326

Chu S, Majumdar A (2012) Opportunities and challenges for a sustainable energy future. Nature 488(7411):294

Daniel DK, Mankidy BD, Ambarish K, Manogari R (2009) Construction and operation of a microbial fuel cell for electricity generation from wastewater. Int J Hydrog Energy 34(17):7555–7560

Deng F, Sun J, Hu Y, Chen J, Li S, Chen J, Zhang Y (2017) Biofilm evolution and viability during in situ preparation of a graphene/exoelectrogen composite biofilm electrode for a high-performance microbial fuel cell. RSC Adv 7(67):42172–42179

Dharmadhikari S, Ghosh P, Ramachandran M (2018) Synthesis of proton exchange membranes for dual-chambered microbial fuel cells. J Serb Chem Soc 83(5):611–623

Dong Y, Feng Y, Qu Y, Du Y, Zhou X, Liu J (2015) A combined system of microbial fuel cell and intermittently aerated biological filter for energy self-sufficient wastewater treatment. Sci Rep 5:18070

Eldridge T (2014) Achieving industry integration with nanomaterials through financial markets. http://www.nanotech-now.com/columns/?article=835. Accessed 21 Nov 2018

El-Naggar MY, Wanger G, Leung KM, Yuzvinsky TD, Southam G, Yang J, Lau WM, Nealson KH, Gorby YA (2010) Electrical transport along bacterial nanowires from *Shewanella oneidensis* MR-1. Proc Natl Acad Sci 107(42):18127–18131

Erable B, Féron D, Bergel A (2012) Microbial catalysis of the oxygen reduction reaction for microbial fuel cells: a review. ChemSusChem 5(6):975–987
Ersan K, Irfan AR, Tukek S (2010) Effect of humidification of gases on first home constructed PEM fuel cell stack potential. Gazi University J Sci 23(1):61–70
Fei K, Song TS, Wang H, Zhang D, Tao R, Xie J (2017) Electrophoretic deposition of carbon nanotube on reticulated vitreous carbon for hexavalent chromium removal in a biocathode microbial fuel cell. R Soc Open Sci 4(10):170798
Feng L, Yan Y, Chen Y, Wang L (2011) Nitrogen-doped carbon nanotubes as efficient and durable metal-free cathodic catalysts for oxygen reduction in microbial fuel cells. Energy Environ Sci 4(5):1892–1899
Gajda I, Greenman J, Melhuish C, Ieropoulos IA (2016) Electricity and disinfectant production from wastewater: microbial fuel cell as a self-powered electrolyser. Sci Rep 6:25571
Ghasemi M, Daud WRW, Hassan SH, Oh SE, Ismail M, Rahimnejad M, Jahim JM (2013) Nano-structured carbon as electrode material in microbial fuel cells: a comprehensive review. J Alloys Compd 580:245–255
Ghasemi M, Halakoo E, Sedighi M, Alam J, Sadeqzadeh M (2015) Performance comparison of three common proton exchange membranes for sustainable bioenergy production in microbial fuel cell. Procedia CIRP 26:162–166
Gorby YA, Yanina S, McLean JS, Rosso KM, Moyles D, Dohnalkova A, Beveridge TJ, Chang IS, Kim BH, Kim KS, Culley DE (2006) Electrically conductive bacterial nanowires produced by Shewanella oneidensis strain MR-1 and other microorganisms. Proc Natl Acad Sci 103(30):11358–11363
Harnisch F, Schröder U (2010) From MFC to MXC: chemical and biological cathodes and their potential for microbial bioelectrochemical systems. Chem Soc Rev 39(11):4433–4448
Hassan SH, Kim YS, Oh SE (2012) Power generation from cellulose using mixed and pure cultures of cellulose-degrading bacteria in a microbial fuel cell. Enzym Microb Technol 51(5):269–273
Hsu WH, Tsai HY, Huang YC (2017) Characteristics of carbon nanotubes/graphene coatings on stainless steel meshes used as electrodes for air-cathode microbial fuel cells. J Nanomater 2017:9875301
Huang Z, Wu Z, Zhou F, Zhou J (1997) Ovulation prediction by monitoring the conductivity of woman's saliva. In: Engineering in medicine and biology society, 1997. Proceedings of the 19th annual international conference of the IEEE, vol 5. IEEE, pp 2344–2346
Hubler AW, Osuagwu O (2010) Digital quantum batteries: energy and information storage in nanovacuum tube arrays. Complexity 15(5):48–55
Humphries AC, Nott KP, Hall LD, Macaskie LE (2004) Continuous removal of Cr (VI) from aqueous solution catalysed by palladised biomass of Desulfovibrio vulgaris. Biotechnol Lett 26(19):1529–1532
Igem.org (2018) Microbial fuel cell. [Online] Available at: http://2013.igem.org/Team:Bielefeld-Germany/Project/MFC. Accessed 21 Nov 2018
Jiang X, Hu J, Lieber AM, Jackan CS, Biffinger JC, Fitzgerald LA, Ringeisen BR, Lieber CM (2014) Nanoparticle facilitated extracellular electron transfer in microbial fuel cells. Nano Lett 14(11):6737–6742
Kaewkannetra P, Imai T, Garcia-Garcia FJ, Chiu TY (2009) Cyanide removal from cassava mill wastewater using Azotobacter vinelandii TISTR 1094 with mixed microorganisms in activated sludge treatment system. J Hazard Mater 172(1):224–228
Khan MD, Abdulateif H, Ismail IM, Sabir S, Khan MZ (2015) Bioelectricity generation and bioremediation of an azo-dye in a microbial fuel cell coupled activated sludge process. PLoS One 10(10):e0138448
Khan N, Clark I, Bolan N, Meier S, Saint CP, Sánchez-Monedero MA, Shea S, Lehmann J, Qiu R (2017) Development of a buried bag technique to study biochars incorporated in a compost or composting medium. J Soils Sediments 17(3):656–664
Khudzari JM, Tartakovsky B, Raghavan GV (2016) Effect of C/N ratio and salinity on power generation in compost microbial fuel cells. Waste Manag 48:135–142
Kim BH, Chang IS, Gil GC, Park HS, Kim HJ (2003) Novel BOD (biological oxygen demand) sensor using mediator-less microbial fuel cell. Biotechnol Lett 25(7):541–545

Kim BH, Chang IS, Gadd GM (2007) Challenges in microbial fuel cell development and operation. Appl Microbiol Biotechnol 76(3):485
Kodali M, Herrera S, Kabir S, Serov A, Santoro C, Ieropoulos I, Atanassov P (2018) Enhancement of microbial fuel cell performance by introducing a nano-composite cathode catalyst. Electrochim Acta 265:56–64
Kumar R, Singh L, Wahid ZA (2015) Role of microorganisms in microbial fuel cells for bioelectricity production. In: Microbial Factories. Springer, New Delhi, pp 135–154
Lal Deeksha (2013) Microbes to generate electricity. Indian J Microbiol 53(1):120–122
Lamberg P, Bren KL (2016) Extracellular Electron transfer on sticky paper electrodes: carbon paste paper anode for microbial fuel cells. ACS Energy Lett 1(5):895–898
Li S, Cheng C, Thomas A (2017) Carbon-based microbial-fuel-cell electrodes: from conductive supports to active catalysts. Adv Mater 29(8):1602547
Lies DP, Hernandez ME, Kappler A, Mielke RE, Gralnick JA, Newman DK (2005) Shewanella oneidensis MR-1 uses overlapping pathways for iron reduction at a distance and by direct contact under conditions relevant for biofilms. Appl Environ Microbiol 71(8):4414–4426
Light SH, Su L, Rivera-Lugo R, Cornejo JA, Louie A, Iavarone AT, Ajo-Franklin CM, Portnoy DA (2018) A flavin-based extracellular electron transfer mechanism in diverse gram-positive bacteria. Nature 562(7725):140. https://doi.org/10.1038/s41586-018-0498-z
Lim DH, Wilcox J (2012) Mechanisms of the oxygen reduction reaction on defective graphene-supported Pt nanoparticles from first-principles. J Phys Chem C 116(5):3653–3660
Logan BE (2010) Scaling up microbial fuel cells and other bioelectrochemical systems. Appl Microbiol Biotechnol 85(6):1665–1671
Logan BE, Regan JM (2006a) Electricity-producing bacterial communities in microbial fuel cells. Trends Microbiol 14(12):512–518
Logan BE, Regan JM (2006b) Microbial fuel cells-challenges and applications. Environ Sci Technol 40(17):5172–5180. https://doi.org/10.1021/es0627592
Ma X, Li Z, Zhou A, Yue X (2017) Energy recovery from tubular microbial electrolysis cell with stainless steel mesh as cathode. R Soc Open Sci 4(12):170967
Malvankar NS, Vargas M, Nevin KP, Franks AE, Leang C, Kim BC, Inoue K, Mester T, Covalla SF, Johnson JP, Rotello VM (2011) Tunable metallic-like conductivity in microbial nanowire networks. Nat Nanotechnol 6(9):573
Mashkour M, Rahimnejad M (2015) Effect of various carbon-based cathode electrodes on the performance of microbial fuel cell. Biofuel Res J 2(4)
McGovern C (2010) Commoditization of nanomaterials. Nanotechnol Perceptions 6(3):155
Mink JE, Hussain MM (2013) Sustainable design of high-performance microsized microbial fuel cell with carbon nanotube anode and air cathode. ACS Nano 7(8):6921–6927
Mink JE, Qaisi RM, Logan BE, Hussain MM (2014) Energy harvesting from organic liquids in micro-sized microbial fuel cells. Npg Asia Mat 6(3):e89
Miran W, Nawaz M, Jang J, Lee DS (2016) Effect of wastewater containing multi-walled carbon nanotubes on dual-chamber microbial fuel cell performance. RSC Adv 6(94):91314–91319
Moqsud MA, Yoshitake J, Bushra QS, Hyodo M, Omine K, Strik D (2015) Compost in plant microbial fuel cell for bioelectricity generation. Waste Manag 36:63–69
Mustakeem M (2015) Electrode materials for microbial fuel cells: nanomaterial approach. Mater Renew Sustain Energy 4:22. https://doi.org/10.1007/s40243-015-0063-8
Nurettin ÇEK (2017) Examination of zinc electrode performance in microbial fuel cells. Gazi University J Sci 30(4):395–402
Omkar S Powar, Lakshminarayana Bhatta KG, Raghavendra Prasad HD, Venkatesh K, Raghu AV (2017) Influence of carbon based electrodes on the performance of the microbial fuel cell. Int J Res-Granthaalayah 5(4: RAST):7–16. https://doi.org/10.5281/zenodo.803412
Park DH, Zeikus JG (2003) Improved fuel cell and electrode designs for producing electricity from microbial degradation. Biotechnol Bioeng 81(3):348–355
Penteado ED, Fernandez-Marchante CM, Zaiat M, Gonzalez ER, Rodrigo MA (2017) Influence of carbon electrode material on energy recovery from winery wastewater using a dual-chamber microbial fuel cell. Environ Technol 38(11):1333–1341

Prasertsung N, Reungsang A, Ratanatamskul C (2012) Alkalinity of cassava wastewater feed in anodic enhance electricity generation by a single chamber microbial fuel cells. Eng J 16(5):17–28

Qiao Y, Li CM, Bao SJ, Bao QL (2007) Carbon nanotube/polyaniline composite as anode material for microbial fuel cells. J Power Sources 170(1):79–84

Reguera G, McCarthy KD, Mehta T, Nicoll JS, Tuominen MT, Lovley DR (2005) Extracellular electron transfer via microbial nanowires. Nature 435(7045):1098

Ren H, Chae J (2015) Microfabricated microbial fuel cells. In Micro Energy Harvesting, Wiley, pp. 347–361. https://doi.org/10.1002/9783527672943

Ren Z, Steinberg LM, Regan JM (2008) Electricity production and microbial biofilm characterization in cellulose-fed microbial fuel cells. Water Sci Technol 58(3):617–622

Reuben Y Tamakloe (2018) PEM-less microbial fuel cells, Proton exchange membrane fuel cell, Tolga Taner, IntechOpen. https://doi.org/10.5772/intechopen.71479. Available from: https://www.intechopen.com/books/proton-exchange-membrane-fuel-cell/pem-less-microbial-fuel-cells

Rezaei F, Xing D, Wagner R, Regan JM, Richard TL, Logan BE (2009) Simultaneous cellulose degradation and electricity production by *Enterobacter cloacae* in a microbial fuel cell. Appl Environ Microbiol 75(11):3673–3678

Rojas JP, Hussain MM (2015) The role of microfabrication and nanotechnology in the development of microbial fuel cells. Energ Technol 3(10):996–1006

Sayed ET, Abdelkareem MA (2017) Yeast as a biocatalyst in microbial fuel cell. In Old yeasts-new questions. InTechOpen. https://doi.org/10.5772/intechopen.70402

Schaetzle O, Barrière F, Schröder U (2009) An improved microbial fuel cell with laccase as the oxygen reduction catalyst. Energy Environ Sci 2(1):96–99

Schroder U (2007) Anodic electron transfer mechanisms in microbial fuel cells and their energy efficiency. Phys Chem Chem Phys 9(21):2619–2629

Siegert M, Sonawane JM, Ezugwu CI, Prasad R (2019) Economic Assessment of Nanomaterials in Bio-Electrical Water Treatment. In: Prasad R, Thirugnanasanbandham K (eds) Advanced Research in Nanosciences for Water Technology. Springer International Publishing AG, Cham, pp 5–23

Sure S, Ackland ML, Torriero AA, Adholeya A, Kochar M (2016) Microbial nanowires: an electrifying tale. Microbiology 162(12):2017–2028

Tan L, Li N, Chen S, Liu ZQ (2016) Self-assembly synthesis of CuSe@ graphene–carbon nanotubes as efficient and robust oxygen reduction electrocatalysts for microbial fuel cells. J Mater Chem A 4(31):12273–12280

Tandukar M, Huber SJ, Onodera T, Pavlostathis SG (2009) Biological chromium (VI) reduction in the cathode of a microbial fuel cell. Environ Sci Technol 43(21):8159–8165

Toczyłowska-Maminska R, Szymona K, Król P, Gliniewicz K, Pielech-Przybylska K, Kloch M, Logan BE (2018) Evolving microbial communities in cellulose-fed microbial fuel cell. Energies 11(1):124

Tommasi T, Salvador GP, Quaglio M (2016) New insights in microbial fuel cells: novel solid phase anolyte. Sci Rep 6:29091

Tsai HY, Hsu WH, Huang YC (2015) Characterization of carbon nanotube/graphene on carbon cloth as an electrode for air-cathode microbial fuel cells. J Nanomater 2015:3

Valipour A, Ayyaru S, Ahn Y (2016) Application of graphene-based nanomaterials as novel cathode catalysts for improving power generation in single chamber microbial fuel cells. J Power Sources 327:548–556

Wang Z, Lee T, Lim B, Choi C, Park J (2014) Microbial community structures differentiated in a single-chamber air-cathode microbial fuel cell fueled with rice straw hydrolysate. Biotechnol Biofuels 7(1):9

Waste-management-world.com (2013) Bacteria that turn waste to energy in microbial fuel cells studied. [Online] Available at: https://waste-management-world.com/a/bacteria-that-turn-waste-to-energy-in-microbial-fuel-cells-studied. Accessed 21 Nov 2018

Xie X, Hu L, Pasta M, Wells GF, Kong D, Criddle CS, Cui Y (2010) Three-dimensional carbon nanotube–textile anode for high-performance microbial fuel cells. Nano Lett 11(1):291–296

Yang Y, Xu M, Guo J, Sun G (2012) Bacterial extracellular electron transfer in bioelectrochemical systems. Process Biochem 47(12):1707–1714

Yang G, Chen D, Lv P, Kong X, Sun Y, Wang Z, Yuan Z, Liu H, Yang J (2016a) Core-shell au-Pd nanoparticles as cathode catalysts for microbial fuel cell applications. Sci Rep 6:35252

Yang N, Ren Y, Li X, Wang X (2016b) Effect of graphene-graphene oxide modified anode on the performance of microbial fuel cell. Nano 6(9):174

Yang Q, Liang S, Liu J, Lv J, Feng Y (2017) Analysis of anodes of microbial fuel cells when carbon brushes are preheated at different temperatures. Catalysts 7(11):312

Yin Y, Huang G, Zhou N, Liu Y, Zhang L (2016) Increasing power generation of microbial fuel cells with a nano-CeO2 modified anode. Energy Sources, Part A 38(9):1212–1218

Yong YC, Dong XC, Chan-Park MB, Song H, Chen P (2012) Macroporous and monolithic anode based on polyaniline hybridized three-dimensional graphene for high-performance microbial fuel cells. ACS Nano 6(3):2394–2400

Yuan Y, Zhou S, Liu Y, Tang J (2013) Nanostructured macroporous bioanode based on polyaniline-modified natural loofah sponge for high-performance microbial fuel cells. Environ Sci Technol 47(24):14525–14532

Zhang L, Liu C, Zhuang L, Li W, Zhou S, Zhang J (2009) Manganese dioxide as an alternative cathodic catalyst to platinum in microbial fuel cells. Biosens Bioelectron 24(9):2825–2829

Zhang B, Jiang Y, Han J (2017a) A flexible nanocomposite membrane based on traditional cotton fabric to enhance performance of microbial fuel cell. Fibers and Polym 18(7):1296–1303

Zhang Y, Jiang J, Zhao Q, Gao Y, Wang K, Ding J, Yu H, Yao Y (2017b) Accelerating anodic biofilms formation and electron transfer in microbial fuel cells: role of anionic biosurfactants and mechanism. Bioelectrochemistry 117:48–56

Zhao Y, Collum S, Phelan M, Goodbody T, Doherty L, Hu Y (2013) Preliminary investigation of constructed wetland incorporating microbial fuel cell: batch and continuous flow trials. Chem Eng J 229:364–370

Zhou M, Wang H, Hassett DJ, Gu T (2013) Recent advances in microbial fuel cells (MFCs) and microbial electrolysis cells (MECs) for wastewater treatment, bioenergy and bioproducts. J Chem Technol Biotechnol 88(4):508–518

Index

R. Prasad (ed.), *Microbial Nanobionics*, Nanotechnology in the Life Sciences,
https://doi.org/10.1007/978-3-030-16383-9

CPSIA information can be obtained
at www.ICGtesting.com
Printed in the USA
LVHW080547091220
673656LV00006B/33